CLIMATE
IN THE
AGE OF EMPIRE

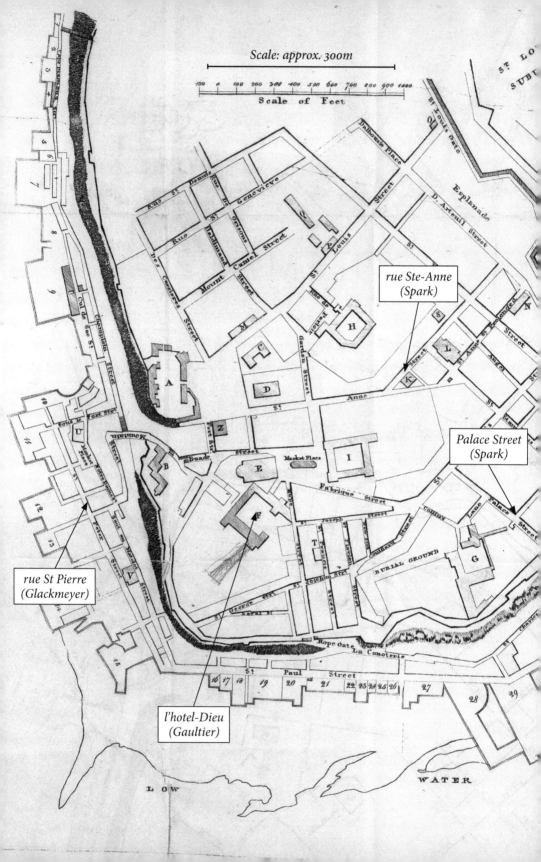

CLIMATE

IN THE

AGE OF EMPIRE

WEATHER OBSERVERS IN COLONIAL CANADA

↔ VICTORIA C. SLONOSKY ↔

AMERICAN METEOROLOGICAL SOCIETY

Climate in the Age of Empire: Weather Observers in Colonial Canada © 2018 by Victoria C. Slonosky. All rights reserved. Permission to use figures, tables, and brief excerpts from this book in scientific and educational works is hereby granted provided the source is acknowledged.

Front cover: *Vue de Montréal à partir du fleuve Saint-Laurent*, c. 1840. Courtesy of Bibliothèque et Archives Canada.

Maps: City of Quebec: Modified from Bennet E. 1822. Collection numérique, cartes et plans, item number 4039528, P600, S4, SS2. *City of Quebec, 1822.* Reprinted with permission from Bibliothèque et Archives nationales du Québec. City of Montreal: Map of the city of Montreal with the latest improvements, showing the locations of the nineteenth-century weather observers. Reprinted with permission from Bibliothèque et Archives nationales du Québec (image number 65542).

Published by the American Meteorological Society
45 Beacon Street, Boston, Massachusetts 02108

The mission of the American Meteorological Society is to advance the atmospheric and related sciences, technologies, applications, and services for the benefit of society. Founded in 1919, the AMS has a membership of more than 13,000 and represents the premier scientific and professional society serving the atmospheric and related sciences. Additional information regarding society activities and membership can be found at www.ametsoc.org.

Print ISBN: 978-1-944970-20-8
eISBN: 978-1-944970-21-5

Library of Congress Cataloging-in-Publication Data

Names: Slonosky, Victoria C., 1972- author. | American Meteorological Society.
Title: Climate in the age of empire : weather observers in colonial Canada / by Victoria C. Slonosky.
Other titles: Weather observers in colonial Canada
Description: Boston, Massachusetts : American Meteorological Society, [2018] | Includes bibliographical references and index.
Identifiers: LCCN 2017048199 (print) | LCCN 2017057374 (ebook) | ISBN 9781944970215 (eBook) | ISBN 9781944970208 (pbk.)
Subjects: LCSH: Meteorology--Canada--History. | Meteorology—Canada—Observations. | Meteorology—Canada. | Meteorological Service of Canada—History. | Meteorologists—Canada. | Climatic changes—Canada. | Canada—History.
Classification: LCC QC985 (ebook) | LCC QC985 .S56 2018 (print) | DDC 551.5097109/032—dc23
LC record available at https://lccn.loc.gov/2017048199

Contents

List of Illustrations vii
Preface ix
Acknowledgments xix

Part I   The Landscape: Scientists, Practices, and Theories

1   Territory, Networks, and Tools   3
2   Dr. Jean-François Gaultier: New France's Climatologist   17
3   Clearing and Cultivation: Eighteenth-Century Climate Improvement Theory   27
4   British-American Weather Observers to 1830   49
5   McCord and the Montreal Natural History Society   65
6   Nineteenth-Century Scientists Question Climate Amelioration   87

Part II   Meteorology Takes Shape

7   Meteorology and the Military   125
8   The Magnetic Crusade and the Founding of the Toronto Observatory   141
9   Medical Meteorology   165

10 The Establishment of the Meteorological Service of Canada   183
11 The McGill Observatory and the Professionalization of Meteorology   197
12 What Do Three Centuries of Observations Tell Us?   207
13 Extraordinary Seasons   243

Biographical Sketches   269
Index   311

Illustrations

Following page 97

North America in the mid-eighteenth century
Photographs of thermometers sent by Joseph-Nicholas Delisle to Anders Celsius
Illustration of a page of Gaultier's records
A view of Quebec City from the Saint Lawrence River
The Hôtel-Dieu hospital at which Gaultier worked in Quebec City
The first page of Gaultier's report for 1754
Alexander Spark
St. Andrew's Presbyterian Church
Memorial to Dr. Spark
Example of Spark's weather journal
Example of the first McCord journal
La Grange aux Pauvres
Thomas McCord
Thermometer from the McCord Museum
Example of John Samuel McCord's second weather diary
Henry William Cotton's *View of Old Montreal from Temple Grove*
John Samuel McCord

Alexander Skakel
John Bethune
Saint James Street, Montreal
View of Montreal from Saint Helen's Island
Modern view of Old Montreal from Saint Helen's Island
Lieutenant (Major General) C. J. Buchanan Riddell
Reconstructed officer's writing desk at Fort York
First Toronto Magnetic and Meteorological Observatory
Portrait of John Henry Lefroy
Reconstructed barracks, Fort York
Sleeping and living arrangements, Fort York
Smallwood's Saint Martin Observatory
Dr. Charles Smallwood
Dr. Archibald Hall
The second observatory building (Toronto)
Professor George Templeman Kingston
Eugene Haberer's *The Public Institutions of the Dominion of Canada*
Clement McLeod and students
Transit telescope
Barometer and clock
Equipment used to relay time signals
The old McGill Observatory in winter
Snow gauge
Sunshine recorder
Current automatic weather station on McTavish Street

Preface

I decided to write this book in part to address two common misunderstandings: first, that climatology is a young science, and second, that Canada has little in the way of a scientific history. Science in Canada, as well as the sciences of climatology and meteorology generally, suffer somewhat from the perception of being relatively new, of being "in their infancy." Historians know this is not so. Historians of meteorology and climatology such as Clarence Glacken (1967), Antonello Gerbi (1973), John Kington (1980), Karen Kupperman (1982), James Fleming (1988, 1990), Theodore Feldman (1990), and Vladimir Janković (2001) provided a foundation for the history of ideas about climate in the last half of the twentieth century. In the past decade a number of new studies about the history of ideas of climate change, and particularly anthropogenic climate change and colonization, have been explored by Jan Golinski (2008), Brant Vogel (2011), and Sam White (2015), while Colin Coates, Dagomar Degroot, and Yvon Desloges (Coates 2000; Coates and Degroot 2015; Desloges 2016) have looked at ideas about climate change in Canada. On the historical climatology side, data rescue groups concerned with historical climate, such as Atmospheric Circulation Reconstructions over the Earth (ACRE), are growing and proliferating (www.met-acre.org).

This book is the story of early scientists—they would have called themselves natural philosophers or natural historians—in colonial Canada who kept instrumental daily weather observations in the tradition of the science that was blossoming in eighteenth-century Europe. These observations were taken in a social and intellectual context in which there was hope that the climate *would* change and that the climate in the colonies in particular would improve through human action. As was usual in those times, the weather observers discussed here were men whose occupations included the clergy, the law, teaching, the military, fur-trading, and gardening. Practitioners of medicine, too, had an abiding interest in the climate and how it related to health and disease, and doctors from the King's Physician Jean-François Gaultier (1706–1756) towards the end of the French regime to Dr. Charles Smallwood (1812–1873), who founded the McGill Observatory, are among the most assiduous of the colonial weather observers.

The aim of this book is neither to provide a complete history of meteorology or climatology in Canada, nor to present a detailed analysis of the climate of the past several centuries. This work stemmed originally from an analysis of instrumental observations recorded in historical weather diaries, and so the focus is on the weather observers who kept daily records in colonial Canada, starting towards the end of the French regime with doctor Jean-François Gaultier in 1742 and ending around the time of the establishment of the Meteorological Service of Canada in the 1870s.[1] What were the ideas and motivations of these scientific forebears of ours, who left us such rich and detailed sources with which to study our climate's past? What inspired them to keep to strict observing schedules for years and sometimes even decades, committing themselves to a task of rigid routine that considerably limits freedom of movement within a given day and demands great concentration and attention to detail? What did they hope to discover? How did they think about climate, its variation, and the relationship between people and climate? Are any of their ideas or motivations similar to ours?

Some were. William Kelly and John Samuel McCord were interested in using observations to evaluate theories of climate change. Too often, we consider climate change, especially climate change brought about by human activity, as a new and modern concern. Climatology's "deeper roots in historical experiences" (White 2015, p. 566) has been buried under the vast amount of data and machinery involved in recording, storing, and producing

---

1. Technical papers discussing the analysis of daily weather observations in eighteenth- and nineteenth-century Canada are available (Slonosky 2003, 2014, 2015); ongoing efforts in Canadian climate data rescue can be seen at the citizen science website (citsci.geog.mcgill.ca).

climatic information from observing systems and computer modelling in recent decades. As it happens, over two centuries ago weather observers were motivated by the same issue that concerns us today: is the climate changing, and is the change caused by human activity?

During much of the eighteenth and nineteenth centuries in Canada, it was thought that the climate improved as colonists cleared forests and turned the swamps of the Saint Lawrence River lowlands into cultivated fields of grain. The hope was that clearing the forests and swamps would moderate the climatic extremes of North America. Mitigating the long, cold winters or abating the heat of the short, mosquito-ridden summers could only be considered beneficial. Cutting down trees to improve the climate was practically a patriotic duty.

In exploring the fascinating topic of historical thinking about climate change, Jean-François Gaultier's comments and John Samuel McCord's scientific notebook led me to William Kelly's work, especially his "Remarks on the climate of Lower Canada";[2] Theodore Mann's works on historical climatology, *Mémoires sur les Grandes Gelées* (*History of Great Frosts*); and then, by way of James Fleming's *Historical Perspectives of Climate Change*, to Clarence Glacken's *Traces on the Rhodian Shore*. It was a captivating journey. References to the improvement of climate through the clearing the forests can be found in letters and scientific documents from both the French (1608–1763) and British (1763–1867) colonial periods. Early Canadian weather observers looked for evidence of climate change both in written documents from earlier periods and in weather observations recorded using scientific instruments.

As industrial civilisation and colonial empires expanded across the globe, so did meteorological observers and their instruments. Herein lies one problem in trying determine how climate changed over time: the location of the observers and the network of observations have also been continually changing. Records are fragmented, some lasting only a few months, others lasting a decade or more, and a very few persisting over centuries in formal institutions such as observatories. Short registers in particular are difficult to evaluate for comparison with later records as the number of observations are too small to be able to detect potential problems such as instrument calibra-

---

2. The full title is the "Abstract of the meteorological journal Kept at Cape Diamond, Québec, from the 1st of January, 1824, to 31st December, 1831, with some remarks on the climate of Lower Canada."

tion or exposure.[3] Climate manifests at all scales from the very local to the global;[4] moving a barometer or thermometer over even short distances—to a place farther up the hill or closer to the river, in the middle of a field or at the edge of the forest, outside a window in a town house or on a shed in the country—all make a difference to the pressure, temperature, and precipitation recorded. These changes are not because of any inherent changes in the regional or global climate but because of changes between the different microclimates of the instruments' environments.

To complicate matters further, those few records that do tend to be stable over time, sometimes over decades, such as those made by the McGill Observatory, or even over centuries, such as those made by the Paris Observatory, were kept in places where the environment has changed over time. The Paris, Toronto, and McGill Observatories were all initially in the countryside; indeed, the Paris Observatory was so far from town that the academicians who were supposed to use it as their principal meeting place complained about the travel and stayed instead in central Paris, leaving the observatory out in the country to the astronomers. Today, these observatories are in the centres of their respective urban conglomerates. Cities produce an additional warming effect as a result of both the heat generated by homes, workplaces, and industries and because of the changes in land surface from grass, trees, and water to roads, brick, and concrete, all of which have a tendency to absorb heat during the day and radiate that heat to the environment at night. Changes in the heat and water-absorbing properties of built environments also affect the energy balance in cities.[5]

These issues—changes in instrumentation, station relocation, irregular

---

3. If a record lasting only a few months shows a particularly unusual climate—or even if it shows a normal one—it's impossible to know if those months really were unusual or if there was a problem with the instrument. Over longer periods of time, warmer months and colder months usually even out, and the reliability of the records can be estimated statistically.

4. This also complicates the discussion of climate change and anthropogenic climate change. At what level are we discussing change? Building a house causes the climate to change, even more so building a city. Are we only counting what can be measured globally when we talk about climate change or are we also thinking regionally or locally?

5. Together, these effects are known as the urban heat island effect. The process of trying to account for these non-climate-related changes to instrumental records, so that we can then compare them across time and space, is today known as homogenization.

or changes in observation procedure—combined with a sparse observation network[6] make the pressing issue of current evaluation of climate change a far from obvious task. Reams of paper (or today, electrons) and countless conference sessions and international meetings are devoted to the issue of climate data quality and homogenization in modern climatology. Imagine, then, my astonished delight when, reading through John Samuel McCord's scientific notebook (ca. 1830s) in the hopes of finding information about his instruments and their location, I came across the following, incredibly familiar sentiments:

> No one but the zealous meteorologists knows how very difficult it is to obtain observations in this science which can be depended upon. During the course of some investigations made several years ago, on the subject of the climate of Canada with a view of ascertaining whether any and what changes had taken place in its temperature, many tables came into my possession, some in MS and others published in the periodicals of the day . . . Knowing from experience how difficult it was to obtain even these scanty data on which to base a comparison, I resolved to [make summaries and publish the results of my work] . . . in order that future students in this interesting, but infant science . . . may be saved all the trouble and research which fell to my lot. (J. McCord, Scientific Notebook 1836)

The scientific concerns expressed by early observers are the same as ours today. They were concerned with issues of instrument calibration and reliability, and we wonder how effective were instruments designed in temperate Europe in capturing the rigours of Canadian climate? How could different observations, taken at different places, with varying instruments and observation times, be compared to each other? How did the local environment, such as town versus country, affect the observations? Overall, how reliable were the data? And were they reliable enough to show empirically whether the climate was changing or not? How common or uncommon are extreme events and unusual seasons? Though the instruments and means of analysis have changed over the past 200 years, these questions have not.

Then as today, climate change was not the only reason for keeping weather records. Those who were interested in evaluating climate change,

---

6. This is a problem that is not improving in our time, as we currently have only a fraction of the number of meteorological stations reporting today that we had in the 1960s.

William Kelly and Theodore Mann, and to some extent John Samuel McCord, relied on observations others had kept. If not an interest in observing change, what had motivated Alexander Spark, John Bethune, or Charles Smallwood to keep such diligent records for such long periods? "A sincere love of science" (McCord 1836, p. 36) cannot be discounted. Scientific curiosity, together with an integrated view of nature alongside the more practical hope of improving navigation, drove governments and institutions such as the British Army and the Royal Society to put enormous resources of time and money into the magnetic crusade of the 1840s. While theories of that time linking geomagnetic phenomena to meteorology fizzled out as understanding of Earth's magnetic core advanced (Courtillot and Le Mouël 2007), the recording of meteorological observations that accompanied magnetic observations persisted.

An additional reason our forebears had for keeping weather records was to aid in the examination of the connection between climate and health, based on the prevailing theories of weather-related epidemics of disease. It is difficult for us to imagine, living in the early twenty-first century, how different the experience and ideas of disease were before the advent of cellular biology and germ theory in the late nineteenth century. The scourges of cholera and typhus that swept the globe in the nineteenth century brought a new urgency to the question of the causes of these pandemics. Jean-François Gaultier and Archibald Hall, both doctors, kept track of weather and disease simultaneously.

Another motivating factor in keeping weather records was religion. To "contemplate the teachings of God in Nature . . . to study the method of God's workings in nature . . . to decipher some new word in the pages of that great book," as described by Charles Smallwood (1866, p. 126), was an act of devotion. Natural theology, the belief that laws of nature "are within the grasp of the human mind; God wanted us to recognise them . . . so that we could share in His own thoughts," as Johannes Kepler put it as far back as the seventeenth century, expressed scientific discovery as following the mind of God (Baumgardt 1951, p. 50; see also Polkinghorne 2011 for a description of natural theology). Nearly all of the observers and writers discussed in this book for whom personal information can be found were practicing Christians, and many were devout. Alexander Spark and John Bethune, who kept the two longest and most complete individual records for the Saint Lawrence valley region, were clergymen. On the other hand, Alexander von Humboldt, arguably the most influential and revered scientist of the early nineteenth century, was a noted atheist. Especially after the

French Revolution, religious faith wasn't something that could be taken as axiomatic in the lives of the intellectual or scientific elite. Nevertheless, a belief in Providence, a search for knowledge, and an understanding of a world perceived to have been created by divinely ordained laws of nature, were both publicly and privately referred to by most of the scientists in this book as a source of inspiration. To overlook this aspect would be to miss a significant influence on their thinking about climate and motivations for observing the weather.

Acknowledging that many of the observations made by other scientists were too sporadic to be useful in determining climate change or deducing general laws of weather, McCord and others, such as Royal Artillery Officer John Henry Lefroy (1817–1890), director of the Toronto Observatory, kept their records with the avowed purpose of being "of use to posterity." They preserved information for the generations to come in the hopes that it would be used in the future to understand the climate. Today, with the advent of modern computers, digital photography, numerical weather models, and increased digital storage and memory, scientists are finally able to rescue and use the vast sets of numbers in paper archives that were so faithfully, determinedly, and hopefully recorded by scientists of the past.

Our interest in climate, and particularly climate change, seems to be as cyclical as the climate itself. Much of the hard-won knowledge gathered by the weather observers in this book was lost or discarded once interest in the climate-related theories, whether of climate improvement, climate-caused diseases, or the relationship between magnetism and climatic change, fell by the wayside. Much patient effort by climatologists, archivists, and citizen scientists around the world has gone into rescuing these historical observations, too often found in fragmentary form with much of their context missing.

In 1853, Lefroy wrote that

> by reference [to the observations] we learn ... whether we can bring about changes in climate through human agency: whether such changes are always beneficial, and therefore in harmony with the design of the Universe: or whether sometimes noxious, and therefore in favour of the opinion that there are pre-ordained bounds to the extension of civilised man over the Globe. (Lefroy 1853, p. 29)

One hundred and sixty years on, we are concerned as ever, though with different hopes and fears, with climate change.

Notes on Terminology

A large part of my interest in past ideas about weather and climate is in seeing which are still held today, which have been proved wrong, and which still hold merit but have been forgotten. This is all within the context of what came before and led up to the climatological studies I do now, and as such is not and cannot be the full picture of all the people concerned with and making observations of climate or with other cultural traditions. The theory of modifying the climate by modifying the environment seems to be recurrent in western culture and philosophy, continually rediscovered in different guises. While western culture and philosophy are the context, by its very nature tje history of scientific inquiry produced ideas that have not stood the test of time, such as the miasma theory of disease (the idea that disease is caused by pollution or unpleasant air; "malaria" translates literally to "bad air") or phlogiston (the idea that heat is a tangible material with measureable mass that can be gained or lost from material entities such as trees). These can be difficult to grasp today by the public who understands (if not by name) the germ theory of disease and the calorie as measurement of energy.

Further, many of the terms we use today—geographical, political, and scientific—didn't exist and wouldn't be recognised by the people about whom this book is written. The country of Canada, as such, didn't exist until Confederation in 1867, although the term "Canada" or "New France" for French North America was used from the time of the earliest explorers in the sixteenth century (and is used as such in this book). Under the French Empire, the Saint Lawrence valley was divided into three separate "governments": Quebec, Trois-Rivières, and Montreal, while "New France" was the ever-changing North American territory that at one point encompassed North America west of the Mississippi and stretched from the Great Lakes to the Gulf of Mexico. With the arrival of the British in 1763, the territory around the Saint Lawrence valley and surrounding regions was named Lower Canada and then Canada East, eventually becoming the province of Quebec; the province of Ontario was known as Upper Canada and Canada West.

Those who did the work of science were not called scientists; prior to the mid-nineteenth century, they were known (as referred to earlier) as either natural philosophers, if they were interested in the physical sciences such as astronomy, chemistry, or physics, or as natural historians, if they studied what we now call the earth sciences or geoscience, including climate. The term "meteorologist" has been in use only since the beginning of the nineteenth century, although "meteorology" dates to ancient Greece.

Similarly, the term "climate change" can refer to a variety of different concepts. Today it's generally understood that the climate has always changed naturally, from the warm environment of the Carboniferous period of 300 million years ago to the depths of the great ice ages, with kilometers of ice in huge continental glaciers covering North America and Europe. The climate change most often discussed today is in part anthropogenic (human caused), related to the release of carbon dioxide and other greenhouse gases into the atmosphere. However, people can change local and possibly even regional climates by other means, for example, by changing the surface environment, as in a city. To be able to attribute any changes in climate to human agency, we must understand how the climate changes both naturally and anthropogenically; this is the biggest challenge in climatology now, as it was in colonial Canada.

References

Baumgardt, C., 1951: *Johannes Kepler: Life and Letters*. Philosophical Library. 209 pp.

Coates, C., 2000: *The Metamorphoses of Landscape and Community in Early Quebec*. McGill-Queen's University Press, 231 pp.

——, and D. Degroot, 2015: Les bois engendrent les frimas et les gelées: Comprendre le climat en Nouvelle-France. *Rev. Hist. Amér. Fr.*, **68**, 197–215.

Courtillot, V., and J.-L. Le Mouël, 2007: The study of Earth's magnetism (1269–1950): A foundation by Peregrinus and subsequent development of geomagnetism and paleomagnetism. *Rev. Geophys.*, **45**, RG3008, https://doi.org/10.1029/ 2006RG000198.

Desloges, Y., 2016: *Sous les cieux de Quebec: Météo et climat, 1534–1831*. Septentrion, 220 pp.

Feldman, T. S., 1990: Late enlightenment meteorology. *The Quantifying Spirit in the Eighteenth Century*, T. Frangmyr, J. L. Heilbron, and R. E. Rider, Eds., University of California Press, 143–177.

Fleming, J. R., 1990: *Meteorology in America, 1800–1870*. Johns Hopkins University Press, 264 pp.

——, 1998: *Historical Perspectives on Climate Change*. Oxford University Press, 194 pp.

Gerbi, A., 1973: *The Dispute of the New World: The History of a Polemic, 1750–1900*. University of Pittsburgh Press, 700 pp.

Glacken, C., 1967: *Traces on the Rhodian Shore*. University of California Press, 763 pp.

Golinski, J. V., 2008: American climate and the civilization of nature. Science and Empire in the Atlantic World. James Delbourgo and Nicholas Dew, Eds., Routledge, 153–174.

Janković, V., 2001: *Reading the Skies: A Cultural History of English Weather, 1650–1820*. University of Chicago Press, 272 pp.

Kington, J., 1980: Daily weather mapping from 1781: A detailed synoptic examination of weather and climate during the French Revolution. *Climatic Change*, **3**, 7–36, https://doi.org/10.1007/BF02423166.

Kupperman, K. O., 1982: The puzzle of the American climate in the early colonial period. *Amer. Hist. Rev.*, **87**, 1262–1289, https://doi.org/10.2307/1856913.

Lefroy, J. H., 1853: Remarks on thermometric registers. *Can. J.*, **1**, 29–31.

McCord, J. S., 1836: Scientific notebook. McCord Family Fonds, Papers P001-825. McCord Museum Archives.

Polkinghorne, J., 2011: *Science and Religion in Quest of Truth*. Yale University Press, 143 pp.

Slonosky, V. C., 2003: The meteorological observations of Jean-François Gaultier, Quebec, Canada: 1742–56. *J. Climate*, **16**, 2232–2247, https://doi.org/10.1175/1520-0442(2003)16<2232:TMOOJG>2.0.CO;2 .

——, 2014: Daily minimum and maximum temperature in the St-Lawrence valley, Quebec: Two centuries of climatic observations from Canada. *Int. J. Climatol.*, **35**, 1662–1681, https://doi.org/10.1002/joc.4085.

——, 2015: Historical climate observations in Canada: 18th and 19th century daily temperature from the St. Lawrence valley, Quebec. *Geosci. Data J.*, **1**, 103–120, https://doi.org/10.1002/gdj3.11.

Smallwood, C., 1866: Address to the members of the Montreal Natural History Society. *Can. Nat.*, **3**, 126–134.

Vogel, B., 2011: The letter from Dublin: Climate change, colonialism, and the Royal Society in the seventeenth century. *Osiris*, **26**, 111–128, https://doi.org/10.1086/661267.

White, S., 2015: Unpuzzling American climate: New World experience and the foundations of a new science. *Isis*, **106**, 544–566, https://doi.org/10.1086/683166.

Acknowledgments

Dedication: To Cynthia V. Wilson, whose pioneering work with early instrumental observations in Canada opened the way for others to follow, for her unfailing encouragement. The overall project of recovering, digitizing, and analysing historical weather and climate observations in Canada and this book, which came out of that work, have been ongoing for over a decade and a half now, and the list of people who have generously given their time and expertise grows longer every year.

I'd like to thank all the archivists who are always so enthusiastic about sharing their knowledge of the documents under their care. Josée Alexandre at the Paris Observatory Library first showed me Gaultier's manuscripts while taking a short break from other research one afternoon in 2000. Gordon Burr and Lori Podolsky at the McGill University Archives, Nora Hague and Stephanie Poisson at the McCord Museum Archives, Sylvie Dauphin at the Stewart Museum, and the archivists at the Galileo Museum were all welcoming and helpful guides to their collections. Anna Deuptsch-Staph, Morley Thomas, Roberta McCarthy, and Maria Latyszewskyj of Environment and Climate Change Canada all provided unstinting help in locating files, with special thanks to Anna for her personal and professional hospitality and encouragement.

Morley Thomas at the Meteorological Service Canada was unfailingly generous in sharing his wealth of material on the history of Canadian meteorology, collected over a lifetime and documented in his numerous works. Bill Hogg, Francis Zwiers, and Val Swail at Environment and Climate Change Canada supported the initial phase of data collection, and WMD Consulting undertook the initial keypunching of the McCord and Spark diaries. Timothy Slonosky undertook detailed research and organization of the information from archives across the country, while Eddie Graham discovered and copied the original logbooks of the Royal Engineers. Xuebin Zhang and Lucie Vincent of Environment and Climate Change Canada provided further support for data rescue in 2013–2014. Federico Ponari generously shared the articles and data he collected on the historical climate of Montreal. Thanks also to all those at Ouranos for their ongoing interest and discussions over the years.

Cary Mock consistently kept in touch, sending news and articles and sharing his finds and results during times when I was unable to keep up with current research. Jürg Luterbacher and his students Thanos Tsikerdekis and Lamprini Dergianli of the University of Giessen contributed much time and labour in the keypunching of the Royal Engineers and other historical diaries. Emeritus Professor André Plante from the Université de Montréal provided invaluable statistical advice. Lorne McKee of Natural Resources Canada provided information on the natural magnetic background of Saint Helen's Island and Toronto. Alan MacEarchen and Liza Piper of the Network in Canadian History and Environment (NiCHE) provided a valuable meeting place with their Canadian Climate History (2008) and The Climate is History (2014) workshops. Paula O'Connor provided the first ideas and encouragement to actually write this book.

The entire McGill Data Rescue: Archives and Weather (DRAW) team, Renée Sieber, Frédéric Fabry, Gordon Burr, Lori Podolsky, and Eun Park, along with student members Rob Smith, John Lindsay, Jeremy Cullen, and Pippa Bartlett continue to inspire research into the climate and climate observers of the past. Renée Sieber gave me an academic home in 2017 as a visiting scholar at McGill's Department of Geography, while Eun Park, Lori Podolsky, and Gordon Burr provided visiting scholar affiliation and working space for 2016 at the School of Information Studies and the McGill University Archives. Stéphan Gervais, MaryAnn Poulaten, Pascal Brissette, and Sherry Olson at the Centre for Interdisciplinary Research on Montreal at McGill University provided an encouraging environment during my year there as researcher in residence in 2015. Everyone listed above spent many hours in helpful discussions.

A huge thank you to all the citizen scientist volunteers who donated their time and effort to transcribing the historical weather records discussed here from digital photographs: Jennifer Dowker, Rose Dlhopolsky, Gilles Paquette, Ray Couture, Pat Fortin, Carolyn Verduzco, Kristin Davoli, Alana Cameron, Kyle Hipwell, Dan Manweiler, Lisa Woodward, Nancy Hagen, and Jason Ferguson. Thanks also to Denis Robillard and Yvan Dutil for pointing out memoirs and other historical works. Thanks especially to Jean-Paul Hacot and everyone who tested the DRAW site, our next step in rescuing historical weather data.

Immense thanks especially to Sarah Jane Shangraw at the American Meteorological Society (AMS), who made this book happen, for her enthusiasm and encouragement during a very long process, as well as to Production Manager Beth Dayton and Copy Editor Jordan Stillman, and to early readers Sam White, Lourdes Avilés, and Charles Schafer, whose comments much improved this book.

And none of this would have been possible without the support of my family: thanks to Tony and Clare Slonosky for babysitting, chauffeuring, and critiquing; Tim Slonosky for advice on historical matters; and the Hollands for cheerfully providing a place to stay in Toronto and the Hacots for the same in France. Most of all, thanks to my children Julien and Eloïse for so patiently visiting all those museums in different cities while their mother was delving into the archives, and to Hervé Hacot, for everything.

PART I

The Landscape

Scientists, Practices, and Theories

CHAPTER ONE

Territory, Networks, and Tools

In the early hours of the morning of January 10, 1859, a mass of cold, dense air swept south and eastward from the frigid continental interior of North America towards the Great Lakes and then spread up the Saint Lawrence valley. The temperature had first started to drop on the night of January 8. By six in the morning on January 9, Dr. Charles Smallwood (1812–1873) recorded a temperature of −29.9°F (−34.3°C) on the outside wall of his homemade observatory on Île Jésus in the Saint Lawrence River, just to the northwest of Montreal Island. The temperature would stay below 0°F (−18°C) for the next five days. At 3:00 p.m., the temperature outside William Skakel's (unknown–1863) house on Saint James Street in the heart of the city of Montreal had only risen a few degrees to −28°F (−33°C), the coldest afternoon reading by a margin of nearly 10°F (5.5°C) since his record began in 1820. By the morning of January 10, it was −43.6°F (−42.0°C) on Île Jésus. At 7:00 a.m., it was −33°F (−36°C) on Saint James Street. There were reports of mercury freezing in 15 minutes outside. As the wave of frigid air travelled eastward from an epicentre to the northwest, Louis-Edouard Glackmeyer (1793–1881) in Quebec City recorded −35°F (−37°C) at sunrise on January 10 and −40°F (−40°C) at sunrise the next day, while Colonel William Ward (dates unknown) and the officers of the Royal Engineers up at the Citadel of Quebec registered a minimum temperature of −38°F (−39°C) on January 10.

These were the coldest temperatures ever recorded in the Saint Lawrence valley.[1]

At Kingston, where the Great Lakes flow into the Saint Lawrence River upstream, and to the southwest of Montreal, Professor of Natural Philosophy James Williamson (1891–1900) and his students recorded a minimum temperature of −21.5°F (−29.7°C) on the ninth and −29.5°F (−34.2°C) on the tenth. Even farther to the southwest, on the northwestern edge of Lake Ontario, Professor of Meteorology George Kingston (1816–1886) at the Toronto Observatory recorded the lowest-ever temperature there: −26.5°F (−32.5°C). The cold continued eastward, reaching Halifax on the eleventh and Saint John's on the thirteenth, though moderated to a mere −9°F (−22.8°C). Astonishingly low temperatures were being recorded across the eastern part of the continent: New York recorded −9°F (−22.8°C), and Boston recorded −14.8°F (−26.0°C).

Only a few years earlier, scientists had been vigorously debating whether humans were warming the Canadian climate. Many people were convinced that the increasing clearing of the forests and cultivation of the Saint Lawrence valley and Great Lakes regions was causing the climate to change. Scientists in Canada with an interest in weather climate set up observatories, collected historical documents and weather diaries, and kept personal weather diaries to settle the debate over climate change that had been ongoing for the better part of a century, if not longer. In the Saint Lawrence valley region, the first long-term measurements of air temperature, winds, weather, rain, and snow started with the arrival of Jean-François Gaultier (1706–1756) and his thermometers in 1742. Measurements of the air temperature and weather have been recovered from locations in the Saint Lawrence valley for nearly every day since the 1790s.

These early Canadian observers kept their weather records during the time of the French Empire in the eighteenth-century Enlightenment, through the rise of the British Empire, to the Canadian Confederation in 1867, and then as part of the Meteorological Service of Canada from 1874 onward. Their observations were taken during the ongoing and increasingly significant transformation of the North American landscape from forest to farmland and then from farmland and orchards to nineteenth-century industrialization and twentieth-century urbanization.

---

1. The next coldest temperature for Quebec City, −36.7°C, was recorded on Wednesday January 15, 2015, at the Quebec Airport (record from 1940 to 2015). On the island of Montreal, −37.8°C was registered at the Dorval Airport on January 15, 1957.

These were, of course, times of great technological transformation. The development of the telegraph in the middle of the nineteenth century brought the rapid communication of current states of the weather in different locations, enabling the large-scale mapping of simultaneous weather observations: a snapshot of the atmosphere. This, in turn, gave rise to the possibility of storm warnings and weather forecasting, as, for the first time, the communication of weather information travelled faster than the weather itself. The revolutionary impact of the telegraph acted to bring meteorologists together and to encourage data exchange through the telegraph. British North America (Canada) became part of a network of cooperation between the United Kingdom, Canada, and the United States in collecting and communicating large-scale weather observations. Conferences on the international sharing of weather observations and the necessary standards to accomplish this were held in the 1870s. Just as weather systems transcend national borders, weather observations needed to become international in scope. Meteorology was no longer a science in which any country, let alone an individual, could stand alone. By the end of the nineteenth century, the era of individual observers communicating within minimal or slow networks had come to a close.

During the eighteenth and most of the nineteenth centuries Canadian weather observers were closely connected with some of the most powerful scientific institutions of their times. Networks were formed and driven by associations, such as the Académie Royale des Sciences and the British and American Associations for the Advancement of Science, the Scottish universities, and the clergy, as well as by military chains of command in the Royal Navy, the Royal Artillery, and the Royal Engineers. Canada during the eighteenth and nineteenth centuries was thus integrated into sophisticated global networks.[2]

After the end of the French regime in Canada, weather records were kept during the period of British military occupation in the second half of the eighteenth century by army officers, army engineers, and naval captains. Indeed, many observatories, such as that of the Citadel of Quebec or the Toronto Magnetic and Meteorological Observatory, were founded by the

---

2. "Canada" was a term employed by Jacques Cartier's Iroquois guides to the Saint Lawrence valley and possibly beyond, starting just downstream of Quebec City at the tip of the island of Orleans. The territory described by the word Canada evolved considerably over time, but the Saint Lawrence valley has been encompassed by the term Canada since the sixteenth century.

military in the nineteenth century. But on the whole, the British tradition of science in the late eighteenth and early nineteenth centuries depended on unpaid amateurs: individuals who funded their interest in science either through personal means or through their earnings in another profession, such as law, medicine, or the clergy.

In this book, we'll meet Jean-François Gaultier, médécin du roi (royal physician) in Quebec City; clergymen Alexander Spark (1762–1819) and John Bethune (1791–1872); teacher Alexander Skakel (1776–1846) and his brother William; Judge John Samuel McCord (1801–1865) and his father Thomas McCord (1750–1824); notary Louis-Edouard Glackmeyer; gardener Robert Cleghorn (1778–1841); Hudson's Bay Company Officer John Siveright (1779–1856); and physicians Charles Smallwood, Archibald Hall (1812–1868), and William Sutherland (1815–1875). In the Great Lakes region near Toronto, Royal Artillery Officers Charles Riddell (1817–1903), Charles Younghusband (1821–1899), John Henry Lefroy (1817–1890), and Professor George Kingston left letters conveying their experience setting up a scientific observatory in an outpost of the British Empire.

New France and British North America
France "claimed" the Saint Lawrence valley in 1534, though it would be more than seventy years before the permanent colony of Quebec City was established in 1608. This region would be a vital part of the French Empire for the next 150 years until the territory was ceded to the British after the Seven Years' War. The first permanent settlement of Quebec City, founded by Samuel de Champlain (ca. 1570–1635),[3] was on a northerly promontory overlooking the Saint Lawrence River approximately 500 kilometers from the Gulf of Saint Lawrence. Here, the Saint Lawrence narrows to less than 1 kilometer across, enabling control of water access to the heart of North America. Montreal, located on a large island archipelago in the middle of the Saint Lawrence River, was founded in 1642 as a religious and missionary settlement. Montreal Island's situation at the beginning (when heading upstream) of the navigable portion of the Saint Lawrence and at the confluence of the Saint Lawrence and Ottawa Rivers made it the hub of the fur-trading industry. If Quebec City controlled access to the continental interior, Montreal was the starting point of all voyages farther west, either southwest towards Lake Ontario, the

---

3. For an account of the intervening century between Jacques Cartier and Samuel de Champlain, see White (2017).

southern Great Lakes, and the Ohio and Mississippi Rivers or northwest towards the northern Great Lakes, Hudson Bay, and the North American prairies. In order to protect these vulnerable outposts, the military would be a major contingent of the population of Canada, at times composing 30% of the population under both the French and the British Empires.

The territory composing New France varied considerably over time, at one point stretching from the Atlantic coast to the Mississippi delta and at times including much of central North America. Particularly in the Saint Lawrence valley, New France was a largely agrarian society, though with 30% of the population living in cities it was much less agrarian than France in the same period. It was also a highly militarised society, with soldiers and military officials as a significant, if sometimes transient, portion of the population. The incessant, three-way endemic warfare between the British, French, and First Nations peoples led to the establishment of a permanent force in New France, the Companie Franche de la Marine (Moore 2012, p. 139). The fur-trade was also a significant economic force, and many young men engaged themselves as canoe guides or trappers with the mercantile fur-trading associations for a few seasons before settling down to farm. Nevertheless, education and science were not neglected in New France; the Jesuit boys' school, the Collège de Québec, was established in 1635 and offered a thorough education in the classics and sciences for the colonial elite. The college eventually became the Université Laval, the oldest university in North America. It was especially known for cartography and hydrography, vital skills for mapping the continent. The colonies were closely connected with the French government and its networks, including the scientific networks maintained by the state, either under the military control, such as the Ministére de La Marine, or the civilian Académie Royale des Sciences.

Three of the wars that pitted New France against New England in the eighteenth century were the War of Spanish Succession from 1701 to 1713 (known as "Queen Anne's War," in North America; Chartrand 2008, p. 28), the War of Austrian Succession, which lasted from 1744 to 1748 in North America, though it started in 1740 in Europe; and the Seven Years' War from 1754 to 1763. The Seven Years' War was officially declared in Europe in 1756, but battles and skirmishes had been endemic along the frontiers of the French and British Empires in North America for decades. The Battle of the Plains of Abraham, where in 1759 the British General James Wolfe's force defeated the defending French Army of General Louis-Joseph de Montcalm, and the surrender of Montreal to British General Jeffery Amherst in 1760 turned New France into a militarily governed area until the formal transfer of New France to Great Britain in the Treaty of Paris in 1763. The military remained important for several decades thereafter.

From 1763 until the Canadian Confederation in 1867 and after, Canada was an integral part of the growing British Empire. During this period, briefly, Canada and the American colonies under British rule were together known as British North America. In 1773, after the American Revolution, this large, conjoined colony was confined to Canada. Continued conflict in North America, first between the American colonists and Great Britain during the American Revolution and later between Canada, as a part of Great Britain, and the United States in the War of 1812, meant the military presence continued to be a dominant factor in Canada.

Enterprising settlers and tradesmen followed the British Army to North America and became, along with retired soldiers, among the first English-speaking settlers in Canada in the 1760s. (John Samuel McCord was descended from a trading family.) The first large influx of English speakers in Canada, however, were the Loyalists, Americans who remained loyal to the British Crown and had to flee their homes in the newly independent United States during and after the American Revolution. John Bethune was from one such refugee family; his father had been a minister and served as a chaplain in the British Army during the American Revolution. A second wave of largely British immigrants arrived in Canada after the end of the Napoleonic Wars in 1815. Charles Smallwood was one of these immigrants, arriving in Canada sometime in the 1830s.

Enlightenment Science
The historical period of these early weather observers spanned the century between Isaac Newton (1642–1727) and Charles Darwin (1809–1882), when the ideas of Edward Jenner (1749–1823) and Louis Pasteur (1822–1895), concepts of viruses and bacteria, were only starting to be understood. Scientists such as James Croll (1821–1890), James Hutton (1726–1797), Charles Lyell (1797–1875), and Louis Agassiz (1807–1873; Imbrie and Imbrie 1979) provoked a revolution in thinking about time and climate on Earth. Concepts of deep time, geological-scale climate change, and the Great Ice Ages were developed. Observing and recording the weather was very much a part of the growing scientific interest and approach to the natural world.

However, despite being referred to as "infant sciences," climatology and meteorology had been pursued for much longer. Regular, quantitative observations of the state of the atmosphere were made possible by the invention of the barometer by Evangelista Torricelli (1608–1647) in 1643, one of the first modern scientific instruments invented. Not only does weather play a significant role in shaping day-to-day life (particularly throughout temperate areas of the globe, typically in the midlatitudes, which host a succession of ever-changing skies, droughts and rains, cold snaps and heat waves), but our drive to understand it has played an integral, if for some time overlooked, role in the history and development of science. Indeed, trying to understand the weather and atmosphere instigated some of the most famous and defining experiments in the history of science, leading the way to discoveries in fluid dynamics by Blaise Pascal (1623–1662), atomic theory by John Dalton (1766–1844), and chaos theory by Edward Lorenz (1917–2008). In fact, Pascal established the scientific principle of verifying theory with measurement through what is widely recognised to be one of the first scientific experiments: the famous 1648 Puy de Dôme exercise in measuring the decrease in air pressure with height. Pascal and his brother-in-law Florin Périer's quantified observations, measured by the barometer, established the precedent that measurements of natural phenomena are required to support a theory. After Pascal, measurement as well as reasoning became part of the scientific method, and quantity and calculation were introduced into scientific experimentation (Plackett 1988, p. 183).

Mid-eighteenth-century France was the heart of the Enlightenment, the *Siècle de Lumières*. Both private and state patronage of science catalysed advances in almost every contemporary field: astronomy, botany, cartography, geology, metallurgy, and zoology. Since the late seventeenth century, the idea of unity of knowledge was in vogue, characterised by attempts to find out how the world worked by relating different phenomena to each other (this was not a time of the extreme specialization of science we know today). By the eighteenth century, scientific research in France was largely state sponsored, with salaried researchers appointed to state institutions, including the Académie Royale des Sciences (the Royal Academy, founded in 1666 by Colbert, Louis XIV's minister of finance), the Jardin du Roi (the King's Gardens), and the Observatoire Royale (Royal Observatory).

In 1699, the academy already had strict rules limiting the number of appointments and stipulating that members must live in Paris. Anyone wishing to publish in the *Mémoires and Histories de l'Académie Royale des Sciences*, the journals of the academy, had to be formally appointed at least as a correspondent of a full member. Technically, each correspondent was supposed to communicate only with their designated member, but this does not appear to have been very strictly adhered to. The resulting web (not a strict hierarchy) of scientific communication includes letters from naturalists in Canada to several different academicians.

The Jardin du Roi

Louis XIV gave over his gardens in Paris for the use of botany and later zoology; this became the Jardin du Roi (Royal Garden), where the connection between weather, medicinal herbs and plants, and disease was investigated. Doctors were interested in climate for two reasons: for the impact climate had on disease and health and for the importance of the climate for growing the plants used for remedies. Many physicians of the era attended lectures given by botanists at the Jardin du Roi, which also provided a point of contact between physicians and the botanists elected to the Académie Royale (Royal Academy). Jean-François Gaultier was described as being "neo-Hippocratic" for his belief in the link between prevailing weather and prevalent diseases. Medicine and botany and, as a subinterest, climate thus fell mostly under the discipline of "natural history." While Gaultier was studying in Paris, the Jardin du Roi was headed by the Jussieu brothers, a family of talented botanists: the eldest, Antoine (1686–1758), was a member of the Royal Academy of Sciences and a Professor of Botany at the Jardins du Roi, with an interest in medical botany. His younger brother, Bernard (1699–1777), also a member of the Academy, held the position of Demonstrator and designed the botanical gardens at Versailles, while the youngest brother Joseph (1704–1779) accompanied the 1739 expedition to Peru to measure the arc of the meridian at the equator; he remained there for the next three decades, sending botanical specimens, including quinine, to his brothers at the Jardin. The boundaries between disciplines among the Parisian scholars were not rigid, however, and many early scientists, such as Réné Antoine Ferchault de Réaumur (1683–1757), worked in both natural philosophy at the observatory and in natural history at the Jardin du Roi.

At the time, climate was also seen as having an influence on individual and national temperament, as climate was both an attribute and determinant of local environments.[4] This led to some concern about the effects on health and character of "transplanting" people from one climatic zone to another, as was happening with the European settlement of the New World (Gerbi 1973;

---

4. As this idea goes back to ancient Greece, there is considerable historical literature examining it. For recent works, see J. Fleming's *Historical Perspectives on Climate Change* (Fleming 1998). For an extended discussion of this "heresy" and how it applied to the Americas, see Gerbi (1973).

Zilberstein 2016). Specimens of plants and animals from around the world were continually being sent to the Jardin du Roi, including specimens from Canada sent by Gaultier's predecessor as king's physician, Michel Sarrazin (1659–1734), to his correspondents Joseph Pitton de Tournefort (1656–1708), Sébastien Vaillant (1669–1722), and Réaumur at the Jardin du Roi and Académie Royale. Sarrazin's work, and later Gaultier's, especially in collecting and identifying botanical specimens, was incorporated into the growing body of knowledge being collected and organised at the Jardin du Roi in Paris.

The Paris Observatory

The Observatoire de Paris (the Paris Observatory), founded in 1672, was initially meant to be the main meeting place and experimental laboratory of the academy. It proved to be too far away from the centre of Paris to be convenient for most academicians, however, and soon became known instead as a centre for instrumentation and the measurement sciences. The observatory measured not only the heavens but Earth as well; the state of the atmosphere with recordings of the temperature, pressure, and precipitation, measured as rain and melted snow, was meticulously recorded from the 1680s onward. State funding for astronomical observations was not only provided for the pure advancement of science, but also to sustain an imperial era dependent on sailing ships for commerce, war, and transportation to distant colonies. Astronomical observations were crucial for obtaining even a rough idea of a ship's position at sea (Barrie 2014). The observatory thus also became a centre for cartography and geodesy as well as a centre for magnetic research, again of considerable importance for navigation. Not only was it important to know the spatial changes in the magnetic deflection from true north, but it was hoped by scientists and navigators alike that measurements of magnetic disturbance might be a means of determining longitude. The more mathematical and observational sciences, such as astronomy, geodesy, and cartography, were thus under the purview of the observatory and generally came under the discipline of "natural philosophy."

Eighteenth-Century Weather and Climate Networks

The attempt to solve the puzzle of the weather and how it relates to the atmosphere was behind the early attempts at systematic, long-term, international coordination of scientific measurement. The scientists connected with Observatoire de Paris and Académie Royale des Sciences, particularly Jean Philippe (Giacomo Filippo) Maraldi (1665–1729) and Philippe de La Hire (1640–1718; Hire 1700), published measurements and observations from Paris beginning in 1688 and deduced general patterns in atmospheric pressure from observations collated from areas ranging from Lancashire to Malacca (Derham 1698; Maraldi 1709). Weather observations and measurements from as far afield as Massachusetts and Calcutta, as well as from around Europe from Uppsala to Danzig, were collected by William Derham (1657–1735) and published along with his own observations in the *Philosophical Transactions of the Royal Society* in the early eighteenth century. Later in the eighteenth century, the Mannheim Society, which was started by the elector of Mannheim in the 1780s and at its height included observations from some thirty-five locations across Europe and the North Atlantic region, was one of the first international efforts to standardise meteorological instruments and measurement practices (Kington 1980). Unfortunately, the society's operations were disrupted by the outbreak of the Napoleonic Wars, during which international communication came under suspicion as cover for espionage. International exchanges of weather observations were not seriously undertaken again for a century. While the Mannheim Society crumbled in the 1790s under the combined effects of bankruptcy and war, it remained an inspiration for future meteorologists over the next century (Anderson 2005).

The Paris Observatory itself was under the direction of successive generations of the Cassini family from its foundation until the French Revolution, starting with Jean Dominique (Giovanni Domenico) Cassini (1625–1712), invited to Paris by Louis XIV to become a member of the Académie des Sciences and director of the Paris Observatory, then still under construction (Connor 1947, p. 147). His nephew, Jean Philippe Maraldi soon joined him, and later so did Maraldi's nephew, Jean Dominique Maraldi (1709–1788). Cassini's son, Jacques Cassini (1677–1756), was best known for his geodesic work. Jacques Cassini's son, César François Cassini de Thury (1714–1784), continued his father's geodisc and geographical working on the map of France (Connor 1947, p. 152). Cassini de Thury's son, Jean Dominique Cassini, known as Cassini IV (1748–1845), had his career cut short by the French Revolution. Many records were lost during the revolution, but the observatory was eventually reestablished as the Observatoire de Paris, while the Jardin du Roi became the Jardin des plantes and the Académie Royal des Sciences was somewhat uneasily incorporated into the Institut de France.

The Development of Thermometers

The Paris Observatory was principally concerned with measurement, and it was here that much of the development of meteorological instruments took place. The astronomers themselves made daily recordings of temperature and pressure in their astronomical notebooks, while measurements of precipitation were made on the observatory terrace from 1688 until 1885. Unfortunately, many of the logbooks and original observations went missing during the revolution.

Experiments on the construction and calibration of thermometers were continued throughout the first half of the eighteenth century at the Paris Observatory. Communications between the King's Physician Gaultier in Canada and the observatory were of importance both as a record of measurement of temperature in Quebec and for furthering the development of thermometers calibrated to measure the coldest temperatures of a Canadian winter.

Delisle's Thermometer
Joseph-Nicholas Delisle (1688–1768) and Réaumur were among those who experimented extensively with the development of the thermometer: how to calibrate it, how to construct it, and how to make sure the readings of individual thermometers could be compared with each other. Throughout the late seventeenth and early eighteenth centuries, the development of a standardised thermometer with a fixed unit of degrees proved elusive. Liquid-in-glass thermometers work on the principle that certain fluids expand linearly with increasing temperature; the value of a degree is a certain fraction of this expansion. However, there was something of a vicious circle in trying to determine how to calibrate a thermometer, that is, in the procedure of finding universal fixed points to ensure that the 0° mark on one thermometer indicates the same temperature as the 0° on another thermometer. It is now known that temperature is constant during a phase change, so the melting point of ice and the boiling point of water can be used as universal fixed points, but it took some time before thermometers had become sufficiently accurate for these points to be established. Fixed points used at the time included the lowest temperature reached in a mixture of ice and various salts: −18° on the centigrade scale or the zero point used by Gabriel Fahrenheit (1686–1736). Another fixed point used for decades was the temperature of the cellars in the observatory, which are part of the catacombs and ancient quarries of Paris. As deep as 28 meters underground, the temperature in the cellars was remarkably constant, varying between 11.8° and 12.8°C over the course of two centuries (Esclangon 1944); for many years, the temperature was assumed to be constant at 12.8°C. As thermometers developed by French scientists based at the observatory, particularly those of Réaumur, became

more common, the value of the temperature of the observatory cellars (often abbreviated as "c.o." for "caves de l'Observatoire") became a widespread benchmark.

Delisle's temperature scale was a reasonably straightforward one, and his thermometers were constructed using mercury, with zero at the freezing mark and 150 at the boiling point of water. Delisle had been called to Russia in 1725 as part of the scientific delegation to Tsar Peter the Great, and much of his work on the thermometer was developed there. According to Middleton (1966), he was one of the first people to see mercury freeze in 1740 during his expedition to Siberia to observe the transit of mercury. Gaultier didn't record mercury freezing, but the cold temperatures he observed in the Quebec winters of the 1740s often went beyond the calibration of his instruments, with the mercury of his Delisle thermometers contracting into the bulb, leaving nothing to measure in the graduated tube. Delisle regretted this circumstance and concluded from this that the cold recorded by Gaultier in Quebec must have been colder than his observations in Russia, as the thermometers Delisle took with him to Russia and those Gaultier took to Quebec had the same scale. Several new thermometers with extended tubes were sent out to Gaultier from Paris throughout the 1740s, until finally in 1754 a specially constructed Réaumur thermometer "designed to capture the great cold of Canada" (Gaultier 1755) was graduated down to 54°Réaumur (R). Several of Gaultier's manuscript treatises and letters are conserved in the Delisle papers of the Paris Observatory, although they must have originally been sent to Henri-Louis Duhamel Dumonceau (1700–1782); they may have been passed on to Delisle in the course of his work on thermometers and so have been conserved in the observatory archives.

Réaumur's Thermometer

In 1730, Réaumur wrote a long treatise, published in the *Mémoires*, on the construction and placement of thermometers such that observations from different places and different instruments would be comparable. During this age of exploration and colonial expansion, the physical description and numerical observations of different climates was becoming of more interest to the natural philosophers, and reports from as far afield as Canada and Russia were being collected at the observatory. Unfortunately, Réaumur's essay is confused, and he seems to have lost track at various points as to whether the liquid in the thermometer he was referring to was water or diluted alcohol (*l'esprit du vin*). Consequently, the composition of the liquid that was actually boiling at his designated boiling point of 80° is unclear;[5] 80°R is now generally taken to be the boiling point of water (i.e., 100°C). However, as his thermometers were constructed with a liquid composed of a mixture of alcohol and water, and as the boiling point of ethanol is 78°C, there is a problem; most of the alcohol would have boiled off by the time the boiling point of water was reached. Therefore, 80°R must refer to the boiling point of the diluted alcohol mixture used, which is probably somewhere between 80° and 100°C, depending on the dilution of the alcohol. Moreover, the thermal expansion of alcohol and mercury aren't the same, making it difficult to calibrate the boiling point of the liquid used in Réaumur's original thermometer using a mercury thermometer. Réaumur's fame and authority nevertheless resulted in the Réaumur scale being widely used throughout Europe, and by the 1770s, the "Réaumur" thermometer seems to have described a mercury thermometer with the melting point of ice as the 0 mark and the boiling point of water (i.e., 100°C) as the 80°R mark.

By Gaultier's time, the use of the freezing point as zero seems to have been in widespread, if unofficial use, as nearly all of his observations are reported with respect to the freezing mark; a few are reported with respect to the temperature of the observatory cellars. Thus, when Gaultier set sail for New France in 1742, he brought with him not only an array of the latest scientific instruments but also connections to the most prestigious scientific network of the eighteenth century. The challenge of Canada's cold temperatures spurred even more developments in thermometers over the course of Gaultier's time there.

References

Anderson, K., 2005: *Predicting the Weather: Victorians and the Science of Meteorology*. University of Chicago Press, 331 pp.

Barrie, D., 2014: *Sextant: A Voyage Guided by the Stars and the Men who Mapped the World's Oceans*. William Collins, 348 pp.

Chartrand, R., 2008: *The Forts of New France in Northeast America 1600–1763*. 2008 Osprey Publishing, 64 pp.

Connor, E., 1947: The Cassini Family and the Paris Observatory. *Astronomical Society of the Pacific Leaflets*, **5**, 146–153. http://adsbit.harvard.edu/full/1947ASPL....5..146C

---

5. See Gauvin (2012). See also the photograph of the Universal Delisle thermometer in Fig. 1.2, where the boiling point of *l'esprit du vin* appears to be at 73°C.

Derham, W., 1698: Part of a letter of Mr. William Derham, Rector of Upminster, dated Dec. 6. 1697. Giving an account of some experiments about the heighth of the mercury in the barometer, at top and bottom of the monument: And about portable barometers. *Philos. Trans. R. Soc. London*, **20**, 2–4, https://doi.org/10.1098/rstl.1698.0002.

Esclangon, E., 1944: Sur les variations de la température des caves de l'Observatoire de Paris, et leur relation éventuelle avec celles de la chaleur interne du Globe. *Ciel Terre*, **60**, 99–101. http://adsabs.harvard.edu/full/1944C%26T....60...99E

Fleming, J. R., 1998: *Historical Perspectives on Climate Change*. Oxford University Press, 194 pp.

Gauvin, J.-F., 2012: The instrument that never was: Inventing, manufacturing, and branding Réaumur's thermometer during the Enlightenment. *Ann. Sci.*, **64**, 515–549, http://dx.doi.org/10.1080/00033790.2011.609073.

Gerbi, A., 1973: *The Dispute of the New World: The History of a Polemic, 1750–1900*. University of Pittsburgh Press, 700 pp.

Hire, P. D., 1700: Observations du baromètre, du thermomètre et de la quantité d'eau de pluie & de neige fondue qui est tombée à Paris dans l'Observatoire Royal pendant l'Année 1699. *Mém. Acad. Roy. Sci.*, **1700**, 6–9.

Imbrie, J., and K. Imbrie, 1979: *Ice Ages: Solving the Mystery*. Harvard University Press, 224 pp.

Kington J, 1980: Daily weather mapping from 1781: A detailed synoptic examination of weather and climate during the decade leading up to the French Revolution. *Clim. Change*, **3**, 7–36.

Maraldi, J.-P., 1709: Comparaison des Observations du Baromètre faites en differens lieux. *Mém. Acad. Roy. Sci.*, **1709**, 233–245.

Middleton, W. E. K., 1966: *A History of the Thermometer and Its Use in Meteorology*. Johns Hopkins Press, 249 pp.

Moore, C., 2012: Colonization and conflict: New France and its rivals, 1600–1760. *The Illustrated History of Canada*, C. Brown, Ed., McGill-Queen's University Press, 96–80.

Plackett, R., 1988: Data analysis before 1750. *Int. Stat. Rev.*, **56**, 181–195, https://doi.org/10.2307/1403641.

Smallwood, C., 1860: Contributions to meteorology reduced from observations taken at St. Martin, Isle Jesus. *Canadian Journal.*, **27**, 308–312.

White, S., 2017: *A Cold Welcome: The Little Ice Age and Europe's Encounter with North America*. Harvard University Press, 376 pp.

Zilberstein, A., 2016: *A Temperate Empire: Making Climate Change in Early America*. Oxford University Press, 280 pp.

CHAPTER TWO

Dr. Jean-François Gaultier

New France's Climatologist

Dr. Jean-François Gaultier was born in 1708 in la Croix-Avranchin, near Mont-Saint-Michel, Normandy, in northern France, a region with a long association with New France (Tesio 2005). Little is known of his early life, making it unlikely that Gaultier's family was one of particular importance or influence. He studied medicine in Paris, where he met and apparently made a good impression on many influential French scientists. He was connected to the Jussieu family, who at that time were prominent in the Jardin du Roi in Paris, where much of the work on natural history and botany in seventeenth-century France was conducted. It was thanks to the influence of the Jussieu brothers, Antoine (1686–1758) and Bernard Jussieu (1699–1777), that Gaultier was selected for the position of royal physician to Quebec City.

Gaultier was the first Crown official in New France to keep instrumental meteorological measurements and daily weather observations. He arrived in Quebec City, New France, in the autumn of 1742 and from then until his death in 1756 kept daily records of the air temperature, wind, and weather. Gaultier, the "perfect correspondent" (Wien 1999, p. 75) sent detailed annual reports of these observations to his correspondent at the Académie des Sciences, Henri-Louis Duhamel Dumonceau. Duhamel Dumonceau had a particular interest in trees and agriculture and kept his own series of weather observations in France as well as being involved in research on the use of wood and fibres in naval construction for the French Navy (Payne 2002; Duhamel 1994; Lamontagne 1960). He was appointed inspector general of the navy in 1739. Duhamel Dumonceau published four of Gaultier's reports in detail in the *Mémoires de L'Académie Royal*, from the period covering November 1742 to September 1743 until the period October 1745 to September 1746. A brief résumé for 1749 was also published. Further, manuscript letters from Gaultier survive in the Joseph-Nicholas Delisle collection of the Paris Observatory Archives for 1747–1748 (Gaultier 1748), while another letter survives in the Houghton Library at Harvard for 1754 (Gaultier 1754). Delisle worked extensively on thermometers and was interested in Gaultier's reports on the extremes of heat and cold to be found in Quebec.

During his time in New France, in the last decades of the North American French Empire, Gaultier was first and foremost a doctor. His primary responsibility was the health of the soldiers and marines sent to fight in New France, and the health of the inhabitants of Quebec City and the surrounding region. He worked closely with religious orders, mainly nuns dedicated to healing and nursing the sick. He was also a naturalist and botanist, the latter being especially important as plants were the principal source of medical remedies.

Gaultier's time as royal physician in Quebec coincided in part with the tenure of Roland-Michel Barrin de La Galissonière (1693–1756) as commandant general (de facto governor general) of New France, from 1747 to 1749. Gaultier greatly admired La Galissonière, a keen naturalist and friend of Duhamel Dumonceau (Chartrand et al. 2008). La Galissonière instituted a system of botanical and zoological specimen collection across New France for which Gaultier wrote a guide giving recommendations for collecting, preserving, and forwarding specimens to him in Quebec City.

In 1749, Gaultier acted as scientific host to Pehr Kalm (1716–1779); "one of the outstanding utilitarian Linnaean botanists" (Jarrell 1979), a friend and student of the great Swedish naturalist Carl Linnaeus (1707–1778), who developed the binomial Linnaean system used to this day to categorise all living things. During his visit to Canada, Kalm, a member of the Royal Swedish Academy of Science, had been commissioned to visit North America and report back on any plants that might prove useful to the agricultural economy of Sweden (Jarrell 1979). Kalm published his travel diary, a meticulous and accurate account of his visit to North America, which became an influential description of Canada in the last days of the

French Empire. Gaultier showed Kalm around his workplace, the hospitals, and led Kalm on botanical and mineralogical collecting expeditions. Kalm, visiting New France at what was perhaps the apogee of the colony's scientific development, attributed the "great zeal for the advancement of Natural History" to be found in New France to the effort of La Galissonière (Kalm 1880b, p. 4), who he describes as "another Linnaeus" (Kalm 1880b, p. 183). In Kalm's travel diaries, in turn, we find much information about Gaultier's daily life and role in the colony.

Medicine

At the time, one of the most prevalent theories of disease was that it resulted from an imbalance in the four humours of phlegm, blood, black bile, and yellow bile. Bleeding and purging to restore the balance of the humours were treatments commonly used by Gaultier. Another common theory arising from a revival of Hippocrates[1] was the idea that certain types of prevailing weather gave rises to certain illnesses. Because of this connection, medical associations undertook some of the earliest organised meteorological activity (Kington 1980). According to Duhamel Dumonceau, "M. Gaultier is inclined to agree with Hippocrates and against Sydenham, that large differences in air temperature could well cause . . . diseases" (Duhamel Dumonceau 1744, p. 137). Indeed, Gaultier often associated colds, bronchitis, and pneumonia to north and northwest winds, and he believed that sudden changes in temperature could bring on illness as the body did not have enough time to adjust the balance of the humours by sweating.[2] As part of his notes on the general character of each month, he added to each meteorological table a summary of the state of agriculture, the surrounding environment, and prevalent illnesses. In his description of the prevalent illnesses for April 1744, he noted that the malignant and putrid epidemic fevers of that month were caused by the alternating excessively cold and very mild weather. As a physician, Gaultier also performed autopsies in the colony, sometimes for criminal cases, as well as during epidemics, to learn more about prevalent diseases. In addition to attributing disease to the prevailing weather, Gaultier also on

---

1. 460–370 BC.

2. Proper sweating was also considered essential to good health, and the frequent changing of linen shirts was a sign of cleanliness and attention to health among the upper-classes who could afford frequent changing and washing.

occasion attributed illness and epidemic diseases to nutrition or, following a poor harvest, to malnutrition.

Kalm first met Gaultier, who he describes as a man of great learning in medicine and botany, when he visited the convent hospital in Quebec City (Kalm 1880b, p. 99). Kalm depicted Gaultier as visiting the hospital once or twice a day, stopping by each bed, and giving his prescription for the patient. Kalm recorded that the hospital was clean and in good order, with two large rooms containing two rows of beds on either side of the room, heated with a stove, and well lit. The beds were separated from each other, with a bedside table between the beds, clean sheets, and separate rooms for the gravely ill who needed special care. The nuns nursed the sick, who were mostly soldiers, especially in July and August when the ships arrived from France as well as in times of war. When space was not needed by the military, other patients were admitted (Kalm 1880b, p. 102).

Meteorology

Gaultier sent lengthy annual letters to Duhamel Dumonceau describing the weather, the harvest, the dates of flowering, sowing, harvesting, river freeze-up and ice breakup, and other weather-related phenomena as well as prevalent diseases. These letters were generally sent with the last ships sailing from Quebec to France before the freezing of the Saint Lawrence River, so each letter covers the weather from October/November of the previous year to September/October of the year the letter was sent.

In the 1740s and 1750s, when Gaultier was registering the temperature in Quebec, thermometers were still undergoing development, with a variety of scales and liquids being used, principally mercury, alcohol, and alcohol mixtures (see chapter 1). Gaultier found Réaumur's early thermometers in the 1740s difficult to use, not least because it wasn't graduated to measure the cold temperatures of Quebec's winter and in extremely cold temperatures the mercury contracted completely into the bulb, leaving no trace to be measured in the graduated stem of the thermometer. Gaultier was left to try to gauge the intensity of the cold by estimating by how much the mercury had contracted into the bulb of his thermometer. Instead, Gaultier's temperatures in the 1740s were mainly recorded using Delisle's mercury thermometer and scale. He described his thermometer and its setting in his letter of 1747–1748. It faced to the north and northwest, as these were, as Gaultier (1748) states:

the directions of the two coldest winds in Canada. The reasons we can give for this is that these winds pass over the lands of Hudson's Bay, and over several mountains which are covered in snow and have always been so.[3]

A new Réaumur thermometer was sent out to Gaultier sometime in the 1750s, which he described in great detail in his 1754 letter, although the page is somewhat water stained and difficult to read. Gaultier explained that this thermometer was specially calibrated for the "great cold of Canada," with 54° from the bulb at the bottom of the thermometer tube and the zero point and another 54° to the top of the tube. The liquid used was *esprit du vin* or an alcohol distillation with red colouring. This thermometer was exposed outside a window of an unheated room and was exposed to "the winds of the south, east, north-west and north" (Gaultier 1755).[4] Réaumur had also sent thermometers to Gaultier for him to use in experimental artificial incubators for hens' eggs, which Gaultier then dispersed among the inhabitants of Quebec for use in raising hens year-round (Gaultier 1930, p. 34, 38). By 1754, Gaultier, in thanking Réaumur yet again for sending out more thermometers, thought that there were now a sufficient quantity in Quebec to last a fair while, and let Réaumur know that there were by this time several communities employing the thermometers to regulate the incubators and provide a much needed source of protein for lean times of the year (p. 41).

In the opening sentences of his letter for 1747–1748, Gaultier explained some of his reasons for keeping a weather record and his hopes of its eventual use to posterity:

> I have no doubt that the few observations made to this day have already shown the great difference we remark in the temperature of the air and the weather of each season and each year is the reason for the differences in the harvests and in the prevalent illnesses, which proves the necessity of these sorts of observations, which will be infinitely more valuable to society than many others which might seem more curious or interesting, in that they would be able to contribute much to the progress of knowledge . . . on

---

3. The original text reads as follows: "qui sont les deux vents les plus froides au Canada. La raison qu'on peut donner c'est que ces vents passent par dessus les terres de la Baye d'Hudson et par dessus plusieurs montagnes qui sont toujours couvertes de glaces et de neige et l'ont toujours été."

4. The original text reads as follows: "t'au grand air et aux vents du sud, d'ouest, d'est, nord-ouest et du nord."

agriculture ... and posterity ... will know which illnesses to expect, and the remedies which best succeeded.[5] (Gaultier 1748)

While impressed by Gaultier's medical and botanical work, Kalm was less enthusiastic about Gaultier's temperature measurements. Kalm recounted that Gaultier had one of La Hire's thermometers (Kalm 1880b, p. 187),[6] which, as has been noted, wasn't graduated far enough to register extremely cold temperatures. Kalm considered that Gaultier's thermometer was badly placed but gives no further indications. In his extract of Gaultier's journal for 1745, Kalm records that the temperature of May 3, 1745, was −4° "according to the thermometer of Celsius or Swedish" (p. 188). Did Gaultier also have a Celsius thermometer? Or was Kalm converting Gaultier's temperatures to his preferred scale? From his travel diary, we know that Kalm did have his own Celsius thermometer with him in 1749 (Kalm 1880a, p. xiii).

Kalm noted that a Monsieur de Pontarron, a priest in Montreal, started a record of temperature observations in 1749 using a Réaumur thermometer, which was placed in an open window and so, in Kalm's opinion, rarely registered the coldest temperatures (Kalm 1880, p. 57–58). Pontarron found the coldest day of that year to have been January 18, 1749, when the thermometer reached −23°R, roughly −29°C. Here, Kalm heard for the first, but surely not the last, time the opinion that the summers had lengthened remarkably in Canada since the land had been cleared for agriculture, with summers starting earlier and finishing later. Winters, while shorter, hadn't lost any of their severity. Kalm also noted the magnetic declination in Montreal, whether using his own instruments or those of his clergy hosts is unclear (Kalm 1880b, p. 59).

---

5. The original text reads as follows: "Je ne doute nullement que les peu d'observations botanico-météorologiques qu'on a fait jusqu'ici n'ait déjà appris que la grande différence qu'on Remarque entre la température de l'air et les météos de chaque saison, et de chaque année ne soit cause de celle qu'on observe dans les différentes productions de la terre et dans les maladies qui règnent; ce qui prouve de plus en plus la nécessité de ces sortes qui seront infiniment plus avantageuse à la société que beaucoup d'autre qui paroissent plus curieux qu'intéressantes qu'elles pourront beaucoup contribuer au progrès des connaissances sur l'agriculture ... et a la postérité ... on saura ... quelles maladies qu'on pourra craindre, es les remèdes qui auront le mieux réussies."

6. Gaultier, however, described his thermometer as being of the Delisle type. Did Kalm confound the two or did Gaultier have a variety of thermometers of different construction and scale?

Gaultier's predecessor as royal physician in Quebec, Michel Sarrazin, did not himself keep meteorological observations, but his work in natural history and his relations with members of the Royal Academy and the Jardin du Roi set an important example of collaboration between naturalists in Canada. New France was often in a state of war, either overtly with the British or covertly in raids and skirmishes between the French settlers in Canada, the British colonists in the thirteen colonies, and their First Nations allies on both sides. This made scientific exploration and collecting specimens a sometimes hazardous occupation, as Sarrazin remarked (Rousseau 2013).

The governor of New France from 1747 to 1749, Roland-Michel Barrin de La Galissonière was himself a keen naturalist and put in place a network of naturalists at military forts across New France, that at the time extended from the Gulf of Saint Lawrence to the Great Lakes and down the Ohio and Mississippi River valleys to the Gulf of Mexico, to collect both botanical, zoological, and mineral specimens, following the guidelines set out by Gaultier at La Galissonière's request.[7] Letters from the commandant at Fort Niagara, Daniel-Hyacinthe Liénard de Beaujeu, to La Galissonière described the difficulty of finding suitable specimens, but Beaujeu assured La Galissoniére that "he would not give up" (Lamontagne 1960, p. 518). Kalm came across Gaultier and La Galissonière's specimen-collecting network as soon as he crossed the border from the British colonies into New France. At Fort Saint-Frederic on Lake Champlain, Kalm was met by the post commander, Paul-Louis de Lusignan, who promptly showed him Gaultier's handbook on the collection of plant, animal, and mineral specimens in New France, with comments not only on the types of specimens to be found in New France but also their distribution. Ordinary soldiers who diligently participated in the collection of plants, minerals and other "curiosities" (Kalm 1880b, p. 5) received promotion or other rewards based on their abilities. Special attention was to be paid to ascertain any medicinal qualities and to discover how the Native Americans used these "productions of Nature" (Kalm 1880b, p. 5). These were forwarded to Gaultier in Quebec City, who analysed them and prepared them for transport to Paris. This military network contributed considerably to the scientific description and cataloguing of flora and fauna at the

---

7. There were some thirty-nine French forts spread between Newfoundland, the Great Lakes and the Ohio river (Chartrand 2008, p. 6), with another forty-five stretching as far west as Lake Winnipeg and as south as New Orleans (Chartrand 2010, p. 7), though not all were operational at the same time, some were very short-lived, and many were fortified mission or trading posts with small garrisons.

Jardin du Roi and to the advances made by naturalists such as Georges-Louis Leclerc, comte de Buffon (1707–1788, known by his title, Buffon) and later Georges (Baron) Cuvier (1769–1832; Chartrand et al. 2008). These specimens were eventually organised into today's taxonomy system by Carl Linnaeus.

Towards the end of August (August 29), Gaultier and Kalm went on an expedition at the request of the colonial authorities to examine mineral deposits; a silver or lead mine was thought to exist near to Baie-Saint-Paul and Les Éboulements. Kalm was happy to join the expedition, as this enabled him to see more of the country and extend his voyage. They found iron; lead had been found in this region earlier on an expedition in 1739 (Kalm 1880a, p. xv). Kalm left Quebec City soon after his return from the trip to Baie-Saint-Paul, on September 11, 1749. Upon his return to Sweden, he and Linnaeus worked on classifying his Canadian samples. Kalm had hoped to publish a treatise on the Canadian flora, but this never came about.

Gaultier's scientific contribution to botany was acknowledged by Kalm and Linnaeus, who named the genus of native Canadian wintergreen berries after him. Gaultier's botanical discoveries in Quebec gave rise to a small scientific vocabulary based on his name, with 508 plant species in the Gaultheria genus (Boivin 1974).[8] Although in this book it's mainly Gaultier's meteorological observations that are of interest, they are far from being his only contribution to science (Holmes 2008).

An Era's Passing

As evidenced by Gaultier's contacts with the era's renowned scientists and by his contributions to the most influential publications of the eighteenth century, the science of New France, far from being marginalised, was integrated into the heart of the most powerful and influential scientific institutions of the time.

Gaultier died in 1756 during a summer typhus epidemic, thought to have been brought by soldiers from France. He had been depicted in a novel by William Kirby, written a hundred and thirty-five years after Gaultier's arrival in Quebec, as being "rich, generous, learned, and likable" (Boivin 2000) and as "the physician and savant *par excellence* of Quebec" (Kirby 1877, p. 112). He is described by Bernard Boivin as being "faithful to his duty, a pleasant personality . . . of varied talents . . . studying various fields at once and contributing to several new ones" (Boivin 2000).

---

8. See online at www.theplantlist.org/1.1/browse/A/Ericaceae/Gaultheria/.

The confusion and stresses of war prevented the installation of a replacement. Quebec City fell to the British in 1759, Montreal fell in 1760, and France permanently ceded the colony to the British in 1763. Many of the aristocrats and social leaders, including military and civil officers, returned to France when the colony became a part of British North America. The French scientific networks and connections to the colony of Canada faded as a new empire established itself along the Saint Lawrence valley.

References

Boivin, B., 1974: Jean-François Gaultier. *Dictionary of Canadian Biography*, Vol. 3, University of Toronto/Université Laval, accessed April 18, 2012, http://www.biographi.ca/en/bio/gaultier_jean_francois_3E.html.

Chartrand, L., R. Duchesne, and Y. Gingras, 2008: *Histoire des Sciences au Québec de la Nouvelle-France à nos jours*. Boréal, 535 pp.

Chartrand, R., 2008: *The Forts of New France in Northeast America 1600–1763*. Osprey Publishing, 64 pp.

——, 2010: *The Forts of New France: the Great Lakes, the Plains and the Gulf Coast, 1600–1763*. Osprey Publishing, 64 pp.

Duhamel, S., 1994: Henri-Louis Duhamel Dumonceau. *Cap-aux-Diamant*, **37**, 83.

Duhamel Dumonceau, H., 1744: Observations botanico-météorologiques faites à Quebec par M. Gautier pendant l'année 1743. *Mém. Acad. Roy. Sci.*, **1744**, 135–155.

Gaultier, J.-F., 1745 unpublished manuscript: Journal des observations météorologiques de M. Gaultier à Kebec. Observatoire de Paris Manuscript, Doc. 64-5-B, 103 pp. [Available from Fonds Joseph-Nicolas Delisle, Observatoire de Paris, 61 Avenue de l'observatoire, Paris 75014, France.]

——, 1748 unpublished manuscript: Journal des observations météorologiques de M. Gaultier à Kebec depuis le 1 octobre 1747 jusqu'au 1 octobre 1748. Observatoire de Paris Manuscript, Doc. 64-6-A, 71 pp. [Available from Fonds Joseph-Nicolas Delisle, Observatoire de Paris, 61 Avenue de l'Observatoire, Paris 75014, France.]

——, 1754 unpublished manuscript: Journal des observations météorologiques de M. Gaultier à Kebec 1754. Meteorological Collection, Houghton Library, MS Can 42(1), 17 pp. [Available from Houghton Library, Harvard University, Cambridge, MA 02138.]

——, 1755 unpublished manuscript: Observations on Quebec, 1744–1754. Meteorological Collection, Houghton Library, MS Can 42(2). [Available from Houghton Library, Harvard University, Cambridge, MA 02138.]

——, 1930. "Cinq lettres inédites de Jean François Gaultier à M. de Rhéaumur de l'Académie des Sciences." Edited by Arthur Vallée. *Mémoires de la Société Royale du Canada* 24: 31–43.

Jarrell, R. A., 1979: Kalm, Pehr. *Dictionary of Canadian Biography*, Vol. 4, University of Toronto/Université Laval, accessed 02/05/2016, http://www.biographi.ca/en/bio/kalm_pehr_4E.html.

Kalm, P., 1880a: *Voyage de Kalm en Amérique*. Vol. 1. Marchand, L.W., trans. Mémoires de la Société Historique de Montréal, septième livrasion. Impr. par T. Berthiaume, 168 pp.

Kalm, P., 1880b: *Voyage de Kalm en Amérique*. Vol. 2. Marchand, L.W., trans. Mémoires de la Société Historique de Montréal, huitième livrasion. Impr. par T. Berthiaume, 256 pp.

Kirby, W., 1877: *Le Chien d'Or, The Golden Dog, a Legend of Quebec*. Lovell, Adam, Wesson and Co, Montreal and New York, 678 pp.

Kington, J., 1980: Daily weather mapping from 1781: A detailed synoptic examination of weather and climate during the French Revolution. *Climatic Change*, **3**, 7–36, https://doi.org/10.1007/BF02423166.

Lamontagne, R., 1960: La contribution scientifique de La Galissonière au Canada. *Rev. Hist. Amer. Fr.*, **134**, 509–524.

Rousseau, J., 2013: Sarrazin, Michel. *Dictionary of Canadian Biography*, Vol. 2, University of Toronto/Université Laval, accessed Oct 8, 2014, http://www.biographi.ca/en/bio/sarrazin_michel_2E.html.

Tésio, S, 2005: De La Croix-Avranchin à Québec, Jean-François Gaultier, médecin du roi, de 1742 à 1756. *Annales de Normandie, 55 année*, **5**, 403–426.

Wien, T., 1999: Jean-François Gaultier (1708–1756) et l'appropriation de la nature canadienne. Edited by Jean Pierre Bardet and Réné Durocher. *Français et Québécois: le regard de l'autre*. Paris. 73–78.

CHAPTER THREE

Clearing and Cultivation

*Eighteenth Century Climate
Improvement Theory*

The idea that people might be changing the climate is one that has a long history in Canada as well as in the United States and Europe. Indeed, Canada was considered, in some respects, an ideal case study, as the changes wrought by European settlers in clearing the forests and turning the land to agricultural uses provided side-by-side examples of the "untouched" natural state of the climate together with the "improved" land and climate.

Jean-François Gaultier mentioned in his letter to Duhamel Dumonceau at the Académie Royale for 1744–1745 that "the inhabitants of Canada claim that the winters are not as cold as they used to be, this they attribute to the large quantity of land which has been cleared" (Duhamel Dumonceau 1746, p. 91). Later in the same letter, he clarified his thoughts:

> We had said that it was remarked in Canada that the spring started earlier, and the winter later than before[1] and that this change in the temperature of the air was attributed to the quantity of wood which had been cut down, and the quantity of land which was now cultivated: the Elders of the country assert also that before, the wheat harvest was only started on the 15th or 16th

---

1. *Anciennement* is the exact term.

of September, and that they were rarely of perfect ripeness; this observation gives rise to the hope that the more the land is cleared in Canada, the more the country will become fertile. (Duhamel Dumonceau, 1746, p. 96)

This is an example of the idea that deforestation and the cultivation of the land would lead to warming. It is presented by Gaultier as a common belief of the people rather than an academic theory stemming from the natural philosophers.

Gaultier wasn't presenting these observations as his own opinion but was careful to state that these came from the "habitants," a term used to refer to the settled farming population of the Saint Lawrence valley descended from the original French colonists. Gaultier distanced himself from this theory and displayed a certain scepticism towards the opinions he was relating: "the inhabitants *claim*," "this they *attribute*," and "the Elders of the country *assert*" (emphasis added). The scientist Gaultier was reporting local knowledge and beliefs without committing himself to a position or endorsing the cultivation–climate improvement theory himself.

Clarence Glacken, in his authoritative work *Traces on the Rhodian Shore*, ascribed the first expression of the question "is it possible for man to change the climate?" to Theophrastus, the pupil and successor of Aristotle writing in the third century BC. Glacken (1967, p. 130) wrote the following:

> These particular ideas seem to have occurred independently to many men in different periods. The effects of clearing on climate were discussed by Albert the Great in the Middle Ages; after the Age of Discovery discussions multiplied because so many travellers, especially in North America, observed or thought they observed, or believed the reminiscences of the old, that the climate became warmer when the woods of newly discovered lands were cleared.

Gaultier's letter of 1745 fits this description perfectly.

These two theories of climate and society described in Gaultier's work, that deforestation and cultivation lead to climatic change and warming, and that weather and climate have a significant influence on disease, were the principal ideas that motivated the eighteenth- and nineteenth-century observers in Lower Canada (present-day Quebec) to keep meticulous daily diaries of the weather over the course of the next 150 years. As can be seen from Gaultier's writings, the idea of anthropogenic (human induced) climate change has been part of Canada's philosophical ideas of nature since the beginning of European colonization.

Climate and Human-Induced Climate Change

The history of the "neo-classical theory of anthropogenic climate change" (Zilberstein 2016, p. 148), climatic improvement through cultivation, especially in the New World, was discussed in the second half of the twentieth century by Clarence Glacken (1967), Karen Kupperman (1982), Kenneth Thompson (1990), and James Fleming (1998). In the last decade, the climate improvement theory has been explored in detail by Brant Vogel (2011), Sam White (2015), Colin Coates and Dagomar Degroot (2015), and Anya Zilberstein (2016).

Kupperman describes the deforestation, cultivation, and climate improvement theory as perceived by early colonists of the New World, including early-seventeenth-century experience in Newfoundland. Kupperman, White, and Coates and Degroot discuss how the climate improvement through cultivation thesis developed over the course of the seventeenth century, with Samuel de Champlain expressing his belief in the early-seventeenth-century that the climate of New France could be improved by cultivation. Jesuit missionaries thought the climate was more variable in Canada because of the large extent of watery surfaces and because the land was uncultivated (Kupperman 1982, p. 1288; White 2015, p. 561; Coates and Degroot 2015, p. 213–214). Kupperman also describes how this idea affected the attempt to found a permanent colony in Newfoundland, citing the works of early colonists Richard Whitehouse and John Mason. Whitehouse, writing in 1620, "believed that cutting down the woods of Newfoundland would clear vapours and raise temperatures by allowing the sun's rays to penetrate more effectively and be retained" (Kupperman 1982, p. 1287), while Mason may have been one of the first to draw attention to the heat given out by homes and buildings in towns and cities as a source of warming climates. Brant Vogel has drawn attention to a refutation of the climate improvement by cultivation theory from Dublin, published in the *Philosophical Transactions of the Royal Society* in 1676, where the author points out that although the amount of cultivation had decreased in Ireland after the wars led to depopulation in the seventeenth century, the climate of Ireland had warmed also (Vogel 2011). Vogel and White also point to the long experience of Spain, where officials and missionaries since the sixteenth century had been gathering observations concerning the different climates in the Old and New Worlds. By the time Gaultier wrote of the beliefs of the Canadians in warming the climate through deforestation and cultivation, the idea had been a part of the colonial mentality for two centuries. Over the course of the eighteenth century, during decades of clearing the forests, draining swamps, and cultivating the land, the act of deforestation became both an ideal of progress through human exertion and intervention, and engendered a sense of ownership over both the land and the climate in the New World, whose improvements had been brought about by the investment of colonists' labour (Golinski 2008; Zilberstein 2016). Zilberstein detailed how the climatic improvement theory influenced ways of thinking about colonialism, cultivation, and climate in New England and Nova Scotia in the eighteenth century, with climatic improvement seen as a by-product of the general improvement brought about to land in northeastern North America by colonists.

In this chapter, three papers discussing the climate improvement theory, the evidence they present for or against climate change, and the physical mechanisms they invoke to explain the climate are examined. Although written in the second half of the eighteenth century, these papers continued to influence climate observers in Canada, particularly John Samuel McCord, in the nineteenth century.

The climate improvement theory appeared once more in the scientific literature two decades after Gaultier's remarks in the *Mémoires* in a letter written by Daines Barrington (ca. 1728–1800) in the *Philosophical Transactions of the Royal Society* in 1768 (Barrington 1768; Glacken 1967; Fleming 1998; Golinski 2008). Barrington wrote that he had "long entertained the notion that the seasons had become infinitely more mild in the Northern latitudes than they were 16 or 17 centuries ago" (Barrington 1768, p. 58) in an early reference to the idea of climate change over centuries. Barrington also used documentary sources as evidence of climate changes over time. The theory already had a long social history, although with Barrington the time scale had become longer as well. Instead of just considering the few decades or centuries since the colonization of the New World, Barrington looked at climate change in the Old World over centuries, based on the classical writings of ancient Greece and Rome.

While using the descriptions of Roman writers to demonstrate that the climate of Europe had warmed since Roman times, Barrington argued against the theory of deforestation and cultivation of the land changing the climate. Comparing the climate of the eighteenth century to that of the Romans around the year zero, Barrington "chiefly relied upon many of Ovid's[2] letters from Pontus[3] . . . in which he describes the effects of cold at Tomos" (Barrington 1768, p. 59).[4] Barrington chose Ovid's exile to Temisware

---

2. Roman poet, 43 BC to AD 17.
3. This is the area to the west of the Black Sea.
4. This is present-day Temiswolle in Romania.

because "the country being precisely in the same state that it was in the time of Ovid ... entirely excludes the common observation that the cultivation of a country will render the climate more temperate" (p. 64). In Barrington's view, the climate Ovid describes at Temisware (latitude 44°N) was similar to that of "the winter of Hudson's Bay" (p. 60), with the Black Sea sufficiently frozen over to walk on and wine distributed in frozen chunks;[5] Barrington contended that Ovid described a normal winter, not a particularly severe one, and that in many places snow never melted over the summer in Roman times. This seems unlikely, although Barrington did try to bring a certain level of critical analysis to the historical descriptions he quoted, acknowledging that "it may be objected that no credit is to be given to a melancholy poet [Ovid], of warm imagination" (p. 60) or to Virgil,[6] "unfortunately also a poet" (p. 62). Strabo, on the other hand,[7] Barrington considered to be more reliable, and his descriptions, along with those of Ovid and Virgil, point to a cold climate in central and northeastern Europe in Roman times (Barrington 1768).[8]

Barrington next considered Italy, where "the country being better cultivated, in the Augustan age, then it is now, should have made the ... air more warm than ... now" (Barrington 1768, p. 64). If the cultivation theory is correct, Italy should have been colder in the eighteenth century than in Roman times. Again turning to Virgil, Barrington was particularly impressed by the uselessness to contemporary Italians of advice on how to catch eels beneath river ice and takes this as proof that the climate of Italy must have been colder in Virgil's time than in Barrington's. Barrington finished his letter with comparisons of Hannibal's difficulties in crossing the Alps, because of snow and ice, compared to the ease with which later travellers, ranging from the armies of Francis I of France to consumptive patients in search of a cure, crossed the Alps in winter. His main conclusions were that 1) the climate had warmed since Roman times, but 2) it wasn't related to people changing the climate by deforestation and cultivation.

---

5. Considering that the freezing point of ethanol alcohol is −114°C, an estimate of the freezing point of wine with 10% alcohol is around −4°C; with 20% alcohol, it is about −10°C.

6. 70–19 BC

7. ~64 BC–AD 24

8. Anya Zilberstein in *A Temperate Climate* points out the irony in this, as Europeans from northwestern Europe would later apply the same arguments to North America as were applied to their countries by the Romans.

Climate Improvement in Pennsylvania:
Hugh Williamson and the Celtic Forests

Dr. Hugh Williamson (1735–1819), writing in Philadelphia a few years later, disagreed with Barrington's scepticism regarding land cultivation leading to climate change. Williamson was, rather, a fervent believer in the theory. On the one hand, he acknowledged that there may well be other causes to climate change than land clearing but went on to say that he couldn't "recollect a single instance of any remarkable change of climates, which may not be fairly deduced from the sole cultivation of the country" (Williamson 1770, p. 340). Williamson's paper, an early work describing in detail the climate improvement theory and the effect it was perceived to have had in northeastern North America, influenced thinkers such as George-Louis Leclerc, Comte de Buffon in France, who himself had a tremendous impact on scientific thinking in the eighteenth and into the nineteenth century. Williamson's paper and its influence is discussed in detail in both Glacken and Fleming's seminal works.

Williamson started off with the following words: "It is generally remarked . . . that within the last forty or fifty years there has been a very observable change of climate, that our winters are not so intensely cold, nor our summers so disagreeably warm as they have been" (Williamson 1770, p. 336). At this time in history, between the British Conquest of Canada in 1763 and the American Revolution in 1775, all of northeastern North America was part of the British Empire, with the thirteen colonies and Canada forming British North America. While similar to Gaultier's observations some thirty years earlier, Williamson now included milder summers as well as milder winters as a consequence of deforestation and cultivation.[9] Referring to Barrington, Williamson conceded that "it is not to be dissembled that their winters in Italy were extremely cold about seventeen hundred years ago" but suggested that it was in Poland, Germany, and France, covered with "wild extensive forest," from which "piercing North winds used to descend in torrents on the shivering Italian" (Williamson 1771, p. 341), that the true reason for the Roman coolness was to be found. The cultivation of northern Europe warmed the source areas of those north winds, rendering them less piercing. Likewise, the cultivation of North America would "prevent or mitigate those winter blasts which are the general origin of cold" (p. 341–342). According to Williamson, this wouldn't lead to correspondingly warmer summers, as

---

9. Kenneth Thompson (1980, p. 48) calls Williamson's reasoning "convoluted . . . difficult to follow and unconvincing."

a combination of air circulation and a "greater degree of heat being reflected by the plains" would temper summer heat (p. 343). The latter mechanism has been partially confirmed by modern radiation climate studies (Wilson 1975). Williamson concluded with the thought that "every friend to humanity must rejoice more in the pleasing prospect of the advantages we may gain in point of health [and] all the additional luxuries we may enjoy" (Williamson 1771, p. 344) if the land was cleared and cultivated enough to modify the climate.

Although Williamson's visions of the future climate as modified by man and hard labour seem somewhat utopian, his grasp of the principles of climatology was sound. He described the basic principles of the Hadley circulation cell and radiation theory, along with some remarks on the differing heat capacity of land and sea and the different absorption properties of different surfaces (today called albedo) in the first pages of his articles. He then dismissed the above remarks as "trite and general reasonings" (Williamson 1771, p. 339) and went on to discuss the impact of the Gulf Stream on the climate of northeastern North America. One of the biggest puzzles to these eighteenth-century scientists was why, given the redistribution of heat from the equator by the atmospheric circulation and the overwhelming influence of the sun's heat on the climate linking climatic zones to latitude, the temperature wasn't similar in all places with the same latitude. The fact that the climate of North America was much colder at more southerly latitudes than the climate of Europe was a problem that needed explanation.[10]

The works of Barrington and Williamson were noted by those interested in the subject in the English-speaking world, and when Canada was transferred to the British Empire in 1763, their ideas had influence in Canada also, although a suggestion of Gaultier's continuing influence will be seen in chapter 6. But even more than Barrington or Williamson, Theodore Mann (1735–1809), originally an Englishman writing in Flanders, would have a profound influence on the nineteenth-century Montreal climatologist John Samuel McCord.[11]

---

10. Benjamin Franklin's discovery of the Gulf Stream was an important step in recognising the role of ocean currents on climate. A discussion of the ideas of climate and latitude in North America is given in Kupperman (1982), White (2015), and Zilberstein (2016).

11. As can be seen in the appendix, Mann had an interesting career as an English Roman Catholic at a time when Catholicism was illegal in England. After leaving England, he became a priest and spent his scientific career on the Continent. When

Cycles of Climate? Theodore Mann and More Romans

An even more extensive late-eighteenth-century discussion on long-term climate change and the possible role of people comes from (Abbé) Theodore Augustin Mann's *Report on Great Frosts and Their Effects, Where We Try to Determine What to Think of Their Periodic Returns, and the Degree of Greater or Lesser Cold of our Globe* (Mann 1792a). Mann's 124-page memoir is divided into two parts, the first having been originally printed in French in 1789 in the *Physique de l'Académie Electorale de Mannheim*, the first international meteorological society organised by Karl Theodore, Prince-elector of Bavaria and Count Palatine (1724–1799) in southern Germany. [For more on the Mannheim Meteorological Society, see Cassidy (1985), Feldman (1990), and Kington (1980).] A handwritten transcript and partial translation was found in John Samuel McCord's scientific notebook.

References to Climate in Classical Literature

Mann's work is a much longer, and more detailed, examination of the classical literature for evidence of climate change than that which was provided by Barrington some twenty years earlier. He started off his treatise with a reference to Williamson's paper in the first paragraph. In the second, he observed that before looking for causes of climate change, it was first necessary to "prove by reliable authorities that these changes actually happened" (Mann 1792b, p. 1).[12] He thus divided his treatise into two parts: the first describing the evidence from classical authors of the temperature and landscape in "Gaul,[13] Germany, Pannonia,[14] Thrace,[15] Moesia,[16] Dacia,[17] and European

---

the British Empire was suddenly augmented by the Catholic territory of Canada, Mann was suggested as a potential bishop to Quebec. Though in the end this didn't happen, it's interesting to wonder how Mann would have applied his scientific scholarship to the newly British province of Canada and if, given Mann's connections in France and continental Europe, he would have managed to maintain or rebuild the scientific networks in Canada developed under the French Empire.

 12. All translations from the original French are by the author, with occasional reference to McCord.
 13. France
 14. Hungary and parts of Croatia and Serbia
 15. Bulgaria, northeastern Greece, and European Turkey
 16. Drobrogea
 17. Romania and Moldova

Scythia"[18] (Mann 1792a, p. 2). Unfortunately for posterity, he dismisses as "superfluous, in an academic treatise, to describe the modern state of the climate in these countries" (p. 2).

Mann started his proofs of climate change with Herodotus,[19] a native of today's Turkey. Herodotus mentions several times that in European Scythia the "insupportable" winters lasted eight months, the remaining four cool months being labelled summer.[20] Caesar,[21] Virgil,[22] Diodorus of Sicily, Ovid, Strabo, Pomponius-Mela,[23] Seneca,[24] Petronius,[25] Pliny the Elder (the naturalist),[26] Statius,[27] Herodian,[28] and Justinian[29] are all cited by Mann as mentioning "insupportable" cold over the various Celtic countries, from France to the Black Sea, between the latitudes of 44–50°N. According to Mann, they all reported such extraordinary and unequivocal severity of the climate as to make them credible witnesses.

Like Barrington, Mann also compared the descriptions of the "Celtic forests" of northern Europe in Roman times to Canada and the area around Hudson Bay: "The Ancients generally talk of the countries north of 55°N as being filled with lakes, marshes, ice, snow and fog, much as we talk of the countries surrounding Hudson's Bay" (Mann 1792a, p. 4). Even eighteenth-century Scandinavia was, he felt, too mild to compare to Roman Germany, although Lapland and Siberia were close analogues to the climate described by the Romans 2,000 years ago around the Rhine, the Danube, and other regions of eastern Europe. The classical writers mention the effects of cold throughout Europe, which were "absolutely unknown" in Mann's day (p. 11).

---

18. Poland and the Ukraine

19. 484–425 BC. Mann lumps all the classical writers together in this essay, without sorting them in any kind of chronology. Dates are given here to give an approximate idea of the time period these writers were referring to.

20. Not unlike a common description of Canada having eight months of winter and four months of tough sledding.

21. 100–44 BC

22. First century BC

23. Circa 112–35 BC

24. Which Seneca is unspecific in the test, presumably Seneca the Younger, 1 BC–AD 65.

25. AD 27–66

26. AD 23–79

27. AD 45–96

28. AD 170–240

29. AD 483–565

According to Mann, the first effect of the cold noticed by these classical writers, from the western shores of the Black Sea through to France, was that all the seas, lakes, and rivers were constantly frozen every winter, such that the barbarian invaders were able to use them as roads. The barbarians, along with their horses, baggage carts, and chariots, took hay along with them to throw on the ice and prevent the horses from slipping. Mann used these precautions as evidence that using frozen rivers as roads was a common enough occurrence that tricks to overcome the deficiencies of using frozen rivers as transport were devised. Diodorus of Sicily, Herodotus, Virgil, Ovid, and Strabo all report such conditions in the east, while Diodorus again, Seneca, Pliny the Younger, and several others report the same frozen lakes and rivers farther west. With a dubious tone, Mann mentioned that in a treatise on rivers attributed to Plutarch (AD 46–120) the rivers in the Ukraine froze even in summer, something that didn't happen in Mann's time even in Greenland. Mann described other incidents that indicated severe cold, such as river ice in the Danube crushing ships or (again) wine being distributed to soldiers in frozen blocks.

Snow, with which the Celtic countries from France to the Black Sea were filled in winter, is next on Mann's list of evidence of cold, as described by Diodorus of Sicily and confirmed by Florut and Petronius. Virgil and Ovid painted equally snow-filled pictures of winters along the Danube, especially from Bulgaria. As had also been mentioned by Barrington, Ovid even described continual snows, which persisted from one year to the next, along the Danube delta:

> When we compare this to the present state of France, Germany, Hungary, Romania, Transylvania, Bulgaria, and the Ukraine! The modern temperature in these regions has no bearing on the state in which they were 2,000 years ago, and these winter circumstances, which were then constant, occur but once a century, and are then regarded as extra-ordinary. (Mann 1792a, p. 6)

The limits of cultivation, as described by olive trees and grapevines, as well as the distribution of Roman fauna, were further proof of climate change. Many of the countries surrounding the Mediterranean, especially Spain, were "rich and fertile" in Roman times but were "rocky and scorched" Mann 1792a, p. 11–12) by Mann's time. Taken all together, these seemed to point to Mann that over the course of the centuries the climate had gone from one of "extreme humidity and cold towards drought and heat, that is, from one opposite to another. An effect so constant and uniform must have a cause which is not less so" (p. 12). Mann quoted Williamson directly:

Dr. Williamson upholds that the climate of America is continually moderating, and proves his assertion with many facts. He says that this change is distinct and tangible, and is one of the surest and most general that is known, as it manifests itself in all the countries of the world. It is, however, in direct opposition to the hypothesis of a famous physicist,[30] that the earth has been continually cooling since it was in the state of fusion which created it. (Mann 1792a, p. 13)

Having proven his point concerning the change in climate at length and to his satisfaction, Mann next turned his attention to the physical causes of climate change. He wrote that "I don't doubt that many different causes each contribute their part, some greater, some lesser, to produce this effect" (Mann 1792b, p. 16). Land-use changes, such as draining water, reducing lakes and marshes, and the removal of forests and cultivation of land, all moderated the temperature of the countries where they were undertaken. However, Mann also had a grander theory, which was that Earth is the combination of two opposing principles, humidity and phlogiston, and phlogiston was slowly destroying humidity.[31]

Mann stated, in the strongest terms possible, that in North America, the more the forests are cut down, the more the swamps are drained, and the more the land is cultivated, the more the climate is moderated: "This is a statement of fact, which cannot be doubted" (Mann 1792b, p. 18). He combined elements of the water cycle, atmospheric circulation, and phlogiston to explain this moderation of the climate. Once the trees are removed, and the soil worked for cultivation, the sun's energy can be absorbed in summer, and the phlogiston can escape in winter. As the forests of France and

---

30. Name not mentioned, but probably Buffon, one of the most influential naturalists of the eighteenth century. Buffon considered that the Earth had been cooling down for millennia, and only the efforts of humankind in cultivating and improving the Earth could temporarily halt this process (Zilberstein 2016, chapter 5).

31. Phlogiston was considered a material substance that consisted of heat, the substance that allowed combustion to happen, the flammable element in a substance. Thus, an object burned not because of the chemical reaction produced by the heat but because the phlogiston inside the object burned. This was an accepted theory of matter and combustion in much of the eighteenth century until Antoine Lavoisier (1743–1794) showed that sometimes materials gained mass when they burned through the process now known as oxidation. Lavoisier is widely considered the father of modern chemistry.

Germany were gradually cut down over the past 1,000 years, and sources of humidity were removed, they became warmer in winter, diminishing the pool of cold air that the north winds blew over Italy and Greece in ancient times. This explained the warming of these southern countries, even as their level of cultivation fell since the height of the Roman Empire. The forests and swamps of northern Europe had produced a pool of cold, dense air, which created a pressure gradient compared to the light, dry air of Italy, which in turn provoked the northerly winds that chilled Italy.

The vast forests and swamps of the interior region of North America also led to the production of cold air, the likely reason, in Mann's view, why similar latitudes in North America were so much colder than in Europe. Following Williamson, Mann considered that the summers should also become milder as the forests were cleared, with fewer intense heat waves as the air became lighter, purer, and more elastic with decreasing humidity.[32]

Climate Change since AD 588: Recurrences of Cold Winters

In what is quite possibly the first publication in the field of historical climatology, Mann's second paper, "Report on Extraordinary Frosts, Which Are Mentioned in History, from the Most Ancient Times until and including That of the Winter 1788 to 1789; Followed by Physical Considerations of Great Frosts in General"[33] (Mann 1792b) investigated the recurrence of harsh winters in Europe, inspired by the extremely cold winter of 1788–1789. He summarised nicely the main points of climate research:

> Our knowledge in Meteorology has not yet allowed us to know in detail the causes and combinations [of factors which contribute to such phenomena as severe winters], even less to determine if there is a constant and uniform pattern which causes regular recurrences of these extraordinary seasons, or if these are random effects of local causes, isolated and transient, which are subject to no return period, not being caused by any uniform and regulated combination of causes. Another object . . . would be to determine if the

---

32. The belief in the humidity of the Americas, and how this affected the construction of social theories of the Americas by Europeans, is described in detail by Gerbi (1973).

33. The original text is as follows: "Mémoire sur les Gelées extraordinaires, dont il est fait mention dans l'Histoire, depuis les temps les plus reculés jusque et compris celle de l'hiver 1788 à 1789; suivi de Considérations Physiques sur les Grandes Gelées en général."

quantity of heat in our sphere increases, gradually decreases, or always stays more or less the same. (Mann 1792b, p. 39)

Mann looked through historical documents, from antiquity to modern times, to determine whether the frequency or severity of cold winters, and by extension the climate, had changed over the space of twenty-three centuries. He wrote that he supposed another would have already compiled such a record of cold winters and searched long and hard without finding such a chronology. As he had previously covered the period from Herodotus to Justinian in the first report, he started the second one at the Justinian era. He didn't seem to apply any sort of differentiation as to time in the first treatise, comparing observations from Roman writers to modern observations without seeming to be overly concerned that the writers from antiquity spanned 1,000 years. In this second part, however, he listed dates of hard winters, starting in AD 588, with descriptions of phenomena showing the winters to have been extraordinarily cold: the freezing of rivers, depth of snow, presence of ice in harbours, and effects of the severe cold on plants, especially various trees. After an exhaustive tour of records of harsh winters in chronicles, histories, and scientific publications from across Europe, Mann concluded that 1) neither hard winters, nor other seasons nor meteorological phenomena, were subject to regular, constant, and periodic behaviour, and 2) none of the hard winters of recent times had attained the degree of cold of winters of classical times.

While most of Mann's collection of historical documents was for the period before the invention of meteorological instruments, especially the thermometer, he included measurements of air temperature as soon as they became available in the published literature. Starting in 1684 with the measurements from the spirit thermometer in Gresham College in London, Mann also used thermometer observations to compare degrees of cold during hard winters over the next 100 years, including the "Long Winter" of 1708–1709 and the "Great Frost" of 1788–1789 (Mann 1792a).[34]

A Modern Evaluation of the Eighteenth-Century Theory

The ideas of these eighteenth-century climatologists are contrary to modern received wisdom. The Roman period is considered to have been warm, even

---

34. William Derham (Derham 1709) had earlier described the winter of 1708–1709 as "The Great Frost."

culminating sometime after the year zero in what is called an optimum, an ideal and benign state of generally warm and pleasant weather. The eighteenth century, on the other hand, was an interlude in the middle of the climatic period known as the Little Ice Age. Was the perception, held more or less throughout the eighteenth century by Gaultier in Quebec in 1744, Barrington in 1768, Williamson in Philadelphia in 1770, and Mann in Brussels around 1790, that the climate was warming, true? From present-day studies of past climates, the eighteenth century is considered to be a slight lull in between the exceptionally severe seventeenth and nineteenth centuries, but it was still colder than the Medieval Warm Period immediately before the Little Ice Age or the twentieth century. Can our present view of the past three centuries and the view of the eighteenth-century writers, basing their ideas on the Roman writers of classical times, be reconciled?

The most comprehensive and detailed survey of past climates from historical sources over the past two millennia remains H. H. Lamb's magisterial work, *Climate: Past, Present, and Future*, particularly the second volume, *Climatic History and the Future*, with some updates in *Climate, History, and the Modern World* (Lamb 1995). Lamb considered the Roman period to have been warming since the first few centuries BC:

> The coldest conditions of the Sub-Atlantic climatic period [were] ... between about 900 and 300 BC ... A number of severe winters had been reported in ancient Rome at that time (with mentions of the Tiber being frozen in 398, 396, 271 and 177 BC) ... The next several centuries seem to have been an easier time ... The glaciers in the Alps seem to have been in retreat from about 300 BC to AD 400 [when] traffic over the Alpine passes continued even in winter ... Reports of winters in central and northwestern Europe ... indicate occasional great frosts that were prolonged and widespread over Europe seem not to come to prominence until between AD 359 and 565 and again between 664 and about AD 1000. (Lamb 1977, p. 424–425)

As Mann's study of great winter frosts only started in AD 558, it's impossible to find out if Mann and Lamb would have agreed in their assessments of the climate over AD 359–565, but Mann did list severe winters between AD 565 and 664: AD 568, 602, and 604.[35] Lamb took his collection of writings of the ancient Roman authors from Rothman (1848; referred to in Lamb 1977), who was, Lamb wrote, "especially impressed by the similarities be-

---

35. This could be a misdating of the winter of AD 602.

tween the descriptions of the classical writers of around 100 BC to AD 100 and the position in the eighteenth and nineteenth centuries AD" (p. 5). This is in stark contrast to those writing in the eighteenth century (including Gibbon; Gibbon 1789), who considered the climate of Roman times to have been appreciably colder than their own. The similarity of the climate of the nineteenth century and that of the Roman writers, as seen by Rothman, may have been in part the reason for the long held belief towards the end of the nineteenth century and early twentieth century that climate was constant since the end of the last Great Ice Age, with no change on the scale of decades to centuries. Lamb's detailed analysis of climatic changes over the past centuries and millennia did much to renew interest in shorter scales of climate change and variability.

Was the Roman Europe Colder than the Eighteenth Century? Results from Modern Studies

Unfortunately, there are few modern studies of the areas of Europe discussed above that include both the Little Ice Age and the Roman Warm Period. An interesting study on the wood used as part of the fortifications of Jerusalem during the revolt against the Romans and subsequent siege in AD 70 has concluded that this period was indeed colder than present in Israel and possibly other parts of the Mediterranean as well (Issar and Yakir 1997; Yakir et al. 1994). It was precisely "this cooler and more humid climate . . . which turned the provinces of the Levant and North Africa into the breadbaskets of Rome" (Issar and Yakir 1997, p. 104). Lamb also thought that a cold period in the beginning of the Roman times brought more winter rains to Greece and North Africa, making Carthaginian and Roman agriculture possible there (Lamb 1995). Lamb's view was that there was a change of climate going on throughout the Roman period, from cooler, wetter conditions in the early Roman years to a warmer, drier climate culminating around AD 400.

A study of marine sediments off the northwestern Iberian Peninsula also suggests a humid climate during the period from about 0 to AD 200, but Bernárdez and co-authors link this to warmer temperatures rather than colder ones, as do Issar and Yakir (1997) and the eighteenth-century authors (Bernárdez et al. 2008). Moving up the Atlantic coast to Ireland, an isotope record taken from a stalagmite suggests warmer temperatures during the Roman Warm Period slightly more than 2,000 years ago (McDermott et al. 2001). Ice core records from Greenland also show warmer temperatures in the first few centuries of the first millennium (Ljungqvist 2009, and references therein), as do marine sediment records from the North Atlantic (Bianchi and McCave 1999; Sicre et al. 2008; Patterson et al. 2010). Most of these records also show cooler temperatures in the eighteenth century, although as many of the North Atlantic temperatures show a warm period in the intervening Middle Ages, usually around AD 1000 or 1100, temperatures indicated by isotope analysis are not always much colder during the eighteenth century than they were at around the year 0.

Continuing northeastward around Europe, analyses of pollen in northern Europe, particularly Fenno-Scandinavia and Estonia, also seem to indicate a long warm and dry period around 2,000 years ago (Seppä et al. 2009). These records, as well as those of McDermott et al. (2001), show a long period of generally warm temperatures beginning as early as 1000 BC and in northern Europe lasting until about AD 1000. Seppä et al. (2009, p. 1531) write that "this period, which seems to peak at around 2,000 [years before present], has not been widely investigated or documented earlier in northern Europe. In central Europe this period appears as a circa 2,000-year long period of relatively high temperature and low humidity." Meanwhile, a colder period can be found from about 500 to 100 years ago: the Little Ice Age. Melting glaciers in the Alps during the warm summer of 2003 uncovered archaeological finds from the Roman period, which seem to be fairly conclusive evidence that the Alpine glaciers were smaller in Roman times than at nearly any other time since. Some medieval artefacts were also uncovered (Grosjean et al. 2007).

An overview of many of the proxy climate records, which go back to the year zero, has been compiled by Ljungqvist (2009). Ljungqvist writes that "The Roman Warm Period, an episode not much discussed in the literature but usually assumed to have occurred from c 300 BC to AD 300, is visible in many, but not all, of the records" (Ljungqvist 2009, p. 17). Apart from the records discussed above from Iberia, the Alps, and the Levant, there are no records from elsewhere in the Roman Empire included in his fairly comprehensive analysis, and none are included from eastern Europe or the Mediterranean region, the area from which Barrington, Williamson, and Mann were collecting the written evidence from Ovid, Virgil, Strabo, and many other classical writers.

It might be possible to reconcile all this conflicting evidence, especially concerning the century around the year 0, by considering the location of the warm records of the Roman period and the atmospheric circulation. From the records collected by Ljungqvist, it seems that those around the North Atlantic and Greenland have the warmest Roman period and a cooler eighteenth century. Cooler temperatures were widespread from the sixteenth to the eighteenth centuries, with the peak period of cold in the seventeenth century; the eighteenth century, while cold, was warmer than the seventeenth century over North America and Eurasia, so a perception of warming in the eighteenth century is justified (Ljungqvist et al. 2012). Moving farther east, tree rings from the Taymyr Peninsula in northeastern Siberia show a colder Roman period, and an eighteenth century that is cold but still warmer than the Roman period. Issar and Yakir, in their discussion of the cold Roman period in Israel, point out that during the Roman and Byzantine eras, northern China suffered from droughts and famine, and they speculate that this could be caused by the strengthening of the Siberian anticyclone (high pressure system) and a southward shift of the polar front and consequently more outbreaks of cold polar air over western Asia. Research in Japan also found cold, drought, and famine, and Sakaguchi, cited by Issar and Yakir (1997, p. 105), "maintained that this [Roman] period was even colder than the one which prevailed during the Little Ice Age." As postulated by Issar and Yakir (1997), a strengthened Siberian high would bring colder air down to eastern Europe and central Asia and possibly the eastern Mediterranean area as well, especially in the winter months, while the clockwise direction of the wind would bring more southeasterly and southerly winds blowing through western Europe and the Atlantic. The northerly winds passing over the Mediterranean and inland seas of Eurasia would account also for the increased moisture and good growing conditions of the Levant and northern Africa. The atmospheric circulation pattern implied by the curve of the circle of pressure isobars surrounding the Siberian high pressure system to the north would bring warm maritime air on the westerly winds from the Atlantic to northern Europe. Many of the proxy series reflect predominately summer conditions, while the climate change theories of the eighteenth century were focused on the changes in the winters, which further complicate the picture.

If we consider the strengthened Siberian high pressure system proposed by Issar and Yakir to be behind the cold of eastern Europe and central Asia, then both theories may be correct: the eighteenth-century scientists in thinking that winters were colder over eastern Europe and the Mediterranean regions during Roman times (though possibly not for western Europe) and the twentieth- and twenty-first-century scientists who discern a Roman Warm Period in the summers around the Atlantic region.

Lamb cites extensively from the work of Christian Pfister, who showed that in Switzerland the years from 1759 to 1763 and 1778 to 1784 had a "warm tendency," although the period from 1764 to 1777 was "notably cold" (Lamb 1995, p 214–216). However, the eighteenth century as whole in Europe seemed to show an erratic warming tendency, with "rather notable warmth of the European summers in and about the later 1740s and 1750s, and around 1780" (p. 194, 223). So while Barrington, Williamson, and Mann may have been writing in the middle of the several-century-long cold climatic period we know as the Little Ice Age, it's quite possible that they saw themselves living in a period of relative warmth, particularly compared to the late seventeenth and early part of the eighteenth century.

Does Clearing and Cultivation Lead to Warming? An Evaluation of the Climate Improvement Theory

The second thing to consider concerning the theories of Barrington, Williamson, and Mann is whether their reasoning was valid, as least as far as we can tell in the light of twenty-first-century knowledge. While large-scale changes in the atmospheric circulation, as yet themselves unexplained, might have been behind the changes during the Roman period, there was certainly not a continued increase in temperature from Roman times until the eighteenth century. During that time Europe had gone through several temperature swings, from the cold Dark Ages to the Medieval Warm Period and then through the Little Ice Age. Nevertheless, what would we expect to happen to the local climate when forests are cut down and replaced by agriculture over large areas?

The best way to try to answer this question is to consider the energy balance at the Earth's surface, particularly the difference in energy balance between a forested area and an agricultural field (Wilson 1975, 1973). Fortunately, some studies were conducted in the 1960s and 1970s at various locations in Quebec, particularly the Saint Lawrence valley area, where extensive deforestation and cultivation was occurring in the eighteenth and nineteenth centuries.[36]

---

36. The energy balance at a particular location on Earth's surface is usually evaluated in climatological studies by estimating what is called the radiation balance. Radiation is a form of energy and for climatological purposes is usually divided into two parts: shortwave radiation, which is essentially sunlight, and longwave radiation, which is heat. Shortwave radiation entering Earth's system is called insolation (an amalgam of incoming solar radiation) and can be reflected away, the amount of reflection depending on the brightness of the surface; the proportion of radiation

According to these studies, it would seem that cutting down forests and replacing them with fields should cool, not warm, the local climate, as forested areas retain more energy from sunlight than open fields do, especially when snow-covered in winter. Moreover, studies in Mont Saint-Hilaire near Montreal show that forests retain more energy, and have a higher gain of radiation, than grass surfaces. But the temperature at the surface doesn't depend only on how much energy is received but also on how it's used. Plants use sunlight to store energy and in the process evaporate water. Evaporating water takes energy that would otherwise go into heating the air. How does this affect the temperature of forests and fields?

If there is enough moisture, nearly all the available energy over grass or fields will go into the evaporation of water, with only 5%–15% heating the air. In the forest, on the other hand, only about two-thirds of the available energy goes into evaporation, with 33% going into heating the air—nearly double the maximum 15% over grass. The field studies show that in winter, forests converted 89% of the available energy into heat, with only 14% going into evaporation. On the other hand, over snow-covered fields only 3% of available energy, which was already only about 40% of that of forest, went into heating the air.[37]

The results of these studies seem to show that, on the local scale, deforestation would tend to cool the climate, especially in winter where there is usually snow cover. Not only is there less energy being absorbed by the surface to begin with, but less of that smaller amount goes into heating the air. The only exception to this would be in dry summers, when the lack of moisture in a field would lead to less evaporation and more energy left over for heating the air.

The History of Climate Theories in the Eighteenth Century
These three men—Williamson, Barrington, and Mann—hold an important place in the development of climatic theories. Though the deforestation theory would appear to be wrong, and only Williamson wholeheartedly

---

reflected away from a surface is known as that surface's albedo. Bright surfaces such as ice, snow, or thick clouds have a high albedo, and as they reflected away a high proportion of the incoming energy, they tend to be colder. Dark surfaces such as coniferous forests tend to have low albedos and retain more energy.

37. Values obtained from Wilson (1975). This effect has been recently confirmed by Bonan (2001), Bernier et al. (2011), and Arora and Montenegro (2011).

embraced it, their works are among the first systematic investigations of the causes of climate change. Mann's theory of climate change was more related to humidity as a cooling agent and phlogiston as a warming one, although he did think the climate was warmed with the removal of humidity in the form of trees and swamps. Barrington argued against the deforestation and warming theory and discussed the concept of long-term climate change over the course of centuries using documentary sources from the past as evidence that climate changed. Williamson outlined possible mechanisms for climate change, including how water and humidity can affect air temperature and how the winds and changes in land cover can cause climatic changes. Finally, Mann developed ways of thinking about climatic changes over long periods of time as being either 1) constant trends; 2) regular and cyclic, and therefore predictable, changes; or 3) the random effect of a variety of causes, which were not predictable. Mann was also the first to systematically collect information about past climates from historical records, introducing the field of historical climatology.

References

Arora, V. K., and A. Montenegro, 2011: Small temperature benefits provided by realistic afforestation efforts. *Nat. Geosci.*, **4**, 514–518, https://doi.org/10.1038/ngeo1182.

Barrington, D., 1768: An investigation of the difference between the present temperature of the air in Italy and some other countries, and what it was seventeen centuries ago: In a letter to William Watson M. D. F. R. S. by the Honorable Daines Barrington F. R. S. *Philos. Trans. R. Soc. London*, **58**, 58–67, https://doi.org/10.1098/rstl.1768.0009.

Bernárdez, P., R. González-Álvarez, G. Francés, R. Prego, M.-A. Bárcena, and O. Romero, 2008: Late Holocene history of the rainfall in the NW Iberian Peninsula—Evidence from a marine record. *J. Mar. Syst.*, **72**, 366–382, https://doi.org/10.1016/j.jmarsys.2007.03.009.

Bernier, P. Y., R. L. Desjardins, Y. Karimi-Zindashty, D. Worth, A. Beaudoin, Y. Luo, and S. Wang, 2011: Boreal lichen woodlands: A possible negative feedback to climate change in eastern North America. *Agri. For. Meteor.*, **151**, 521–528, https://doi.org/10.1016/j.agrformet.2010.12.013.

Bianchi, G. G., and I. N. McCave, 1999: Holocene periodicity in North Atlantic climate and deep-ocean flow south of Iceland. *Nature*, **397**, 515–517, https://doi.org/10.1038/17362.

Bonan, G. B., 2001: Observational evidence for reduction of daily maximum temperature by croplands in the Midwest United States. *J. Climate*, **14**, 2430–2442, https://doi.org/10.1175/1520-0442(2001)014<2430:OEFROD>2.0.CO;2.

Cassidy, D. C., 1985: Meterology in Mannheim: The Palatine Meterological Society, 1780–1795. *Sudhoffs Arch.*, **69**, 8–25.

Coates, C., and D. Degroot, 2015: Les bois engendrent les frimas et les gelées: Comprendre le climat en Nouvelle-France. *Rev. Hist. Amér. Fr.*, **68**, 197–215.

Derham, W., 1709: The history of the great frost in the last winter 1703 and 1708/9 by the Reverend Mr. W. Derham, Rector of Upminster, F. R. S. *Philos. Trans. R. Soc. London*, **26**, 454–478, https://doi.org/10.1098/rstl.1708.0073.

Duhamel Dumonceau, H. L., 1746: Observations botanico-météorologiques faites à Québec par M. Gaultier pendant l'année 1745. *Mém. Acad. Roy. Sci.*, **1746**, 88–97.

Feldman, T. S., 1990: Late enlightenment meteorology. *The Quantifying Spirit in the Eighteenth Century*, T. Frangmyr, J. L. Heilbron, and R. E. Rider, Eds., University of California Press, 143–177.

Fleming, J. R., 1998: *Historical Perspectives on Climate Change*. Oxford University Press, 194 pp.

Gerbi, A., 1973: *The Dispute of the New World: The History of a Polemic, 1750–1900*. University of Piuttsburgh Press, 700 pp.

Gibbon, E., 1789: *History of the Decline and Fall of the Roman Empire*. Vol 1. Strahan and Cadell, 456 pp.

Glacken, C., 1967: *Traces on the Rhodian Shore*. University of California Press, 763 pp.

Golinski, J., 2008: American climate and the civilization of nature. *Science and Empire in the Atlantic World*, J. Delbourgo and N. Dew, Eds., Routledge, 153–174.

Grosjean, M., P. J. Martin, M. Trachsel, and H. Wanner, 2007: Ice-borne prehistoric finds in the Swiss Alps reflect Holocene glacier fluctuations. *J. Quat. Sci.*, **22**, 203–207, https://doi.org/10.1002/jqs.1111.

Issar, A., and D. Yakir, 1997: Isotopes from wood buried in the Roman siege ramp of Masada : The Roman period's colder climate. *Biblical Archaeol.*, **60**, 101–106, https://doi.org/10.2307/3210599.

Kington, J. A., 1980: Daily weather mapping from 1781: A detailed synoptic examination of weather and climate during the French Revolution. *Climatic Change*, **3**, 7–36, https://doi.org/10.1007/BF02423166.

Kupperman, K. O., 1982: The puzzle of the American climate in the early colonial period. *Amer. Hist. Rev.*, **87**, 1262–1289, https://doi.org/10.2307/1856913.

Lamb, H. H., 1977: *Climatic History and the Future*. Vol. 2, *Climate: Present, Past, and Future*, Methuen, 835 pp.

——, 1995: *Climate, History, and the Modern World*. Routledge, 433 pp.

Ljungqvist, F. C., 2009: Temperature proxy records covering the last two millennia: A tabular and visual overview. *Geogr. Ann.*, **91A**, 11–29, doi:10.1111/j.1468-0459.2009.00350.x.

——, P. J. Krusic, G. Brattström, and H. S. Sundqvist, 2012: Northern Hemisphere temperature patterns in the last 12 centuries. *Climate Past*, **8**, 227–249, https://doi.org/10.5194/cp-8-227-2012.

Mann, T., 1792a: *Sur les Grandes Gelées I: Mémoire sur le changement successif de la température et du terroir des climates, avec des researches sur les causes de ce changement.* P. F. de Goefin, 1–38.

——, 1792b: *Sur les Grandes Gelées II: Mémoires sur les Gelées extraordinaires, dont il est fait mention dans l'Histoire, depuis les tems les plus reculés jusque & comprise celle de l'hiver 1788 à 1789; suivi de Considérations physiques sur les grandes Gelées en général.* P. F. de Goefin, 39–124.

McDermott, F., D. P. Mattey, and C. Hawkesworth, 2001: Centennial-scale Holocene climate variability revealed by a high-resolution speleothem $\delta^{18}O$ record from SW Ireland. *Science*, **294**, 1328–1331, https://doi.org/10.1126/science.1063678.

Patterson, W. P., K. A. Dietrich, C. Holmden, and J. T. Andrews, 2010: Two millennia of North Atlantic seasonality and implications for Norse colonies. *Proc. Natl. Acad. Sci. USA*, **107**, 5306–5310, https://doi.org/10.1073/pnas.0902522107.

Seppä, H., A. E. Bjune, R. J. Telford, H. J. B. Birks, and S. Veski. 2009: Last nine-thousand years of temperature variability in northern Europe. *Climate Past*, **5**, 523–535, https://doi.org/10.5194/cp-5-523-2009.

Sicre, M.-A., and Coauthors, 2008: Decadal variability of sea surface temperatures off North Iceland over the last 2000 years. *Earth Planet. Sci. Lett.*, **268**, 137–142, https://doi.org/10.1016/j.epsl.2008.01.011.

Thompson, K., 1980: Forests and climate change in America: Some early views. *Climatic Change*, **3**, 47–64, https://doi.org/10.1007/BF02423168.

Vogel, B., 2011: The letter from Dublin: Climate change, colonialism, and the Royal Society in the seventeenth century. *Osiris*, **26**, 111–128, https://doi.org/10.1086/661267.

White, S., 2015: Unpuzzling American climate: New World experience and the foundations of a new science. *Isis*, **106**, 544–566, https://doi.org/10.1086/683166.

Williamson, H., 1770: An attempt to account for the change of climate, which has been observed in the middle colonies in North-America. *Trans. Amer. Philos. Soc.*, **1**, 272–280, https://doi.org/10.2307/1005036.

Wilson, C. V., 1973: *The Climate of Québec: Part 2. The Application of Climatic Information.* Climatological Studies Series, No. 11, Environment Canada, 110 pp.

——, 1975: *The Climate of Québec: Energy Considerations.* Climatolgical Studies Series, No. 23, Environment Canada, 120 pp.

Yakir, D., A. Issar, J. Gat, E. Adar, P. Trimborn, and J. Lipp, 1994: ^{13}C and ^{18}O of wood from the Roman siege rampart in Masada, Israel (AD 70–73): Evidence for a less arid climate for the region. *Geochim. Cosmochim. Acta*, **58**, 3535–3539, https://doi.org/10.1016/0016-7037(94)90106-6.

Zilberstein, A., 2016: *A Temperate Empire: Making Climate Change in Early America.* Oxford University Press, 280 pp.

CHAPTER FOUR

British American Weather Observers to 1830

The end of the French Empire in Canada came soon after Gaultier's death; the Battle of the Plains of Abraham in 1759 put Quebec City under British control. The fall of Montreal in 1760 and the ceding of the Canadian territories to the British in the Treaty of Paris of 1763 brought an end to the French North American Empire and a new wave of military officers, colonial officials, and immigrants to the Saint Lawrence valley. The American Revolution and War of 1812 brought further waves of refugees, immigrants, soldiers, and sailors. The period immediately after the British Conquest was a difficult period in Canada. The British armies under General James Wolfe had devastated the countryside and reduced much of Quebec City to rubble during the months-long bombardment of the city over the summer of 1759. Many of the French elite and government officials returned to France by British decree, and the country was under military occupation for some time, leaving little opportunity for scientific endeavours outside of the British military. Years of poor harvests, blockades, and the necessity of supplying the ever-increasing number of troops had strained the colonies' food supplies even before Wolfe's destructive summer campaign of 1759. No new *médécin du roi* had been sent to Quebec to replace Gaultier after his death in 1756.

A number of British officers kept records referred to by contemporary writers, and there are references also to records kept by Louis de Bougainville (1729–1811), aide-de-camp to French General Montcalm during the Seven Years' War. Few original records remain from that period, and few systematic, long-term, land-based weather diaries for the Saint Lawrence valley have been discovered for the period between 1754 and the early 1790s (Lambert 1810; Landmann 1852; Lefroy 1853). There are some fragmentary observations for 1765–1766 by a Captain Alex Rose (Rose and Murdoch 1766), and daily temperature and weather in Montreal during the winters months between 1776 and 1778 recorded by a Mr. Barr (Barr and Saunders 1778) were published in the *Philosophical Transactions of the Royal Society*. Weather observations were certainly made, some by military officers, in the second half of the eighteenth century.[1] It's not until nearly the end of the eighteenth century, however, that Scottish Presbyterian minister Alexander Spark left a reliable record of past temperature, pressure, wind, and weather from 1798 to 1819. At around the same time, from 1813 to 1826, the McCord family also kept weather and temperature diaries in Montreal. These early British American weather observers, Alexander Spark, Thomas McCord, and his son John Samuel McCord, started a new era in Canadian climatology.

The late eighteenth century was a time of explosive growth in new knowledge, with the assimilation of information from the botanical to the geological leading the rapid developments in natural history. With the mid-eighteenth-century wars over, France and Britain turned their attention to exploration and discovery. Several military officers who were in Canada during the Seven Years' War went on to become famous in the history of exploration and science, particulalry James Cook and de Bougainville. Cook's surveys of the Saint Lawrence River were of critical importance in helping the British invasion fleet navigate the difficult Saint Lawrence River. This was the era of the "voyages of discovery," with Cook, Joseph Banks, and Bougainville opening up the South Pacific to European discovery (Holmes 2008). Towards the end of the century Alexander von Humboldt (1769–1859), inspired by Cook and Bougainville, set off on his famous voyage to South America. The results of Humboldt's scientific explorations would, in

---

1. George Landmann (Landmann 1852) made reference to weather observations kept in Quebec City, and James Thompson, a military engineer, also kept a weather diary (Chapman and McCulloch 2010; Desloges 2016). Copies of daily temperature kept in Quebec City in the early 1790s are conserved in John Samuel McCord's scientfic papers.

turn, influence an entire generation of scientists in the nineteenth century, when Humboldt had a stature and reputation similar to that of Einstein in the twentieth century.

Humboldt's concept of a unifying principle in nature would help fuel the nineteenth-century "magnetic crusade," which led to the founding of several observatories of geoscience, including meteorological observatories, across the British Empire (see chapter 7). Humboldt pioneered the graphical representations of geographical relationships between various elements, such as elevation and vegetation, or temperature and bodies of water. His innovative use of isotherms, lines linking together locations with the same temperature on a map of the world, also had a profound influence on the development of nineteenth-century climatology. Most of all, his unifying vision of the "connection existing among all phenomena" (Humboldt 1997, p. 50) and the "portions [of knowledge] which have long stood isolated becoming gradually connected" (p. 163) provided a new way of pursuing scientific investigations, especially in the geosciences, which was both detailed and quantitative yet ordered into a coherent whole. The increasing use of maps and other figures, such as the wind rose developed by Robert FitzRoy (1805–1865), marked a new step in the study of the weather by compressing large amounts of otherwise indigestible observations into beautifully summarised and easily assimilated images. Indeed, most technical science writing today seems to consist largely of explaining the information behind the figures.

Some differences emerge in how science was carried out under the French and British regimes. France, as we saw, had institutionalised many aspects of science and research, with the Royal Academy of Science (Académie Royale des Sciences) appointing state-funded positions, and the Royal Observatory (Observatoire Royal de Paris) and the Royal Garden (Jardin du Roi) also supporting scientific research. Scientific endeavours during the French Ancien Régime, before the Revolution of 1789, were part of the colonial enterprise and were encouraged and supported by state funding (McClellan and Regourd 2000). The French were systematic in surveying their new colonies, including the landscape and weather (Kupperman 1982, p. 1271). Towards the end of the Ancien Regime, the state founded the Société Royale de Médecine (the Royal Society of Medicine) to collect and analyse meteorological records kept by doctors (Kington 1980).

In Britain, on the other hand, most scientific research and data collection was private. In many cases, the practice of natural history and recording of observations was restricted to those with both the leisure time and financial means to invest in instruments, often gentleman amateurs or those with professions such as military officers, clergymen, or lawyers (Golinski 2007; Zilberstein 2016; Janković 2001; Shapin and Schaffer 2011). One of the few publicly supported scientific positions in the United Kingdom was that of the astronomer royal. The structure of the scientific community in the British Empire was centred around the Royal Society of London, with the *Transactions of the Royal Society* as the main venue for exchanging ideas and publishing articles. Details of the links to be found in New England and Nova Scotia between recently arrived settlers, many of them Loyalists who were forced to leave their home in what became the United States after the American Revolution, and the naturalist circles in Great Britain, particularly the Royal Society, are described by Zilberstein (2016).

Meteorologists in Britain at the beginning of the nineteenth century, however, were not happy with the poor quality of the Royal Society weather observations, which were considered by pioneering meteorologists Luke Howard (1772–1864), who kept observations in London and was the first to systematically classify clouds, and John Frederic Daniell (1790–1845), "arguably the leading meteorologist of the day" (Janković 1998, p. 24), as a disgrace in 1823. A short-lived attempt at a meteorological society was organised by Howard, Daniell, and others.

The Hudson's Bay Company and the Royal Society

The Hudson's Bay Company (HBC) and the Royal Society (RS) of London cooperated to organise scientific expeditions on several occasions. The Royal Society Archives contain a number of short records, usually lasting a season or two, from the explorers and factors[2] of the Hudson's Bay Company, such as the weather diaries kept by David Thompson and Peter Fidler (C. Wilson 2009, personal communication; Binnema 2014; Wilson 1985c). The Royal Society and the HBC worked together on the scientific exploration of northern North America, during William Wales's (1734–1798) sojourn at Fort York as part of the worldwide, internationally coordinated scientific expedition to observe the transit of Venus across the face of the sun in 1768 from as many locations around the world as possible in order to calculate the distance from Earth to the sun (Wulf 2012). John Henry Lefroy's magnetic expedition to the Canadian Northwest of 1843 was a similar expedition of scientific exploration undertaken with the help of the HBC (Binnema 2014). Other eighteenth-century naturalists and weather

---

2. Factors were the post managers in the Hudson's Bay Company structure.

observers associated with the HBC include Thomas Hutchins and Samuel Hearne (Houston et al. 2003).

In what is now Canadian territory, explorers and fur-traders for the commercial empires of the HBC and Montreal's North West Company were pushing ever farther north and west towards the Rocky Mountains and the Pacific coast in the late eighteenth century. Alexander Mackenzie reached the Beaufort Sea in the Arctic Ocean in 1789 and the Pacific Coast by way of the Bella Coola River in 1793 in the first crossing of the North American continent north of Mexico. Between the land and sea explorations, the boundaries of North America were becoming known and charted.

The HBC was interested in having meteorological observations recorded by the factors, fur-traders, and explorers; the HBC hoped it could keep down the costs of supplying the outposts in northern Canada with food by cultivating staples and garden produce at the fur-trading posts. Again, there was some confusion caused by the idea that climate was primarily dependent on latitude: York Factory, on the western shore of Hudson Bay, was at 57°N, about the same latitude as Dundee in Scotland, but their climates were very different (Wilson 1985c). According to C. V. Wilson, after 1814 "a high priority was now to be given to the careful observing and recording of the weather and its impact during the active season, with a view to increasing local food production at the Posts" (Wilson 1985b, p. 195). The Royal Society advised the HBC on meteorological instruments, observations, and experiments from the second half of the eighteenth century onward. "A tradition of careful meteorological observations was established [during this] period when the RS was interested in the design and calibration of thermometers, and in regulating, observing and recording procedures, and the HBC was actively involved" (Wilson 1983, p. 154).

Dozens of meteorological manuscripts from explorers and others in connection with the HBC are held in the Royal Society Archives (C. Wilson 2003, personal communication). Some of these records found their way into McCord's collection of weather records, most notably that of John Siveright, who kept a detailed record at Saint Mary's Falls (Sault Sainte Marie) and Fort Coulonge on the Ottawa River until 1833. Many of the archival weather journals, exploration diaries, and daybooks of the Hudson's Bay Company's explorers, fur-traders, and post factors have been carefully examined for evidence of climate change, from the pioneering work of Cynthia Wilson on the records of the Hudson's Bay Company posts on the eastern shore of Hudson Bay to Heather Tomkins's content analysis of the descriptive HBC journals in the southern Yukon (Wilson 1985a, 1988; Tompkins 2009).

British North America in the Saint Lawrence Valley

The new British government was primarily concerned with surveying and mapping the region, building supply lines and roads, and rebuilding the cities. Twenty years after the Conquest of Quebec, the continent erupted once more into war with the American Revolution, and Montreal was held under siege by the American Army. Tensions between British North America and the United States rose again during the Napoleonic Wars, culminating in the War of 1812 between Britain and the United States. The war ended with a stalemate and an imposed settlement in 1815. All this led to a considerable military presence in Canada, with major fortifications in Quebec City and south of Montreal along the Richelieu River, leading to Lake Champlain, and in the Chateaugay River valley. A major battle was fought at Chateaugay, where an invasion force from the United States was repulsed on October 25–26, 1813. On May 31, 1814, McCord recorded in his diary "Cloudy NE fresh rain PM. Troops arrived in 21 hours from Quebec," a reference to the British reinforcements arriving from Europe on their way to the battle arena of Lake Champlain (McCord and McCord 1826).

There was considerable interaction in the colonies of British North America between the professional classes, such as clergymen, lawyers, and doctors interested in natural history, and the military officers stationed in the new British colony. The officers and professional elite of Quebec City organised the Literary and Historical Society of Quebec in 1824, whose transactions have proved to be a valuable source of information about the climate in the early nineteenth century, while the Natural History Society of Montreal was established in 1827 (chapter 5). Both of these institutions would play an important role in organizing and transmitting meteorological observations in early nineteenth-century Canada. Before the establishment of organised science, however, personal diaries kept by individuals with an interest in the weather remain our main source of climatic information. One of those individuals was Alexander Spark.

Alexander Spark and the Influence of the Scottish Enlightenment

Alexander Spark was a clergyman of the Church of Scotland. He first came to Canada in 1780 and worked as a tutor in a Quebec City school. Spark had been recommended for the post as he was "an excellent mathematician, and a sensible, discreet young man with much practice in teaching" (Campbell 1887, p. 241). After three years of teaching classics, geography, and mathematics in Quebec City, Spark decided to become a Presbyterian minister and returned to Scotland for training in 1783.

The time when Spark was studying in Edinburgh was an intense period of the Scottish Enlightenment. Edinburgh had become one of the intellectual centres of Europe, known particularly for its medical schools, development of science, and philosophy. Figures such as David Hume (1711–1776) and Adam Smith (1723–1790) had been active in Edinburgh over the previous few decades, while James Hutton, father of geology, was at that time developing the ideas of geological epochs later published in his *Theory of Earth*. Hume is considered to have contributed to the philosophy of the scientific method and was himself interested in some of the ideas concerning climate and climate change, particularly the association between climate and character (Hume 1752). The medical school in Edinburgh, considered to be the best in the British Empire, provided the training for many Canadian physicians of the nineteenth century, including Archibald Hall. Alexander Skakel, who as one of the principal early-nineteenth-century educators in Montreal had considerable influence over the intellectual development of the city, also had a late-eighteenth-century Scots education.

Quebec City, 1798–1819: Alexander Spark's Meteorological Diary

Spark returned to Quebec City in either 1784 or 1786, where he was assistant to the incumbent minister, George Henry, in the Scottish Presbyterian Church, until Henry died in 1796 and Spark formally succeeded him (Campbell 1887; Lambert 2000). The church Spark helped establish in Quebec City, Saint Andrew's Presbyterian, describes him as "eccentric but exceptionally brilliant" and that "a study of his life reveals a much-loved man of God, consistently dedicated to the good of his congregation and tireless in his service to the church".[3]

Spark was a formative figure in the establishment of Presbyterianism in Canada, where the state religion under the British Empire was Anglicanism, with exceptional accommodation made for the majority francophone and Catholic population in Lower Canada. Spark was a moderate, rather than evangelical, minister of his nonconformist denomination. He was described as both a fine mathematician and as having a fine turn for business, dealing with matters relating to the Presbyterian Church and the Colonial Government with "tact and energy" (Campbell 1887, p. 172). His sermons were written with "mathematical concision," (Lambert 1984, p. 9) and Lambert described Spark as "the practical man of action, the businessman, the

---

3. www.standrewsquebec.ca/en/whoweare.html

experimental scientist, the practical joker, the writer of poetry . . . and the practical theologian" (p. 13).

Spark's diary, which was never formally published, is one of the longest continuous individual meteorological records in Canada. The diaries for 1798–1819 are held at the McGill University Archives, but it's thought that there may exist earlier volumes of his observations for the period before 1798 (Lambert 2000). Although Spark was remarkably meticulous and dedicated in keeping his observations—his rate of missing observations is only 3.5%—he didn't seem to have any particularly pointed interest in meteorological or climatic theory. A few notes on the climate or weather are listed in the margins of his journal, such as the "theory of winds," which was that the winds were "restoring the equilibrium when one part of the atmosphere is more rarefied [less dense] than another which is generally affected by a greater degree of heat on one place than another" (Spark 1819; Lambert 1984).[4] This is an interesting entry in light of the fierce controversies about the nature of storms that would dominate meteorological theory in the United States for much of the nineteenth century (Fleming 1990). He further noted, from the occurrence of thick smoke during a fire in the city, that a volcanic eruption was the likely cause of the infamous "dark day" of October 16, 1785. His diaries generally contain the meteorological observations on one side of the page and notes on the other. These notes range in subject matter from the domestic, botanical, philosophical, and theological, such as the life of Burke or extracts from Rousseau, to the literary, including Romantic poems and either an original manuscript or a copy of what appears to be parts of a novel. He also recorded on occasion his dealings with the members of his church, his finances and payment of salaries to his servants, and political news, including remarks on the Napoleonic Wars, the War of 1812, and Canadian parliamentary matters. None of his comments on the books he was reading or his intellectual life are related to his meteorological observations, although he sometimes included descriptions of local weather events. He performed experiments on the relationship between the volume of freshly fallen snow and its equivalent liquid water content. The value he obtained was a ratio of 11 to 1, very close to today's rule of thumb of 10 centimetres of snow corresponding to 1 centimetre of rain.

A general interest in natural philosophy, rather than a particular theoretical or practical motivation, was probably behind Spark's interest in the weather, as he made few calculations based on his observations, and there are only a handful of notes and comments related to theories of the weather and climate

---

4. This is essentially correct.

in his extensive weather records. He was rather a typical eighteenth-century polymath, writing poetry and comments on philosophy and theology as well as having a lifelong interest in the education of the young and relief of the poor.

The impression left by his biographers, chiefly Lambert, is that of a tolerant, generous man, with wide-ranging interests in science, medicine, and the humanities but without an intense focus on any particular discipline outside his well-regarded theology. It is somewhat ironic, then, that despite his apparently cursory interest in climate theories, he left one of the most complete and consistent records of the weather before the establishment of a professional meteorological service, second only to that of his fellow clergyman, John Bethune.[5]

Early Nineteenth-Century Montreal: Thomas and John Samuel McCord
In the same containers in the McGill University Archives as Spark's ledgers another notebook can be found, with the first few pages written in a blotchy and wavering handwriting, replete with ink blots. This weather diary, which appears to have started off as the project of a school boy, marks the beginning of continuous weather recording in Montreal, up to the present day. This makes Montreal one of the few locations with over 200 years of nearly continuous daily weather observations in North America. But who was this first observer?

The notebook covers the years 1813–1826 and contains references to the daily lives of the McCord family in Montreal. It's clearly a record from the McCord family, but to discover which McCord kept the weather diary, we have to do some detective work to trace the history and location of the members of the McCord family. Fortunately, the McCords became one of the most prominent families of nineteenth-century Montreal, and today the archives of the McCord Museum, founded by David Ross McCord at the end of the nineteenth century, hold many of the personal papers of his father, John Samuel McCord, and his grandfather, Thomas McCord.

The McCord Family in Quebec
The McCord family's history in Canada starts soon after the Seven Years' War, when John McCord (1711–1793) emigrated to Quebec from Ireland the

---

5. John Bethune's father had been well-acquainted with Spark (see Campbell 1887).

summer after the Conquest (1760) to provision the British troops now installed in Canada. John McCord's son Thomas McCord (1750–1824) settled on the island of Montreal. During the 1790s, Thomas McCord lived as a gentleman farmer in a stone farmhouse known as La Grange aux Pauvres in the Nazareth fief (estate). In order to sort out some financial difficulties, he returned to Ireland from 1796 to 1805. His sons John Samuel and William King were born in Ireland, but the family moved back to Montreal in 1805, when John Samuel was four (Fyson and Young 1992).

The First McCord Weather Diary: 1813–1826
The first months of 1813 are written in an immature handwriting, very plausibly that of the eleven-year-old John Samuel. At this time, John Samuel had been attending Alexander Skakel's school in Montreal. In September 1813, John Samuel left for Quebec City to attend his first term at the school of Daniel Wilkie (1777–1851). Examining the manuscript diary, the handwriting appears to change around August 25, 1813. From then until December 1816, the handwriting is more mature.[6]

There are references to local landmarks, such as the gate to the convent of the Grey Nuns, the Saint Lawrence River, and the winter ice roads to Longueuil, Saint-Lambert, and La Prairie on the southern shore of the Saint Lawrence River, which John Samuel couldn't have made while at boarding school in Quebec City. The dock for the summer ferry across the Saint Lawrence River to Longueuil on the southern mainland was on the Nazareth fief property, and the starting point of the winter ice roads across the river were probably also nearby.

On occasion, when a journey was recorded and John Samuel was known to be home from school, the handwriting changed. This suggests that at this point the usual observer was Thomas McCord, but John Samuel took over when his father was away on business. In 1817, there are no receipts for John Samuel's schooling, and his handwriting becomes more frequent. The mixed handwriting continues, with entire months sometimes written in one hand

---

6. One of the more distinctive characteristics of Thomas McCord's presumed handwriting is a way of writing the number 4 with it looking like little more than a triangle. This way of writing the number 4, along with the presumed John Samuel McCord's more traditional 4's, are mixed in together from the beginning of the diary. The triangle 4 isn't seen after 1824, the date of Thomas McCord's death, which is recorded in the diary.

or the other until Thomas McCord's death in 1824. The weather diary then continues sporadically until the summer of 1826. It seems likely that this diary was a joint effort between Thomas and John Samuel McCord and kept at their family home, La Grange aux Pauvres.[7]

La Grange aux Pauvres was originally a farmhouse located outside the city boundaries, in what was considered to be the country. The area had become considerably urbanised by the time of Thomas McCord's death in 1824.[8] In the first quarter of the nineteenth century, Thomas McCord's property developed into Canada's nascent industrial sector, becoming increasingly urbanised as the population of Montreal grew from 9,000 in 1805 to about 15,000 in 1815 and 22,000 in 1824 (Fyson and Young 1992). Construction of the Lachine Canal on the northern boundary of this property began in 1821, and by 1825 the area had "become the very core of Montreal's industrial development" (Fyson and Young 1992, p. 37).

John Samuel McCord's Education: Connections to Alexander Skakel and Alexander Spark

From receipts kept by Thomas McCord, we can see that John Samuel was educated at Alexander Skakel's grammar school in Montreal from 1811 to 1812. Alexander Skakel and his brother, William, kept weather observations over an extended period (1816–1868), although only two fragments of this long record have been identified so far. Miller and Young (1992, p. 63) mention that "Thomas McCord sent his sons to schools that emphasised science . . . and bought scientific instruments, apparently for the boys' use in school." It is very likely, then, that a thermometer was bought for John Samuel's use while he was at Alexander Skakel's school. John Samuel was then sent to Daniel Wilkie's boarding school in Quebec City from September 1813 until the end of December 1816 (McCord Family Papers P001-0286). Wilkie was a close friend of Spark's and wrote a memoir of Spark's life that was published in 1837. Wilkie was also a friend of Alexander Skakel, John Samuel's Montreal teacher, the Scots community in Quebec City and Montreal being closely linked through the Presbyterian churches. The last receipt for John Samuel's

---

7. Throughout his life John Samuel alternated between two kinds of handwriting.

8. Although outside the scope of this work, Young (2014) gives an account of how the McCords managed the transition from agrarian gentlemen farmers in Thomas McCord's generation to ruling elite in John Samuel's, financed in large part through their rental income from the insalubrious dwellings in working-class Griffintown.

board and tuition at Wilkie's school in Quebec City dates to December 31, 1816, and in 1818 John Samuel was enrolled in Le Petit Séminaire, as the Collège de Montréal was then known (Miller and Young 1992). No receipts remain for the year 1817, the year in which John Samuel's writing becomes more frequent in the weather diary.

John Samuel's career as a lawyer and later a judge, his lifelong interest in the natural sciences and meteorology, and his ease with English, French, and the classical languages all suggest his education was certainly thorough. The McCords seemed to move easily between cultures and religions, between the British and French Canadian communities and Anglican, Presbyterian, and Catholic surroundings. John Samuel McCord also kept in touch with his Jewish cousins on his mother's side; the communities of British immigrants in Montreal and Quebec City were small and closely knit.[9]

Influence of Spark and McCord
While we know today that the earliest systematic meteorological records in Canada date back to the 1740s, knowledge of Gaultier's records seems to have largely disappeared by the nineteenth century in Canada, although there is one unmistakable reference to them, published in London in 1810. Spark's record thus became the earliest reference for many of the later meteorologists of the nineteenth century. These two registers, while important documents containing valuable information and scrupulously kept meteorological observations, also appear as somewhat casual journals. Spark in particular interspersed temperature, pressure, wind, and weather remarks and readings with personal comments, details of financial transactions, and some fascinating marginalia that are not connected to the weather observations. While there are numerous comments on any weather events taking place overnight, there is very little to be found in the diaries concerning scientific analysis, details of instruments, comparison of results, or communication with other observers.

McCord's journal deals extensively with observations not only of the weather but also of the influence of the weather and climate on their surroundings: the dates of the freeze-up and ice breakup of the Saint Lawrence

---

9. Many of these relationships, especially with the francophone community, would shatter after McCord's role in suppressing the Rebellions of 1837–1838 as a leader of the volunteer militia, and McCord would become cemented in the Anglican elite (Young 2014).

River, the dates of fruit trees blooming, birds arriving for the summer, and other phenomenological observations. There are occasional notes on events of importance to the family or to the city, such as the departure and arrival of John Samuel and William King to and from school and troop movements during the War of 1812.

Spark's example of keeping regular weather observations may have encouraged his friend Alexander Skakel to also keep observations. Skakel, as a prominent Montreal teacher, influenced, in turn, a younger generation, including twelve-year-old John Samuel McCord. Skakel's records for 1826–1835, though the originals have yet to be located,[10] would be used as the basic record to provide information for the climate of Montreal for much of the nineteenth century (Dove 1840; Schott 1876). John Samuel McCord started a second, more complete weather register in 1831 and opened yet another chapter in Canadian climatology: comparative climatology.

References

Barr, and R. Saunders, 1778: Journal of the weather at Montreal. By Mr. Barr. Communicated by Richard Saunders, M. D. F. R. S. *Philos. Trans. R. Soc. London*, **68**, 559–563, http://www.jstor.org/stable/106332.

Binnema, T., 2014: *"Enlightened Zeal": The Hudson's Bay Company and Scientific Networks, 1670–1870.* University of Toronto Press, 458 pp.

Campbell, R., 1887: *A History of the Scotch Presbyterian Church, St. Gabriel Street, Montreal.* W. Drysdale, 807 pp.

Chapman, E. J., and I. M. McCulloch, Eds., 2010: *A Bard of Wolfe's Army: James Thompson, Gentleman Volunteer, 1733–1830.* R. Brass Studio, 361 pp.

Desloges, Y., 2016: *Sous les cieux de Québec: Météo et climat, 1534–1831.* Septentrion, 220 pp.

Dove, H. W., 1840: *Über die nicht periodischen Änderungen der Temperaturvertheilung auf der Oberfläche der Erde in dem Zeitraume von 1789 bis 1838.* G. E. Reimer, 131 pp.

Fleming, J. R., 1990: *Meteorology in America, 1800–1870.* Johns Hopkins University Press, 264 pp.

Fyson, D., and B. Young, 1992: Origins, wealth and work. *The McCord Family: A Passionate Vision,* P. Miller et al., Eds., McCord Museum of Canadian History, 26–53.

Golinski, J., 2007: *British Weather and the Climate of the Enlightenment.* University of Chicago Press, 284 pp.

---

10. Many of Skakel's observations for 1816–1828 have been recently retrieved from newspaper clippings by Federico Ponari (2017, personal communication).

Holmes, R., 2008: *The Age of Wonder: How the Romantic Generation Discovered the Beauty and Terror of Science.* Harper, 554 pp.

Houston, S., T. Ball, and M. Houston, 2003: *Eighteenth-Century Naturalists of Hudson Bay.* McGill-Queen's University Press, 338 pp.

Humboldt, A. V., 1997: *Cosmos: A Sketch of the Physical Description of the Universe.* Vol. 1. E. C. Otté, Trans., Johns Hopkins University Press, 375 pp.

Hume, D., 1752 (1824): Of the populousness of ancient Nations. *The Philosophical Works of David Hume, Volume III Essays: Moral, Political, and Literary,* London, 421–508.

Janković, V., 1998: Ideological crests versus empirical troughs: John Herschel's and William Radcliffe Birt's research on atmospheric waves, 1843–50. *British J. Hist. Sci.*, **31**, 21–40, https://doi.org/10.1017/S0007087497003178.

——, 2001: *Reading the Skies: A Cultural History of English Weather, 1650–1820.* University of Chicago Press, 272 pp.

Kington, J., 1980: Daily weather mapping from 1781: A detailed synoptic examination of weather and climate during the French Revolution. *Climatic Change*, **3**, 7–36, https://doi.org/10.1007/BF02423166.

Kupperman, K. O., 1982: The puzzle of the American climate in the early colonial period. *Amer. Hist. Rev.*, **87**, 1262–1289, https://doi.org/10.2307/1856913.

Lambert, J., 1810: *Travels through Lower Canada and the United States in the Years 1806, 1807, and 1808.* Volume 1. Richard Philips, 496 pp.

Lambert, J. H., 1983: Alexander Spark. *Dictionary of Canadian Biography*, Vol. 5, University of Toronto/Université Laval, accessed 08-10-2014, http://www.biographi.ca/en/bio/spark_alexander_5E.html.

——, 1984: *One Man's Contribution: Alexander Spark and The Establishment of Presbyterianism in Quebec 1784–1819.* St. Andrew's Presbyterian Church, 26 pp.

Landmann, G., 1852: *Adventures and Recollections of Colonel Landmann, Late of the Corps of Royal Engineers.* Vol. 1. Colburn and Co., 347 pp.

Lefroy, J. H., 1853: Remarks on thermometric registers. *Can. J.*, **1**, 29–31.

McClellan, J. E., III, and F. Regourd, 2000: The colonial machine: French science and colonization in the ancien regime. *Osiris*, **15**, 31–50, http://www.jstor.org/stable/301939.

McCord, T., and J. S. McCord, 1826 unpublished manuscript: McCord's meteorological register. Meteorological Records, 1798–1972, Container 1039, Envelope 320. Faculty of Arts and Sciences, Department of Meteorology, McGill University Archives.

McCord Family Papers P001-0286, unpublished. The Archives and Documentation Centre of the McCord Museum. Montreal, Canada.

Miller, P., and B. Young, 1992: Private, family and community life. *The McCord Family: A Passionate Vision*, P. Miller et al., Eds., McCord Museum of Canadian History, 55–83.

Rose, A., and P. Murdoch, 1766: Abstract of a journal of the weather in Quebec, between the 1st of April 1765, and 30th of April 1766. By Cap. Alex. Rose, of the

52d regiment; communicated by Rev. P. Murdoch, D. D. R. R. S. *Philos. Trans. R. Soc. London,* **56**, 291–295, https://doi.org/10.1098/rstl.1766.0037.

Schott, C. A., 1876: *Tables, Distribution and Variations of the Atmospheric Temperature in the United States, and Some Adjacent Parts of America.* Smithsonian Institution, 345 pp.

Shapin, S., and S. Schaffer, 2011: *Leviathan and the Air-Pump: Hobbes, Boyle, and the Experimental Life.* Princeton University Press, 391 pp.

Spark, A., 1819 unpublished manuscript: Diary 1798–1819. Meteorological Records, 1798–1972, Container 1039, Envelope 318–319. Faculty of Arts and Sciences, Department of Meteorology, McGill University Archives.

Tompkins, H., 2009: A seasonal warm/cold index for the southern Yukon Territory: 1842–1852. *Historical Climate Variability and Impacts in North America,* L.-A. Dupigny-Giroux and C. J. Mock, Eds., Springer, 209–229.

Wilson, C.V., 1983: Some aspects of the calibration of early Canadian temperature records in the Hudson's Bay Company Archives: A case study for the summer season, eastern Hudson/James Bay, 1814 to 1821. *Climatic Change in Canada: 3,* R. Harrington, Ed., Syllogeus Series, No. 49, Canadian Museum of Nature, 144–201.

——, 1985a: The summer season along the East Coast of Hudson Bay during the nineteenth century: A sample study of small-scale historical climatology based on the Hudson's Bay Company and Royal Society Archives. Atmospheric Environment Service, Environment Canada, Canadian Climate Centre Rep. 85-3, 38 pp.

——, 1985b: Daily weather maps for Canada, summers 1816 to 1818—A pilot study. *Climatic Change in Canada: 5,* C. R. Harrington, Ed., Syllogeus Series, No. 55, Canadian Museum of Nature, 191–218.

——, 1985c: The Little Ice Age on eastern Hudson/James Bay: The summer weather and climate at Great Whale, Fort George and Eastmain 1814–1821, as derived from Hudson's Bay Company records. *Climatic Change in Canada: 5,* C. R. Harrington, Ed., Syllogeus Series, No. 55, Canadian Museum of Nature, 147–190.

——, 1988: The summer season along the East Coast of Hudson Bay during the nineteenth century. Part III: Summer thermal and wetness indices. B: The indices, 1800 to 1900. Atmospheric Environment Service, Environment Canada, Canadian Climate Centre Rep. 88-3, 42 pp.

Wulf, A., 2012: *Chasing Venus: The Race to Measure the Heavens.* Alfred A. Knopf, 304 pp.

Young, B., 2014: *Patrician Families and the Making of Quebec.* McGill-Queen's University Press, 452 pp.

Zilberstein, A., 2016: *A Temperate Empire: Making Climate Change in Early America.* Oxford University Press, 280 pp.

CHAPTER FIVE

McCord and the Montreal Natural History Society

Canada was in a nearly continuous state of war for much of the second half of the eighteenth century and the beginning of the nineteenth. The Seven Years' War, the Conquest of Canada by the British and the assimilation into British North America, the American Revolution, the Napoleonic Wars, and the War of 1812 all occupied much of the attention of governors and the governed alike. Indeed, for most its existence, Canada was as much a military outpost as a settler colony and was governed by military commanders as well as, and sometimes instead of, civil authorities.

With the end of the War of 1812 and the Napoleonic Wars, of which the War of 1812 was a regional manifestation, Canada entered a period of cautious peace. The close of the generations-long wars, poor harvests following the eruption of Tambora in 1815–1816 (which entered history as "the year without a summer"), and otherwise difficult times in Europe led to increased immigration to Canada. Peace gave military officers stationed in Canada and others in the colony, such as clergymen, doctors, and lawyers, more time for cultural and social activities, including natural history. Similar societies and associations were formed in many countries around this time, reflecting a renewed and more widespread interest in the arts and sciences (Zilberstein 2016, chapter 2). The first Mechanics' Institutes, which provided further education to working adults in mainly technical and scientific areas, were established in the 1820s. In the United Kingdom, the first meteorological society and geological society were also established in the 1820s, with the British Association for the Advancement of Science (BAAS) being formed in 1831.

As a part of this early-nineteenth-century movement, two societies for intellectual development and the promotion of knowledge sprang up in the colony of Canada: the Literary and Historical Society of Quebec (LHSQ) in Quebec City and the science-oriented Natural History Society (NHS) in Montreal. Both would prove venues for publishing observations, results, and investigations on weather, climate, and climate change in Canada. As Quebec City was the capital of the colonial civil government and military headquarters, the LHSQ tended to have more military and naval officers among its members and contributors of articles to its publications. Men and women published articles on a variety of topics. Papers on geology and zoology were contributed by Naval Commander Henry Bayfield (1795–1885; Bayfield 1824; Bayfield 1831), while meteorology was addressed by Captains Richard Bonnycastle (1791–1847; Bonnycastle 1824) and William Kelly (dates unknown; Kelly 1837). Articles on natural history were published by a Mrs. Sheppard (1824), a catalogue of plants by the Countess of Dalhousie (1824), and an article on ethnography by Daniel Wilkie (1831). Montreal, the commercial centre, had merchants, lawyers, doctors, and clergymen as members of the NHS. It was through his position as corresponding secretary of the NHS that McCord could get military collaboration for extending meteorological observations at the Montreal garrison from a few times a day to once every 2 hours. McCord also used his position at the NHS to collaborate with the Royal Society in London, with John Herschel (1792–1871) through the Royal Meteorological Society, and with Professor Charles Daubeny (1795–1867) of Oxford University.

The Establishment of the Montreal Natural History Society

The NHS of Montreal was founded from James Somerville's natural history collection of mineralogical and some botanical specimens. Somerville was a close friend of Daniel Wilkie's and the minister of the Saint Gabriel Street Church, Montreal's Presbyterian community. Somerville had been recommended to the position by Alexander Skakel and was ordained by Spark and John Bethune, father of the meteorologist John Bethune discussed here (Campbell 1887, p. 157). Somerville was an ardent naturalist who spent much of his time exploring the countryside; he also kept a diary and daily weather record since the age of twenty-two. Robert Campbell gives the following account of the founding of the NHS from the Scots Presbyterian community of the Saint Gabriel Street Church:

> [Somerville's] practice of rambling in the fields in quest of objects suitable for the study of Natural History [and] attractive conversation naturally drew to his society others who possessed similar tastes . . . One gentleman, especially, of highly scientific attainments, (known to be A. Skakel, a teacher in this city) assisted to give accuracy and order to their observations. A considerable collection of natural objects was, in consequence, formed; a place was found [to be] necessary for their reception, the assistance of others was solicited and obtained, and out of these humble endeavours arose, on the 16th May, 1827, the 'Natural History Society of Montreal.' (Campbell 1887, p. 159–160)

Here, again, the influence of the Presbyterian Church and the Scottish Enlightenment on science in Canada during the early period of the British Empire can be seen, from Spark and Wilkie in Quebec City to Somerville and Skakel in Montreal and from there to Skakel's many pupils.

Most of the "amateur" observers, and some of the military officers, presented papers and published results in the *Transactions of the LHSQ* and in publications sponsored by the NHS of Montreal and other Canadian journals. Some results and abstracts were published in Europe: Hall published the results from Skakel's observations in the *Edinburgh Philosophical Journal*, which were later used by Heinrich Willhelm Dove (1803–1879) in his *Handbook of Climatology*. McCord himself published only a handful of papers on weather and climate, although his personal papers in the McCord Museum Archives show an extensive interest in climatology. It is from his collection of records kept by Spark, Siveright, Cleghorn, and Bethune (see below), that climate of the Saint Lawrence valley could be reconstructed for much of the nineteenth century (Slonosky 2014, 2015). McCord, using the prestige and authority of his role of secretary of the Montreal NHS, advocated as early as 1836 for the military to involve themselves in regular meteorological observations for climatological purposes, such as to determine the mean annual temperature. The U.S. military had been keeping regular meteorological observations since the War of 1812, with an interest primarily in maintaining the health of their soldiers; more lives had been lost to disease during the War of 1812 than to warfare.

The NHS is most well known as the Montreal institute that promoted the work of pioneering geologists such as William Logan (1798–1875), another of Alexander Skakel's students, and John William Dawson (1820–1899) and for the contributions of its members to botany and zoology (Zeller 2009; Sheets-Pyenson 1996; Kuntz 2010). The society was also an important meeting place to bring together members of the meteorological community. All the weather observers in nineteenth-century Montreal were members of the society, and Alexander Skakel was one of its founders. According to Kuntz, "several of the more active members of the society had been educated by Daniel Wilkie or Alexander Skakel," with John Samuel McCord and Archibald Hall among them (Kuntz 2010, p. 269). Robert Cleghorn was a member who contributed plant specimens as well as meteorological reports. The NHS, with Smallwood as one of the delegates, played a leading role in securing Montreal as the venue for the American Association for the Advancement of Science (AAAS) annual meeting in 1857, as well as for a second meeting later in the century in 1882 and for the BAAS meeting of 1884. The 1857 AAAS meeting was the first time the association meeting had taken place outside of the United States, and was hailed as an occasion to mark the close ties between American scientists and those of British North America. That the members of the NHS were able to convince the British members of the BAAS to cross the ocean and hold their first meeting outside the British Isles in 1884 points to the high level of Canadian science in general and that of Montreal in particular. These meetings in Montreal provided an important venue of scientific and cultural exchange in an Anglo-American triangle involving Great Britain, Canada, and the United States.

**The Second McCord Weather Journal
and J. S. McCord's Contributions to Science**
John Samuel McCord was a seminal figure in the history of Canadian climatology. Although his contributions to meteorology are not well known today, he was recognised in his lifetime as the "Pioneer of Canadian Meteorology" (Smallwood 1860, p. 309). A member and corresponding secretary of the NHSM, a circuit court judge, a military officer in the Canadian Militia who participated on the side of the government in the Rebellions of 1837–1838, and an amateur scientist who corresponded with such scientific luminaries of the nineteenth century as John Herschel of the Royal Society (one of the most influential voices in nineteenth-century geoscience) and Charles Lyell (Charles Darwin's mentor), John Samuel McCord was largely forgotten until a recent work by Brian Young included a biography of John Samuel McCord in a chapter on a larger work on nineteenth-century Quebec (Young 2014). It is only thanks to the fact that his son, David Ross McCord, founded the

McCord Museum of Canadian History that most of McCord's work survives as a part of the McCord Family Papers.[1] John Samuel McCord's varied correspondence related to meteorological affairs, his ability to persuade the military to instigate regular meteorological observations, and his prominent role in the NHS as the organization's secretary point towards an energetic and affable personality, active in creating scientific alliances and in promoting the science of meteorology and climatology in Canada. Most of the information about him comes from his notebooks, biographical information in the "Family Papers" collection of the McCord Museum Archives, and publications of the McCord Museum.

After a break of several years in the late 1820s, McCord started a second set of meteorological observations in April 1831. From 1826 to 1831, he appears to have been occupied in settling his father's complicated financial affairs, looking after his younger brother William King McCord (whose student gambling debts at Oxford University also led to financial complications), and establishing his own career as a lawyer. From 1819, when he received his first commission as an ensign, to 1837, when, as a lieutenant colonel, he was involved in quelling the Patriot Rebellions, he served in the militia. In an 1838 letter, he mentioned his planned project to start a series of hourly observations at the garrison on Saint Helen's Island, an island in the Saint Lawrence River between Montreal and Longueuil, "if Peace be granted to this distracted Province" (McCord 1838).

It is not completely clear where McCord lived in the years after 1826 or what happened to the Grange. In 1832, John Samuel McCord married Anne Ross (Alexander Spark's wife's niece) and lived on Little Saint James Street, the same street as the Grange's later address. They may have been living in the same house, which was now part of the city rather than an isolated farmhouse, or they might have sold the Grange and moved to a different house on the same street. In any case, in 1832 they moved to Great Saint James Street. In 1836 McCord bought a second property on the southwestern side of Mount Royal, where he built his country home, Temple Grove. From 1839 onward, his weather diaries specify "Town" during the winter months or "Country" during the summer months.

Apart from his involvement in the establishment of the NHS, not much

---

1. Fortunately for historical climatology in Canada, J. S. McCord's son David Ross McCord was a collector of artefacts relating to the history of Canada and founded a museum, which preserved the family papers. Among the papers are John Samuel McCord's scientific notebooks and weather observations.

else is known about McCord's scientific or meteorological activities until 1831, the starting date of the manuscript journal of daily weather records found among his scientific papers. The meteorological observations continue until April 1842, at which time McCord was appointed a circuit court judge. This new duty meant frequent travel of several days' duration around the district south of Montreal, travel that left him unable to continue his daily registers at the level of accuracy and punctuality he felt they required. He subsequently sold most of his equipment to McGill University, although he continued to note the temperature and pressure along with weather descriptions in his personal diary until shortly before his death in 1865.

McCord's Second Series of Observations

The period from 1832 to 1842 was when McCord was the most active in his climatological and meteorological research. During this decade, he held the post of secretary of the NHS of Montreal and memberships in the London Meteorological Society (Kuntz 2010, p. 298), the LHSQ, and the Albany Institute of New York, a society similar to the Natural History Society of Montreal or the Literary and Historical Society of Quebec. The Albany Institute was at this time largely devoted to natural history in the state of New York and had a young Joseph Henry (1797–1878) as curator. Henry delivered his first papers on electromagnetism at the Albany Institute before leaving for Princeton and eventually the Smithsonian Institute. He exchanged letters and meteorological data with others interested in climate within Canada, including naval physician Lieutenant William Kelly. Kelly was a member of Captain Bayfield's decades-long nautical surveying mission and investigated the climate of Quebec City and the water temperature of the Saint Lawrence while stationed in Canada. McCord's meteorological observations were also considered extremely valuable by Lieutenant Charles Riddell, sent by the Royal Artillery to establish a magnetic and meteorological observatory in Canada.

McCord spared no expense on his instruments, buying many of them from the best instrument-makers of the day in London and Edinburgh. He generally recorded his observations twice a day, around 8:00 a.m. and 4:00 p.m., although the times varied as he experimented with the times of day mostly likely to give an unbiased daily mean temperature. He started recording maximum and minimum temperatures and pressure to three decimal places in September 1835. McCord used a two letter code modelled on Francis Beaufort's recommendations to record weather conditions: "sn" for snow, "cl" for clear, "rn" for rain, and so on (Hamblyn 2001, p. 197).

In September 1837, Oxford professor Charles Daubeny met with McCord on his visit to North America and performed several experiments on solar radiation with McCord on McCord's summer estate of Temple Grove, on the western slope of Mount Royal. On September 16, 1837, McCord noted that he had "compared my actinometer[2] with one owned and brought to Temple Grove by Prof. Daubeny of Oxford, England, and found it to correspond within a fractional distance" (McCord 1843, p. 90; see also McCord 1839; Young 2014, p. 198). This result pleased McCord as it confirmed the accuracy of his instruments and observations with those of a prestigious authority.

On his return to Britain in 1838 Daubeny presented his findings on the climate of North America, based in part on McCord's records, to the BAAS, a summary of which was printed in the *Journal of the Franklin Institute* in Philadelphia. He "observed that the general fact was admitted that the eastern portion of the New World possessed a lower temperature than the western portion of the old, yet that much remained to be done" (Jones 1838, p. 394) to determine the relative temperatures of North America and Europe. In his opinion, most of the observations were "very little to be relied upon as to accuracy" (p. 394), though an exception was made for McCord, praised in Daubeny's lecture as having carried out "the best observations made in Canada" (Daubeny 1838, p. 29). Daubeny also mentioned the "promptitude" with which Herschel's ideas for hourly observations to be made worldwide on the solstices and equinoxes were taken up in both Canada and the United States (Jones 1838, p. 394).

Herschel, speaking after Daubeny at the 1838 BAAS meeting, said that, "of all the sciences which now engross the attention of the thinking part of mankind," none required greater international cooperation than meteorology, and promoted his scheme for 24 hours of observations to be made simultaneously across the globe for four days a year: the solstices and equinoxes (Jones 1838, p. 395). McCord had been participating in Herschel's hourly observation campaign since 1836; in 1840, Herschel thanked McCord for his "valuable series" (Herschel 1840). Professor Alexander Bache (1806–1867), in attendance from Philadelphia, urged further cooperation between the United States and British North America "in connecting observations" (Franklin Institute 1838, p. 396).

McCord remained in contact with Lieutenant Riddell and Captain Henry Lefroy as they went to work on setting up the Magnetic and Meteorological Observatory in Toronto. He was requested on several occasions to send

---

2. An instrument devised by John Herschel for measuring solar radiation.

copies of his meteorological observations to Herschel at the Royal Society in London as well as to Lefroy at the Toronto Observatory.[3] By 1840, McCord was an associate member of the London Meteorological Society (McCord 1840).

As secretary of the NHS of Montreal, he undertook several scientific experiments and the organization of meteorological projects. One was the recording of hourly observations at the Saint Helen's Island garrison, with the view of obtaining a measure of the variation of temperature during a 24-hour period over the course of a year and determining which method was most suitable for calculating a daily mean temperature.[4] McCord was the driving force behind this project and was in correspondence with Sir John Colborne (1778–1863), commander in chief of the army in British North America, concerning the initiation of meteorological record keeping by the army. His long-delayed project would finally be undertaken in 1839, in one of the earliest instances of land-based[5] systematic weather observations kept as part of regular military duties. McCord also tried to instigate a second project to take regular meteorological observations at the Hudson's Bay Company's post, enlisting the help of Company Governor George Simpson (Kuntz 2010, p. 289). A circular with instructions was sent out to the HBC post and NHS members, but what came of the initiative remains unknown. Not many records appear to have made their way back to McCord, though one proved extremely valuable: the observations of John Siveright from 1821 to 1833. Siveright's earliest records were from Saint Mary's Falls (Sault Sainte Marie). From October 1823 onward he was stationed at Fort Coulonge on the Ottawa River.

McCord experimented with determining snow and liquid water equiva-

---

3. McCord letters.

4. At the time, scientists were interested in determining absolute values of temperature at various places across the globe and in comparing the temperatures of different locations. Today, climatologists are generally more interested in changes in temperature and recognise the difficulty of obtaining an absolute value given the errors in instrumentation and the small-scale changes in temperature over short distances. Most climate studies today look at the departures from an average value, or anomalies, and this is how the well-known graphs of global temperature are represented. This is why the graphs are usually centred around zero rather than 15°C, which is about the mean temperature of Earth.

5. Meteorological information had been recorded in naval logs since the eighteenth century.

lents at different temperatures after reading Spark's notes on his earlier experiments. He also measured solar radiation. In 1836, he spent a June day on Mount Royal making observations every 15 minutes, again with a variety of instruments. He used blackened thermometer bulbs to make blackbody measurements of solar energy as well as the usual normal thermometer for measuring air and ground temperature.

McCord was also in contact with James Espy (1785–1860), in Espy's quest to collect information about the May 1833 storm. The "storm controversy" in the United States was then at its height, with several actors proposing different physical mechanisms to account for North American storms. Espy held that the heat arising from convection was the principal driver of storms, while William Redfield (1789–1857) held that storms were gravitationally induced and, in part, caused by Earth's rotation (see also chapter 9; Fleming 1990; Moore 2015).[6] While McCord was away during the relevant period and consequently missed observing the storm, he collected journals and registers from William Robertson (M.D.) for Montreal and John Siveright at Fort Coulonge and passed them on to Espy, with McCord's assurances as to their accuracy as observers.

McCord on Instruments

McCord was a meticulous observer who was keenly interested in the developments and limitation of instruments. Instrument standards and calibration are the key to providing accurate and reliable observations. On several occasions he carefully compared up to nine different thermometers over a range of temperatures. Like Gaultier, McCord was able to repeatedly test thermometers manufactured in Europe in cold conditions; also, like Gaultier, he was frustrated by their poor performance. Although the thermometers were now graduated to measure colder temperatures, so that McCord wasn't obliged to try to guess how much the mercury or spirit had contracted into the bulb as Gaultier did, thermometers were still poorly calibrated at these low temperatures.

---

6. Fleming (1990), Lockett (2012), and Moore (2015) provide extensive discussion of the nineteenth-century storm controversy. Elements of both theories are correct: the latent heat given off in condensation fuels the ascension of air and the formation of clouds while the rotation of Earth leads to the Coriolis effect, whereby moving parcels of air are deflected by the rotation of Earth, maintaining pressure imbalances which drive winds and atmospheric circulation.

In the winter of 1836–37, he compared nine thermometers mainly constructed by Adie and Sons [Alexander Adie (1775–1858) and son John Adie (1805–1857)] in Edinburgh and John Newman (ca. 1783–1860) in London. These were a variety of standard and register thermometers. At around 0°F, the readings varied between −4.00° and +3.50°F; at about −16.5°F, the seven thermometers that survived the first trial read between −13° and −17°F. Two thermometers had been "withdrawn as differing too much after the second experiment," one of which was described as "old, in case" (McCord 1843). Another two were withdrawn "as being untrue" after the third experiment, an Adie's and a Newman's spirit register thermometer. McCord writes, "Therefore, only No 1, 2, 4, 6 and 7 are to be depended upon when the mercury is below 0" (McCord 1843).

Another comparison between thermometers in the winter of 1841–1842 resulted in a letter to instrument-maker John Newman, which begins: "After the pains in preparing the standard thermometer which you sent out last year . . . I feel some anxiety in communicating to you." The differences between thermometer readings became "considerable," up to 4°F as the temperature decreased. Among the thermometers McCord used for this comparison was another instrument made by Newman for the Toronto Observatory and given to McCord by Lieutenant Riddell for comparison purposes. This one did not fare well, however, showing differences of nearly 4°F at temperatures between −6° and −12°F. McCord was concerned as well about changes in the contraction coefficient of alcohol with temperature, asking Newman whether "below zero is any allowance made for the diminishing contraction of spirit" (McCord 1843). McCord also kept an eye on the registers of those around him as a means of checking his instruments. On May 3, 1831, a few weeks after starting his second weather diary, he "sent my Bar[ometer] to be repaired, finding it too low when compared with Mr Skakel's table published in the Herald [newspaper] & during those days copied from it" (McCord 1835).

McCord on Climate

One of McCord's main interests, however, was climate change, and he set himself the task of determining whether the climate had changed in Canada with the advent of agriculture and the clearing of the forests, as had been first suggested by Gaultier in 1745. With the aim of finding out if the climate had changed using instrumental observations of temperature, McCord collected a considerable number of copies of manuscript weather registers and diaries from colonial Canada, including copies of Spark's records. It was from Wilkie that he later obtained "a communication of the Tables kept in Quebec City by the late Revd. Dr. Sparks" (McCord 1838a).

By the 1840s, McCord had collected an impressive number of weather records, started analysing them, and published the results. He published his analysis of Spark's record in the *British American Journal* in 1845, announcing at the same time his long term project of determining climate change from long-term records:

> But the object of my present communication is to discuss a subject which has been so ably treated by my friend Dr. Kelly[7] of the Royal Navy ... [and] to request a place in your next number for some of those tables which, with much difficulty, I have collected at different periods; they may prove useful to some future student, if published, whilst at present they are liable to be lost or destroyed. (McCord 1845, p. 35)

McCord also took the opportunity to enumerate the problems he encountered with many of the "tables without number which have been submitted to my inspection." These included the "very imperfect instruments used;" the lack of basic statistics to summarise the masses of daily observations ("the meteorologist who desires to form a comparison between the climate of this Province and that of other countries is compelled to wade through an enormous mass of figures"); and the lack of a standard observing time, for without the adoption of "any one daily fixed hour for observation," it was impossible to use the individual observations to calculate mean values (McCord 1845, p. 35). These were issues which were to preoccupy all future attempts to set up networks of regular weather observations and to compare values across both time and space (Fleming 1990).

McCord found a note in Spark's journal stating that the instruments were defective and only published Spark's observations for 1803 and 1810–1818. This could refer to Spark's comment in January of 1801 that the thermometer hung between double windows was too much sheltered; recent analysis, however, suggests there is no discernible effect on the distribution of temperatures. Spark's move from Sainte-Anne Street to Sainte-Helene Street in 1807 did give rise to a break in the temperature records, which was statistically corrected for in a recent analysis (Slonosky 2015).

---

7. Further discussion of this topic of climate change and Kelly's paper is in chapter 6.

McCord continued to publish the results of the meteorological tables he had collected, giving a detailed account of snowfall measured by Dr. William Belin of Long Point for the winters of 1830–1831 to 1835–1836, along with monthly mean calculations taken from Siveright's long series of observations at Fort Coulonge from 1825 to 1831 (McCord 1846). Tables from René Boileau in Chambly for the 1820s appeared in 1847 (McCord 1847), while McCord's example inspired Dr. W. Marsden to compile and publish the observations from 1838 to 1846 kept by Reverend Francois Desauniers, professor at Nicolet College (Marsden 1847).

In 1860, Dr. Charles Smallwood, the first professor of meteorology at McGill University, paid tribute to McCord's immense efforts at organising weather records, instigating regular observations, and analysing climate change by naming McCord "the Pioneer of Canadian Meteorology" (Smallwood 1860, p. 309).

Other Influences on Temperature

There appears to have been a polite dispute between McCord and Kelly on the true value of the average temperature in Montreal and Quebec City. Calculating the "true" average temperature of a location was of particular concern, both to establish baseline statistics and to be able to compare temperatures across space in Humboldt's synthesizing manner, as discussed in the next chapter (Humboldt 1817). Determining the best time of the day to take observations in order to approximate the daily mean temperature of 24 hourly observations was one of the reasons McCord enlisted the army at Saint Helen's Island to record bihourly observations in 1839 (McCord 1842). It was widely recognised that readings taken predominantly during the daylight hours would bias the daily mean values. Kelly thought that McCord's average for Montreal would be too high, because of the influence of buildings and urban heating, while McCord considered that the fact that the observations on which Kelly was basing his estimates were taken at the Quebec Citadel, at a higher elevation than the town of Quebec (see Fig. 5.1), made Kelly's estimates for Quebec City too low. The influence of urbanization on temperatures, still a hotly contested issue today, was being considered in 1830s Montreal as it had been by Luke Howard in 1810s London (Howard 1833; Hamblyn 2001).

Concerning the effect of elevation on the temperatures of Quebec City, McCord (1838) reasoned as follows:

> My friend Dr. Kelly assigns from tables prepared by Mr. Watts on Cape Diamond a mean temperature to Quebec of 36.68. Atho' Mr. Watts has doubtless bestowed the greatest attention in the construction of the above register, I am of the opinion that his mean temperature is rather too low but this is a point we can scarcely determine accurately till a series of hourly observations have been executed, an object I have taken measures to get accomplished and which if Peace be granted to our distracted Province, I hope to see commenced during the current year.[8] To the mean temperature as above there ought at least to be added a correction for the elevation, which according to Pouillet's scale would be 1.25°, making therefore the mean temperature of Quebec 37.98°, say 38°.

McCord fully appreciated the difficulties of obtaining an exact value of mean temperature.

Kelly, for his part, seemed to have some reservations about urban influences in Montreal, which McCord addressed as follows:

> Dr. Kelly is of the opinion that the mean temperature which I assign to Montreal is a trifle too high, from the fact that the winter observations are made in the city; this is true, but the great pains I have taken to negate every species of reflection or radiation of heat make me feel confident that, if too high, it is only very fractionally so; from Tables kept by my Father in the Country, in the years 1815, 16, 17' 18' & '19[9] I find the mean of the five years to be 41.96°, differing from the mean of the last three years [by] only .86°. (McCord 1838)

Considering that the years mentioned in the "Tables from my Father" consist of some of the coldest years of the past two centuries, it's scarcely surprising they were nearly 1°F (~0.5°C) colder than the 1830s. However, the issue of climate change, at least on a scale of two decades, doesn't seem to be an issue here; instead, McCord was trying to find evidence of long-term, permanent climate improvement.

---

8. This refers to McCord's project of hourly temperature observations to be undertaken by the military garrison on Saint Helen's Island in order to determine correction factors that would enable temperature registers kept at varying times of day to be compared with each other.

9. Presumably, this was taken from the McCord weather record of 1813–1826.

Other Nineteenth-Century Montreal Observers: The Skakels and John Bethune

With McCord taken up with his judiciary duties, there appears to have been a lull in the organizing, communicating, and publishing of meteorological observations in Montreal, although with the establishment of the Toronto Observatory in 1840 there were, for the first time since Gaultier, officially appointed government-sponsored meteorological observations in Canada. In Toronto, they were taken by the military. Individual observers in Montreal still kept their records; the Skakel brothers in particular had kept records since at least 1816, and John Bethune's weather record is the longest yet discovered in Canada by a single individual, covering the period 1838–1869.

The Skakel Brothers: Alexander and William

Alexander Skakel, as mentioned previously, was the principal of the Grammar School in Montreal. Born in Scotland and educated in Aberdeen in the 1790s, he arrived in Quebec in 1798, where he crossed paths and became good friends with Spark and Wilkie (Kuntz 2010, p. 85). He founded the Classical and Mathematical School in Montreal in 1799 that was, in 1811, equipped through parental and citizen fundraising with "philosophical apparatus so that Skakel's teaching might include demonstrative experiments" (Frost 1988). Two of his pupils, John Samuel McCord and Archibald Hall, went on to keep weather records of their own and were instrumental in publishing meteorological observations and developing international connections. They placed Montreal, and the weather observations taken there, in an international context.

Two letters, one published in the *Edinburgh Philosophical Journal* in 1837 by Archibald Hall (Hall 1836) and the second a letter to the editor of the *Montreal Star* in 1859 (McCord 1847), both suggest that "Mr. Skakel" had been keeping meteorological records since at least 1820. Alexander Skakel had died in 1846, so it's assumed that his brother William kept the record after his death or that the brothers had been jointly responsible for the observations. These observations, and especially the synopsis printed in the *Edinburgh Philosophical Journal*, had a long term of service in the nineteenth century: they were reprinted in Dove's 1840 publication on the temperature of Earth, published again in Lorin Blodget (1823–1901)'s somewhat unauthorised version of the Smithsonian oeuvre in 1857, and finally in Charles Schott's authorised tables printed under the aegis of Joseph Henry and the Smithsonian Institution nearly forty years later in 1876 (Dove

1840; Blodget 1857; Fleming 1990; Schott 1876). Schott's 1876 compilation of international meteorological observations listed the information for Montreal as coming from a "W. S. Kakel" as well as listing McCord, Bethune, and Hall as contributors, though by the 1881 edition Bethune and Skakel seem to have disappeared. In an interesting addition, Blodget included Gaultier's 1743–1744 observations from the 1745 volume of the *Mémoirs de l'Académie Royale des Sciences*. McCord also had the 1744 volume of the *Mémoires* included among his works to consult in a page outlining his basic research questions:

> Has any change occurred
> > What signified by change
> > > What was state of climate by earliest accounts

in his scientific notebook (McCord 1836).

The short article submitted to the *Edinburgh Philosophical Journal* by Hall, Skakel's student, in 1836 states that

> the mean temperature may be relied on as substantially correct, being deduced from the meteorological tables of [Alexander Skakel] who has been in the habit of making daily observations for the last fifteen or twenty years; and moreover his thermometer, having never during that time been removed from the station it at present occupies, indicates a temperature upon which the same modifying powers have always exerted a corresponding degree of influence. The means of the months are deduced from the series of two daily observations, the one at 7 o'clock A.M., when it may be assumed that the temperature is at a minimum, the other at 3 o'clock P.M. when it may be stated to be at maximum. (Hall 1836, p. 236)

This reference to scientific instruments that were purchased by the parents of the students in 1811, the fact that the young John Samuel McCord and his father started their meteorological journal in 1813, and the fact that Archibald Hall would be in a position to remember the meteorological instruments from his time as a pupil, all combine to suggest that the Skakel record could plausibly have started as early as 1811. If the original records were ever recovered, they would provide a fifty-year-long, continuous daily weather register for Montreal: a gold mine of information. Some fragments have been recovered from newspaper clippings between 1816 and 1828 by citizen science climatologist Federico Ponari.

From Hall's article, we can see that the positioning of the thermometer and any external influences on the reading of the thermometer were carefully considered, and changes in the location or surroundings of the thermometer were recognised as potentially introducing uncertainty or inaccuracies to the temperature readings. The timing of the observations was also deliberately scheduled to try to record the highest and lowest temperatures of the day.

Hall further adds that the annual average temperature had decreased between 1830 and 1835, "a circumstance not very consonant with the almost universally received opinion, that countries become gradually warmer in the ratio of their cultivation population, etc. The last year (1835) holds a conspicuous place in point of lowness of temperature, its mean being 2.8°F [1.5°C] lower than the mean of the [previous] ten years" (Hall 1836, p. 238).

The letter to the *Montreal Star* in 1859 referred to one of the coldest spells ever recorded in Montreal, the four days from January 9 to 12, 1859. During this period, the absolute coldest temperatures ever recorded in the Saint Lawrence valley of −42.3°F (−41.3°C) were observed by Charles Smallwood on the island of Laval to the north of Montreal, and −40°F (−40°C) were observed by Edouard-Louis Glackmeyer in Quebec City. The letter to the *Star*, written by Archibald Ferguson, lists the 7:00 a.m. and 3:00 p.m. temperatures of the coldest days of every year since 1820 as extracted from "examining my friend Mr. Skakel's meteorological table for the last 40 years." (McCord 1843). Again, this must refer to William Skakel, not Alexander. This list of the values and dates of the coldest temperatures of each year was key to the positive attribution to the Skakel brothers of two unidentified manuscript ledgers of observations in the McGill University Archives.

John Bethune

John Bethune was the son of the chaplain of the Royal Highland Emigrant Regiment. His father, a Loyalist refugee from the American War of Independence, was briefly a Presbyterian minister in Montreal in the late eighteenth century before moving to Upper Canada. His son John was educated by John Strachan, renowned for his establishment of education in Upper Canada and his role in the founding of the University of Toronto. It was thought to be Strachan's influence that turned the young John Bethune, who had the distinction of being the first Canadian trained as a minister for the Church of England in Canada, towards the Anglican Church, rather than the Presbyterian denomination of his father (Campbell 1887). He was sent to Montreal as a minister of the Anglican Diocese of Christ Church in 1818. The close social links between the weather observers can be seen in their religious practices. John Bethune baptised John Samuel McCord's children, for example, while John Bethune's father had worked with Alexander Spark in establishing the Presbyterian Church in Montreal.

Bethune was appointed principal of McGill University in 1835 and was instrumental in starting the construction of the university. The funds left by James McGill for the founding of a university had been in litigation during the twenty years since McGill's death[10] so that by the mid-1830s, little, if anything, had been accomplished towards the development of the university. Bethune's principalship was controversial, however, not least because he wanted the university to become an exclusively Anglican institution, not a popular proposal at the time in multidenominational Montreal. He was dismissed in 1846. Although he certainly seems to have made himself unpopular and would appear to have been a difficult person to get along with, it is thanks to his tenacity under difficult circumstances that McGill University was eventually established. The account he had printed tells of his years of lonely struggle on limited budgets with scanty funds to hire staff. Bethune paid for basic heating and repairs to the college from his own pocket; a theme that will recur in the struggle to establish the Toronto Observatory as well (Bethune 1846). Bethune was appointed rector of Christ Church in 1850 and dean in 1854, a position he retained until his death. Curiously, his weather journal, which continued until 1869, was conserved in McCord's scientific papers. Since McCord himself died in 1865, Bethune's journal must have been donated after McCord's death to a family member, possibly his daughter Anne, who placed it among McCord's papers. McCord's reputation as a meteorologist and his interest in collecting historical weather records survived even after his death.

It is not always easy, nearly 200 years later, to enter into the thoughts and motivations of these nineteenth-century weather observers, but it is clear that their weather records formed an integral part of their daily routines. Bethune and McCord certainly show characteristics of determination and even tenacity in difficult circumstances. While clearly fascinated by the weather and meteorological observation and instrumentation, McCord, like many of his nineteenth-century contemporary scientists, also saw his scientific activities as a duty to society, motivated at least in part by religious devotion. In this era of clergymen scientists such as Bethune and Spark, understanding

---

10. James McGill's death was recorded in McCord's register on December 19, 1813: "Fair, NW, Mr. McGill died" (McCord & McCord 1826).

"God's Second Book," the book of nature (his first book being the Bible), was complementary to, and even an aspect of, religion. In 1842, McCord was assigned as a circuit court judge, which involved constantly changing courthouses in locations that were several days' travel from Montreal. He was forced to abandon his detailed meteorological observations and research, for which a stable observing location is of paramount importance. If, as suggested by Young, McCord's elevation to circuit court judge was effectively an exile from Montreal, as punishment for his participation as an officer in the militia on the government side of the Rebellions of 1837–1838, this political reprisal had damaging consequences for the development of meteorology and climatology in Canada (Young 2014). McCord never returned to his analysis of weather or his compilation of historical climatic records. Both his historical and personal meteorological records remained unpublished.

Despite this, McCord's role in Canadian climatology is a crucial one. He was the first to systematically collect and analyse instrumental observations from Canada. Where Spark and Skakel appear to have been content to record their observations, McCord analysed both his own and other's records, calculating means, comparing instruments, and setting rigorous standards for instrument calibration. He was the first to try to empirically investigate with instrumental observations whether the climatic changes that had been speculated about for centuries (chapters 3 and 6) were actually taking place. The lack of standardised instruments and observations scheduling in the data he collected hampered his investigations, and this work remained unpublished. During his most active period in the 1830s and early 1840s, he was a member in the Natural History Society, corresponded with other scientists interested in meteorology in the United Kingdom and the United States, and provided some of the first accurate climatological observations for Montreal. He instigated the systematic keeping of weather records by the military and helped to establish standard observing times. Although McCord felt he was unable to adequately analyse the records he collected from observers around the province, he left both his own original daily records and those he had collected for future researchers. The historical climate records left to us by McCord form the core documents around which today's understanding of past weather and climate in the Saint Lawrence valley is shaped.

References

Bayfield, H. W., 1824: On the geology of Lake Superior. *Trans. Lit. Hist. Soc. Que.*, **1**, 1–42.

———. 1831: On the coral animals in the Gulf of St Lawrence. *Trans. Lit. Hist. Soc. Que.*, **2**, 1–7.

Bethune, J., 1846 unpublished: A narrative of the connection of the Rev. J. Bethune, D. D. with McGill College, as Principal of the Institution. McGill University Archives, RG 2, Container 604, Office of Principal.

Blodget, L., 1857: *Climatology of the United States, and of the Temperate Latitudes of the North American Continent.* Lippincott, 536 pp.

Bonnycastle, R., 1824: On meteorological phenomena observed in Canada. *Trans. Lit. Hist. Soc. Que.*, **1**, 47–51.

Campbell, R., 1887: *A History of the Scotch Presbyterian Church, St. Gabriel Street Montreal.* W. Drysdale, 807 pp.

Dalhousie, Countess, 1824: Catalogue of Canadian plants. *Trans. Lit. Hist. Soc. Que.*, **1**, 255–262

Daubeny, C., 1838: On the climate of North America. Report of the Eighth annual meeting of the British Association for the Advancement of Science, p 29–32.

Dove, H. W., 1840: *Ueber die nicht periodischen Änderungen der Temperaturvertheilung auf der Oberfläche der Erde in dem Zeitraume von 1789 bis 1838.* G. E. Reimer, 131 pp.

Fleming, J. R., 1990: *Meteorology in America, 1800–1870.* Johns Hopkins University Press, 264 pp.

Frost, S. B., 1988: Skakel, Alexander. *Dictionary of Canadian Biography*, Vol. 7, University of Toronto/Université Laval, accessed 04/18/2012, http://www.biographi.ca/en/bio/skakel_alexander_7E.html.

Hall, A., 1836: Mean temperature of Montreal, Lower Canada, for the period of ten years, viz. from 1826 to 1835 inclusive. *Edinburgh New Philos. J.*, **21**, 236–239.

Hamblyn, R., 2001: *The Invention of Clouds: How an Amateur Meteorologist Forged the Language of the Skies.* Picador, 403 pp.

Herschel, J. F. W., 1840. Letter to Mr. McCord, December 14th 1840. McCord Family Fonds, Papers P001-820. McCord Museum Archives.

Howard, L., 1833: *The Climate of London.* Vol. 1. Harvey and Dauton, 348 pp.

Humboldt, A. V., 1817: *Des lignes isothermes et de la distribution de la chaleur sur le globe.* Perronneau, 145 pp.

Jones, T. P., Ed, 1838. Meteorology—Meeting of the British Association. *Journal of the Franklin Institute*, **12**, 394–396.

Kelly, W., 1837: On the temperature, fogs and mirages of the River St-Lawrence, *Trans. Lit. Hist. Soc. Que.*, **3**, 1–43.

Kuntz, H., 2010: Science culture in English-speaking Montreal, 1815–1842. Ph.D. dissertation, Concordia University, 391 pp.

Marsden, W., 1847: Meteorological observations at Nicolet. *Br. Amer. J. Med. Phys. Sci.*, **3**, 88–89.

McCord, J. S., 1835 unpublished manuscript: Meteorological tables. McCord Family Fonds, Papers P001-829. McCord Museum Archives.

——, 1836 unpublished manuscript: Scientific notebook. McCord Family Fonds, Papers P001-825. McCord Museum Archives.

——, 1838 unpublished manuscript: Letter to Dr. Skey. McCord Family Fonds, Papers P001-827. McCord Museum Archives.

——, 1839: Meteorological register for 1838 kept at Montreal, Lower Canada. *Amer. J. Sci. Arts*, **36**, 180–183.

——, 1840: Meteorological summary of the weather at Montreal, province of Canada (from 1836 to 1840). *Amer. J. Sci. Arts*, **41**, 330–331.

——, 1842: Report of the meteorological observations made on the island of St. Helen's. Natural History Society of Montreal, Ed., Lovell, 18 pp.

——, 1843 unpublished manuscript: Notebook on climate and meteorology of North America. McCord Family Fonds, Papers P001-828. McCord Museum Archives.

——, 1845: Meteorological observations on the mean temperature of Quebec. *Br. Amer. J. Med. Phys. Sci.*, **1**, 35–36.

——, 1846: Meteorological observations. *Br. Amer. J. Med. Phys. Sci.*, **1**, 89.

——, 1847: Observations on meteorology. *Br. Amer. J. Med. Phys. Sci.*, **3**, 228–229.

McCord, T., and J. S. McCord, 1826 unpublished manuscript: McCord's meteorological register. Meteorological Records, 1798–1972, Container 1039, Envelope 320. Faculty of Arts and Sciences, Department of Meteorology, McGill University Archives.

Moore, P., 2015: *The Weather Experiment: The Pioneers Who Sought to See the Future.* Random House, 416 pp.

Schott, C. A., 1876: *Tables, Distribution and Variations of the Atmospheric Temperature in the United States, and Some Adjacent Parts of America.* Smithsonian Institution, 345 pp.

Sheets-Pyenson, S., 1996: *John William Dawson: Faith, Hope, and Science.* McGill-Queen's University Press, 304 pp.

Sheppard, Mrs., 1824: On the recent shells which characterise Quebec and its environs. *Trans. Lit. Hist. Soc. Que.*, **1**, 188–197.

Slonosky, V. C., 2014: Daily minimum and maximum temperature in the St-Lawrence valley, Quebec: Two centuries of climatic observations from Canada. *Int. J. Climatol.*, **35**, 1662–1681, https://doi.org/10.1002/joc.4085.

——, 2015: Historical climate observations in Canada: 18th and 19th century daily temperature from the St. Lawrence valley, Quebec. *Geosci. Data J.*, **1**, 103–120, https://doi.org/10.1002/gdj3.11.

Smallwood, C., 1860: Contributions to meteorology reduced from observations taken at St. Martin, Isle Jesus, C.E. *Can. J.*, **27**, 308–312.

Wilkie, D., 1831: Grammar of the Huron Language, *Trans. Lit. Hist. Soc. Que.*, **2**, 94–197).

Young, B., 2014: *Patrician Families and the Making of Quebec.* McGill-Queen's University Press, 452 pp.

Zeller, S., 2009: *Inventing Canada: Early Victorian Science and the Idea of a Transcontinental Nation.* McGill-Queen's University Press, 372 pp.

Zilberstein, A., 2016: *A Temperate Empire: Making Climate Change in Early America.* Oxford University Press, 280 pp.

CHAPTER SIX

Nineteenth-Century Scientists Question Climate Amelioration

As European settlement continued and forests gave way for farmlands across eastern Canada, the question of people changing the climate through deforestation and cultivation continued to be debated into the early nineteenth century. The settling of the Loyalist refugees from the American Revolution in the 1780s, and later increasing immigration from the British Isles after the end of the Napoleonic Wars in 1815, continually pushed the frontier farther west and north. The question of the suitability of the climate for British settlement was one that provoked considerable anxiety as the other half of coin of the climatic improvement theory, climatic determinism, raised concerns. Climatic determinism is the belief that human character, and sometimes even appearance, is shaped by climate. Anya Zilberstein has described in detail the political and social implications of climatic determinism in British North America in her recent account, *A Temperate Empire*.

In an introduction to a meteorological summary for 1826 published in a Halifax newspaper, the *Novascotian*, the editors wrote as follows:

> We have long regarded it as a subject of regret, that so little attention has been hitherto paid to the collection of facts such as the following; for the climate of a country is one of the most useful and interesting enquiries that can be

> instituted. In a new one like Nova Scotia it is particularly so. It is a theory which now almost universally obtains that the climate of a young country is undergoing, with the progress of improvement, a slow but gradual amelioration. By some it is contended that the winters of the Province, within the last 30 years have become infinitely more mild, and that our past experience affords new evidence of the theory; while others maintain that they are quite as severe as formerly and that tho' there be some amelioration observed in particular seasons the change is only periodical and temporary. We are satisfied ourselves, that the former opinion is correct; but at the same time it would be more satisfactory if we could refer to actual observations and found our reasonings upon meteorological tables. (Young 1827)

Considering that this was written not quite a decade after the brutally cold 1810s, the universal belief in "infinitely milder" winters again shows a contradiction between our current understanding of past climatic changes and contemporary experience, similar to the contradiction between our understanding of the Little Ice Age and the Enlightenment discussion of climate change in the eighteenth century (chapter 3).

By the time this was written in 1826, Spark had been keeping methodical records for more than twenty years, which could have been used to furnish those very tables whose absence is lamented in the *Novascotian*. Spark's records could also have been compared to Gaultier's observations from eighty years earlier. John Lambert (1775–ca. 1811) did just that in 1810.

John Lambert: A Comparison with Gaultier

John Lambert visited Canada from 1806 to 1808 in the hopes of setting up a hemp business. The hemp industry failed but his travel memoir, published in 1810, persists. He devoted a chapter to the Canadian climate, including a description of the extraordinarily warm summer of 1807 and the popular idea of climatic improvement. He effectively dismissed the idea that there had been permanent, long-term climatic warming by referring to "an old journal" (Lambert 1810, p. 119) kept in 1745. Comparing Lambert's analysis with Gaultier's letters, we can see that Lambert was using Gaultier's records to date events such as the last ice on the river or the first strawberries of the season. It is worth quoting Lambert's comparison between the climate of the early nineteenth century and Gaultier's observations:

> It is the general opinion of the inhabitants that the winters are milder, and

that less snow falls now than formerly. That the summers are also hotter. This might easily be accounted for, by the improved state of the country. The clearing of the woods, and cultivation of the land, together with the increased population, must naturally have a considerable effect upon the climate. The immense forest, which before interposed their thick foliage between the sun and the earth, and prevented the latter from receiving that genial warmth which was necessary to quality its rigorous atmosphere, are now considerably thinned, or entirely destroyed in various parts of the country . . . Added to this, the exhalations arising from so many thousands of men and cattle, together with the burning of so many combustibles, must greatly contribute to soften the severity of the climate. (Lambert 1810, p. 118–119)

Lambert continued, however, as follows:

Yet with all these truths, which amount nearly to a demonstration of the fact, and apparently substantiated by the opinion of the inhabitants, I do not find, upon reference to an old meteorological journal, that so great an alteration has taken place, at least within the last sixty years. (Lambert 1810, p. 119)

Lambert then gives point-by-point comparisons:

For the year 1745, it is observed, that on the 29th of January of that year, the river St. Lawrence near Quebec, was covered with ice, but that in preceding years, it had frequently been covered in the beginning of that month, or about the end of December. During my stay at Quebec in 1806, the river was covered with ice by the *first* week in December. . . .

In March, 1745, the journal mentions, that it had been a very mild winter, that the snow was only two feet deep, and the ice in the river, of the same thickness. In 1806 the snow was upon an average . . . four feet in depth. . . .

On the 20th of April, 1745, the ice in the river broke near Quebec, and went down. It is observed, however, in the journal, that it seldom happened so soon, for the river opposite Quebec was sometimes covered with ice on the 10th of May. On the 7th of April that year, the gardeners began to make hot beds, and on the 25th many of the farmers had begun to sow their corn.

In April 1807, the ice began to break up about the third week. . . .

Strawberries were to be had at Quebec on the 22nd June, 1745. But in 1807 we could not procure them till about the 15 or 20th of July . . .

On 22nd August, 1745, the harvest began in the vicinity of Quebec. In 1807 and 1808 it was above a week or ten days later, though the summer of that latter year was remarkably hot. An observation in the old journal, states, that the corn was never ripe in the years preceding 1745 till about the 15th of September. . . .

The Habitans [sic] continued to plough in 1745 till the 10th November. As late as the 18th, the cattle went out of doors . . . From these statements it appears evident, that an *improvement* in the climate of Canada, is extremely doubtful. . . . The winters sometimes differ so materially from each other, as well as the summers, that no accurate estimate can be formed, sufficient to ascertain whether the changes that take place, are occasioned by any increase or diminution of the severity of the climate. (Lambert 1810, p. 119–122)

As we saw in chapter 2, 1745 and 1746 were extremely warm years in Quebec, so if Lambert had more years of Gaultier's records than 1745, it's possible he chose a warm year from the past for his comparison to 1807 and 1808 if he was looking to disprove the climate improvement theory. Lambert doesn't seem to have had a high opinion of Canadians generally, describing them in his next paragraph as "hot-house plants" (Lambert 1810, p. 123) unable to stand the rigours of the cold as well as Europeans, as the Canadians had become used to practices such as indoor heating and dressing appropriately for the weather. Nevertheless, it is clear from Gaultier's and Spark's records that, as Lambert described, the weather changed considerably from one year to the next, making long-term trends, if any, difficult to distinguish.

It's also possible that Lambert found the references to Gaultier's observations of 1745 in Pehr Kalm's travel memoir. Gaultier's descriptions and observations of the climate in Quebec in the 1740s seem to have been unknown by the 1830s, although McCord made a note in his 1838 notebook to obtain or consult the 1744 volume of the *Mémoires de l'Académie Royale des Sciences*. It's not known if he managed to do so, but no further remarks about Gaultier's eighteenth-century observations appear in McCord's notebooks.

McCord, Kelly, and the Climate Improvement Theory

In 1832, William Kelly, a surgeon with the Royal Navy, presented a paper to the Literary and Historical Society of Quebec (LHSQ), which was later published in the society's transactions in 1837. Kelly's paper criticised the "clearing and cultivation" theory of climate change with some vigour. He remarked, somewhat astringently, that

> The general opinion is, that the climate of a country becomes milder, as it is cleared of forests, and cultivated. The application of this doctrine to Canada[1] is not at all new, for I find it entered into the speculation of a writer on this country towards the end of the 17th century.[2] It is founded on the generally admitted fact, that the climate of Europe is much milder now, than at the commencement of the Christian era [as] inferred from the ... Roman writers ... This explanation seems to have been generally received without a very strict examination into the facts connected with it, and kept for want of a better. (Kelly 1837, p. 58–59)

Kelly then proceeded, like Mann some forty years earlier, to look through historical documents to see what had been written about the climate of Quebec in the past. He examined the writings of Samuel de Champlain, which described the climate of Quebec from the year 1608, and noted the dates on which wheat and rye were sown, the first snow fell, whether the river froze in winter, when the snow melted in spring, the dates of sugaring off, and of the maturity of different fruit and crops. Kelly concluded from this that, if anything, the springs of the 1820s and 1830s were later than in Champlain's time in Quebec at the beginning of the seventeenth century: "From the little evidence we have here ... the conclusion that the climate has undergone no change, may I think be safely drawn" (Kelly 1837, p. 63).

Kelly (1837, p. 65) felt that his historical review of the literature on climate in Canada

> conveys little satisfactory information regarding the climate generally ... it seems probable that no material change has taken place in the climate of Canada for the last 200 years at least ... we have little amelioration to expect in the future, however disagreeable ... it may be to admit it.

He noted that many of the indicators often used to demonstrate climate

---

1. By "Canada," these writers are generally referring to the region around the Saint Lawrence valley and the northern shores of Lake Ontario and Lake Erie, not the entire country called Canada today.

2. Unfortunately, Kelly does not give a reference.

change, especially the dates of the first arrival of the ships in spring, or the date of the latest departure in autumn, were not particularly reliable, as they often depended on factors other than the weather. Kelly was also writing during a renewed phase of the Little Ice Age, the generally cold period from about the fourteenth century to the nineteenth century (chapter 3), which was considered to have been at a severe phase in the early part of the nineteenth century in North America. Analysis of historical temperature using Gaultier, Spark, McCord, and others' observations shows the 1810s to have been particularly cold and the 1830s to have been volatile (Slonosky 2014).

It's interesting here to see that Kelly only had one major document to compare the climate of the 1820s and 1830s to that of 200 years earlier; he seemed to be unaware not only of Gaultier's writings but also of Spark's meteorological observations over the period from 1798 to 1819. This is particularly odd as the LHSQ's library and meeting rooms were next door to Spark's church. McCord mentioned that his old school teacher, Daniel Wilkes, sent some of Spark's observations to him after Spark's death, and a number of pages of recopied tables appear in McCord's notebook, while Spark's original diaries are conserved in the McGill University Archives in Montreal. Kelly was in contact with McCord and the two met in May 1836 when McCord went to Quebec to compare and calibrate instruments.[3]

Around this time McCord was translating extensive passages from Mann's Belgian papers, so he was clearly able to obtain some information from eighteenth-century Europe, and his extensive education in French at the Jesuit Collège de Montréal meant language was no barrier to reading works written in French. He referred to publications from the *Mémoires de l'Académie Royale des Sciences* in his scientific notebooks, and both McCord and Spark were familiar with Réaumur's work and suggestions on the exposure of thermometers. It seems quite odd, from a twenty-first-century point of view (and with unprecedented access to scanned documents and online archives), that neither of them appear to have known about Gaultier's papers, although, as we saw before, McCord did try to get a copy of the 1745 *Mémoirs*. Gaultier's observations are roughly contemporaneous with Réaumur's, whose work on thermometer exposure both McCord and Spark had read.

---

3. May 14, 1836: "Visited Quebec City to compare Newman's Bar[ometer] with Dr Kelly's (by Jones) and Mr. Watt's instrument on Cape Diamond. Dr. K's Bar was 0.02 higher than mine with appt heights Mr. Watt's .11- neither had the neut [neutral] point, inverted height of Watt's bar on Cape 330 feet" (McCord 1842).

A letter written by McCord and addressed to Dr. Shey, inspector general of hospitals in Quebec City and vice president of the LHSQ, contained remarks on a letter written by Kelly titled "Notes on the Climate of Lower Canada," and also had a description of McCord's view on changes in climate, or lack thereof, in Lower Canada:

> Most fully do my own observations, supported by many curious and ancient records of the climate of Canada, corroborate the Dr.'s [Kelly's] opinion that there has been no essential amelioration in its climate since the period of its first discovery, and after a careful comparison, in the first instance of these records exhibiting the period of the opening of the navigation, the melting of the snow, the first vegetation and the flowering of trees and plants at the times of earliest settlement of the country with similar phenomena at the present day; —and secondly various old thermometrical registers with the more modern, I am distinctly of the opinion that if the extremes are less intense and long (which I believe) the mean temperature of these Provinces has not materially changed. (McCord 1838)

This view was finally published, in a more succinct form, in the *British American Medical and Physical Journal* in 1847 (McCord 1847, p. 228).

Around 1838, McCord wrote in his scientific journal what appears to be a draft for possible publication:

> All my attention for recent years has been directed to that branch of meteorology connected with ascertaining the mean temperature of Lower Canada. I had not made up my mind as to the fact of the climate having materially altered but was impressed generally with the belief that the climate had improved, or become warmer, as it had increased in population and cultivation ... I resolved to make diligent search for such records as might be found in the Province, in order to answer this interesting question. The difficulty was even greater than I had anticipated. Few individuals had turned their attention to this (even now) infant science, at a period sufficiently removed to bear comparison with the present, and of those which I was fortunate enough to discover, many were unsuitable from the irregular manner in which they were kept. (McCord 1836)

Did McCord include his own earlier and his father's observations from 1813 to 1826 in the "unsuitable" category?

Later, McCord added, with the exasperation expressed by Mann, Kelly, and every climatologist since, a pithy description of the problems he encountered in trying to find and use historical observations to determine whether the climate has changed in the past:

> During the course of some investigations made several years ago, on the subject of the climate of Canada with a view to ascertaining whether any and what changes had taken place in its temperature, many tables came into my possession . . . Knowing from experience how difficult it is to obtain even these scanty data on which to base a comparison, I have taken the liberty of selecting some of them for insertion in the _____[4] in order that future students of this interesting but infant science, and who may not have access to several means, may be saved all the trouble and research which fell to my lot. No one but the zealous meteorologist knows how very difficult it is to obtain observations in this science which can be depended upon. (McCord 1836)

At around the same time as McCord was collecting records in Montreal, Richard Bonnycastle of the Royal Engineers, in his *History of Newfoundland*, alluded to the "various attempts [which] have been made to account for the . . . anomaly between the climates of the old and new world" (Bonnycastle 1842, p. 330), referring to the problem of why the climate of similar latitudes were colder in North America than in Europe. He mentioned the shape of the continents and the fact that the westerly winds over eastern North America came from the cold continental interior and to the "mantle of forest . . . the covering of trees prevents the access of solar heat" (p. 330–331). Lefroy, director of the Toronto Magnetic and Meteorological Observatory, wrote in 1853 that "we ought to learn whether we can bring about changes of climate by human agency" (Lefroy 1853, p. 29), although he didn't specifically mention clearing and cultivation.

Alexander von Humboldt, whose influence on nineteenth-century geoscience was immense (see chapters 7 and 8), was also dismissive of claims of climate change, describing theories of "disturbances in the climatic relations of our globe" and "phantoms of the imagination [which] are so much more injurious as they derive their source from dogmatic pretensions to true science. The history of the atmosphere, and of the annual variations of its temperature, extends already sufficiently far back to . . . afford sufficient guarantee against the exaggerated apprehension of a general and progressive deterioration of the climates" (Humboldt 1997, p. 51). Here, Humboldt is addressing a fear of a

---

4. Presumably McCord meant to publish the results of his research in a journal, but this doesn't appear to have ever happened.

cooling climate, deterioration at that time being associated with cooling and improvement being associated with warmth. Concerning "the questions so often agitated, whether the mean temperature has experienced any considerable differences in the course of centuries . . . whether the winters have not become milder and the summers cooler," Humboldt considered the instrumental record, as it stood in the mid-nineteenth century, to be too short, as the issue "can only be solved by means of the thermometer . . . and its scientific application hardly dates back 120 years" (p. 175).

Finally, in 1867, John Disturnell (1801–1877) devoted four pages in his influential book, *Influence of Climate in North and South America*, to ending the theory that human clearing and cultivation of land in North America altered the climate. He quoted extensively from Dr. Lillie's *Essay on Canada*, which examined more thoroughly the documentary records of Canadian climate since the sixteenth century:

> The historian Charlevoix . . . thinks that, even in his time, the clearings had made some difference in the temperature, and rendered it less cold than in the first years of the colony's existence. In our day, the enquiry is sometimes made, whether the destruction of the great forests which were on the banks of the St. Lawrence, has caused any improvement in the climate of the country—and the question is a very interesting one, which it is worthwhile examining. (Disturnell 1867, p. 71)

Disturnell, quoting Lillie, started with Jacques Cartier's (1491–1557) descriptions of the climate from his voyage in 1535–1536, noting, like Kelly, the dates of the freezing of the Saint Lawrence and the melting of the snow in spring. Lillie next examined Champlain's journals as well as journals from the Jesuits in New France in the 1640s and 1650s.[5]

Following a now well-worn path, Lillie then considered similar observations of the timing of snowfall, river freezing and ice breakup, and snow melt for the 1850s and, as Kelly did in the 1830s, concluded that

> it seems safe to infer that the mean annual temperature of Canada has not

---

5. Although there may be some confusion with dates in Cartier's account because of the Julian calendar that was then in use, France adopted the Gregorian calendar in 1582, so the dates from the French regime correspond to today's dates. The United Kingdom didn't adopt the Gregorian calendar until 1752, a few years before the Conquest, so Canada has always used the Gregorian calendar.

materially changed during the past three centuries ... and the clearings made until now have had very little influence in the present temperature of Canada. It is, then to other influences besides the existence of forests, that the great cold of the winter of our country is due. They are to be looked for in the dryness of the northern atmosphere; the neighbourhood of Hudson Bay, which is covered with ice during a great portion of the year; in the frequency of the northwest winds, which carry away from America the heated moisture produced by the warm current of the Gulf of Mexico. (Disturnell 1867, p. 70–76)

This appears to mark the end of the deforestation and climate improvement theory in Canada. The discussion of the influence of forests on climate, especially in terms of the potential effect on precipitation, continued in Europe and the United States (Stehr and von Storch 2000; Thompson 1990). Research on the influence of forests on climate continues to our day (e.g., Montenegro et al. 2009).

References

Bonnycastle, R. H, 1842: *Newfoundland in 1842: a sequel to "The Canadas in 1841"* H. Colbourn, 367 pp.

Disturnell, J., 1867: *Influence of Climate in North and South America.* D. Van Nostrand, 334 pp.

Humboldt, A. V., 1997: *Cosmos: A Sketch of the Physical Description of the Universe.* Vol. 1. E. C. Otté, Trans., Johns Hopkins University Press, 375 pp.

Kelly, W., 1837: Abstract of the meteorological journal kept at Cape Diamond, Quebec, from the 1st of January, 1824, to 31st December, 1831, with some remarks on the climate of Lower Canada. *Trans. Lit. Hist. Soc. Que.*, **3**, 46–71.

Lambert, J., 1810: *Travels through Lower Canada and the United States in the Years 1806, 1807, and 1808.* Vol. 1. Richard Philips, 496 pp.

Lefroy, J. H., 1848 unpublished manuscript: Letter to Mr. Justice McCord. McCord Family Fonds, Papers P001-820. McCord Museum Archives.

McCord, J. S., 1836 unpublished manuscript: Scientific notebook. McCord Family Fonds, Papers P001-825. McCord Museum Archives.

——, 1838 unpublished manuscript: Letter to Dr. Skey. McCord Family Fonds, Papers P001-827. McCord Museum Archives.

——, 1843 unpublished manuscript: Notebook on climate and meteorology of North America. McCord Family Fonds, Papers P001-828. McCord Museum Archives.

——, 1847: Observations on meteorology. *Br. Amer. J. Med. Phys. Sci.*, **3**, 228–229.

Montenegro, A., M. Eby, Q. Mu, M. Mulligan, A. J. Weaver, E. C. Wiebe, and M. Zhao, 2009: The net carbon drawdown of small scale afforestation from satellite observations. *Glob. Planet. Change*, **69**, 195–204, https://doi.org/10.1016/j.gloplacha.2009.08.005.

Slonosky, V. C., 2014: Historical climate observations in Canada: 18th and 19th century daily temperature from the St. Lawrence valley, Quebec. *Geosci. Data J.*, **1**, 103–120, https://doi.org/10.1002/gdj3.11.

Stehr, N., and H. von Storch, Eds., 2000: *Eduard Brückner—The Sources and Consequences of Climate Change and Climate Variability in Historical Times*. Springer, 338 pp.

Thompson, K., 1990: Forests and climate change in America: Some early views. *Clim. Change*, **3**, 47–64, https://doi.org/10.1007/BF02423168.

Young, G., 1827: Meteorology. *The Novascotian*.

North America in the mid-eighteenth century. "Canada, Louisiane, possessions anglais, par le S. Robert de Vaugondy, géographe ordinaire du roi." The territory under French control is shaded in green while land under British control is shaded in red (Courtesy of Rare Books and Special Collections, McGill University Library, G3300 1762 R62 RBD map.)

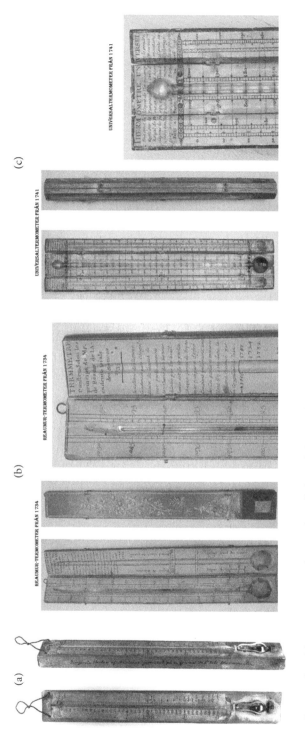

Photographs of thermometers sent by Joseph-Nicholas Delisle to Anders Celsius, currently kept at the Department of Earth Sciences, Uppsala University, reproduced here with the kind permission of Hans Bergström. The various instruments used by Gaultier in Quebec would have resembled these. (a) This thermometer was sent to Celsius by Delisle and was used by Celsius in his research on temperature scales and the influence of pressure on the boiling point of water. The scale is inverted compared to modern usage, with the boiling point of water indicated at the top as 0° and the freezing point (première gelée) at (right) 100° Celsius or (left) 152° Delisle (H. Bergström 2017, personal communication). Photograph taken by Professor Gösta H. Liljequist, Department of Meteorology, Uppsala University. (b) Thermometer constructed according to Réaumur's principles (1734). The freezing point is marked at 0°, and the temperature of the cellars of the Paris Observatory are marked near the 10° line. (c) Universal thermometer, constructed by Delisle at the Paris Observatory in April 1741. Photographs of (b) and (c) courtesy of Hans Bergström, Departments of Earth Sciences, Uppsala University.

OCTOBRE 1743.

Jours.	DEGRÉS du Thermomètre.	VARIATION DU TEMPS.	Situation du vent.
1	10. + 0	très-beau & fort chaud.	nord-ouest.
2	M. 10 + 0	fort beau..........	
Id.	S. 13 + 0	vent violent........	nord-est.
3	M. 9 = 0	grand vent & froid....	
Id.	S. 17 + 0	assez beau.........	sud-ouest.
4	M. 16 + 0	beau & chaud.......	
Id.	S. 2 + les caves.	Idem.............	
5	M. 15 + 0	fort beau & fort chaud...	
Id.	S. 9 + les caves.	Idem.............	
6	M. 12 + 0	fort beau..........	
Id.	S. 17 + 0	Idem.............	sud-ouest.
7	9 − 0	il gela pendant la nuit...	
8	M. 9 = 0	il gela pendant la nuit...	
Id.	S. = les caves.	il fit beau & chaud le jour.	
9	M. 8 + 0.	fort chaud tout le jour...	
Id.	S. 3½ + les caves.	pluie abondante le soir...	
10	M. 1 − les caves.	brouillard épais......	nord-est.
Id.	S. . + les caves.	fort beau..........	
11	M. 15 + 0	le temps fut couvert &	
Id.	S. = les caves.	fort chaud........	nord-ouest.
12	16 + 0	pluvieux...........	
13	8 + 0	neige & pluie.......	
14	..... = 0	beau & froid........	
15	M. . , = 0	très-beau..........	
Id.	S. 12 + 0	Idem.............	
16	M. 8 + 0	pluvieux...........	
Id.	S. 12 + 0	Idem.............	sud-ouest.
17	M. 11 + 0	fort beau..........	
Id.	S. 16 + 0	Idem.............	
18	M. .. = 0	gelée & beau.......	
Id.	S. 15 + 0.	Idem.............	
19	11 + 0	très-beau..........	

Illustration of a page of Gaultier's records with "caves" used as a reference point for temperature (e.g., Oct 4).

A view of Quebec City from the Saint Lawrence River, with the Quebec Citadel and Cap Diamant to the left. Painting titled *Crossing the Saint Lawrence River in winter, at Quebec City, Lower Canada*, by George St. Vincent Whitmore (1836; Library and Archives Canada, Acc. 1970-188-1051, W. H. Coverdale Collection of Canadiana).

The Hôtel-Dieu hospital at which Gaultier worked in Quebec City. *Vue à vol d'oiseau de l'Hôtel-Dieu de Quebec* by Eugen Haberer. [L'Opinion Publique, Vol. 8, No. 35 (30 août 1877), p. 411.] PER O-104. Reproduced with permission from the Bibliothèque et Archives nationales du Québec.

The first page of Gaultier's report for 1754 gives a description of his thermometer, specially designed by Réaumur, and its placement [Gaultier, 1754; Ms Can 42(1); courtesy of the Houghton Library, Harvard University].

Alexander Spark. [Extract from Bibliothèque et Archives nationales du Québec collection P1000, S4, D83, PS36/Collection Centre d'archives de Québec/ Alexander Spark, ministre anglican (version 1950, original avant 1819).]

St. Andrew's Presbyterian Church, Quebec City (H. Hacot).

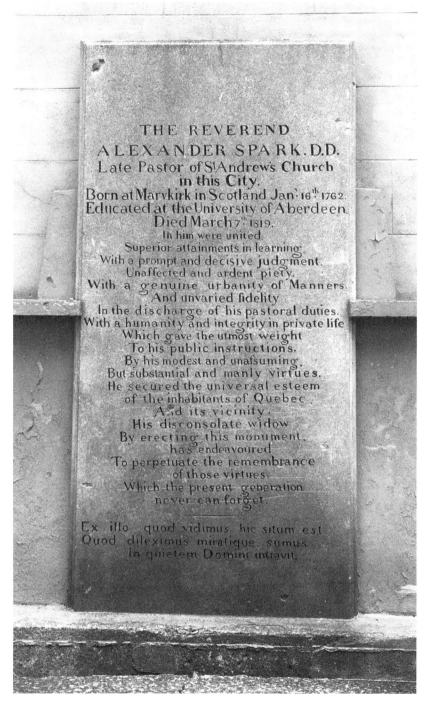

Memorial to Dr. Spark. Photo credit: H. Hacot.

2. Morning, feezes [sun]shine sometimes.
2. Morning [illegible] sometimes.

American States

June 18, 1812. America declared war with England

July 12. Genl. Hull having crossed from Detroit published a proclamation announcing his being in possession of Canada &c—

July 17. Machulimakinac surrendered to the British.

On the 18, 19, & 20 July Genl. Hull made three attempts of [Amentotberg] — & was repulsed

Augt. 15 Genl. Hull, with all his Army 2,500 surrendered to Genl. Brock, together with the City of Detroit.

Octr. 13. General Brock was killed in the commencement of an engagement with a party of Americans who crossed over to Queenstown near Fort George. the rebels who crossed over, about 15.00, were [nearly all] killed or taken prisoners —

September 1812

Days	8 A.M. Ther. Bar. weather	9 P.M. Ther. Bar. weather
1	46 29.7 fine	61 67 29.7 fine
2	52 29.5 overcast	55 53 29.4 rain all night E
3	54 29.5 cloudy E	52 55 29.7 cloudy E
4	54 29.7 overcast B	62 60 30.0 fine
5	52 29.9 fine	66 70 29.8 fine
6	54 29.9 fine	64 66 29.8 fine
7	55 29.8 fine	70 72 29.3 fine
8	56 29.7 overcast	62 64 29.6 cloudy
9	60 29.6 cloudy	66 70 29.5 fine
10	54 29.7 fine	62 67 29.8 fine
11	54 29. fine	68 70 29.9 fine — rain all night
12	56 29.8 fine	72 70 29.9 fine — rain all night E
13	48 29.8 rain E	50 52 29.8 [bleak] E
14	52 29.8 cloudy E	54 54 29.8 rain E
15	51 29.8 fine	54 57 29.8 overcast
16	47 29.9 fine	50 65 29.9 fine
17	55 29.9 fine	65 69 29.7 overcast — rain night
18	58 29.5 cloudy	60 62 29.5 fine
19	48 29.5 fine	54 57 29.5 fine
20	48 29.3 [bleak] E	52 54 29.5 [bleak]
21	46 29.4 cloudy	52 53 29.4 fine
22	41 29.4 cloudy E	50 54 29.4 fine S.E.
23	39 29.7 fine	52 56 29.7 fine
24	44 29.7 overcast	52 56 29.7 overcast
25	48 29.8 fine	52 57 29.3 fine
26	40 29.7 cloudy	54 57 29.5 cloudy — rain every night
27	45 29.5 fine	46 48 29.6 fine
28	39 30.0 fine	50 52 30.0 fine
29	38 30.0 fine	48 60 29.9 fine
30	47 29.8 fine	58 67 29.8 fine

Example of Spark's weather journal for September 1812 with temperature, atmospheric pressure, and the state of the weather and wind direction recorded at 8:00 a.m. and 3:00 p.m. Temperature is also recorded at noon. On the opposite (left hand) page, Spark noted several dates relating to the War of 1812, including the day the Americans declared war on Britain and the death of General Brock at Queenston Heights in Niagara. Image courtesy of McGill University Archives (RG 2 C1039 319-001).

Example of the first McCord journal for April 1818. Alongside the temperature recorded in the morning and the afternoon, detailed descriptions of the weather and the state of the river ice (chapter 12) are noted. Pencil marks in the lower right-hand side show where John Samuel McCord calculated mean temperatures from these diaries in the 1830s. Image courtesy of the McGill University Archives (RG 2, C1039, 320-004).

La Grange aux Pauvres, the farmhouse where Thomas McCord and his sons John Samuel and William King lived in the early nineteenth century (photograph by Alexander Henderson, circa 1872; McCord Museum Archives MP-0000.33.7).

Portrait of Thomas McCord, by Louis Dulongpré (1816; McCord Museum M8354).

Thermometer from the McCord Museum. This relatively simple thermometer, graduated only to 2°F and with a Réaumur scale on the right, is likely one used in the earlier part of the century (McCord Museum M8461).

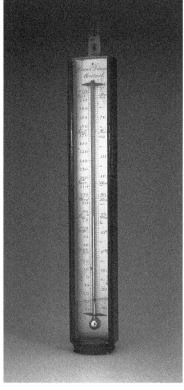

Example of John Samuel McCord's second weather diary from 1831 to 1842. This page from January 1835 shows McCord used several instruments to record, including two barometers and three thermometers (McCord Museum (P001-829).

View of Old Montreal from Temple Grove, by Henry William Cotton, circa 1847. Temple Grove, the McCord's summer house, was built on an estate on the southwestern slope of Mount Royal. McCord's experiments on solar radiation with Daubeny were performed in the garden of Temple Grove (McCord Museum M1859).

Portrait of John Samuel McCord by Frederick William Lock (1847; McCord Museum M8415).

Alexander Skakel, painted photograph by William Notman (1870; McCord Museum N-0000.2.4).

Portrait of John Bethune by William Sawyer (ca. 1845; McCord Museum M986X.137).

Print of Saint James Street, Montreal, by Robert Auchmuty Sproule, 1830, engraved by William Satchwell Leney (McGill Rare Books Lande Collection 32; courtesy of the Rare Books and Special Collections, McGill University Library). Several of the observers, including Alexander and William Skakel, William Sutherland (chapter 9), and John Samuel McCord, lived on Saint James Street.

View of Montreal from Saint Helen's Island by Robert Auchmuty Sproule, 1830, engraved by William Satchwell Leney (McGill Rare Books Lande Collection 35; courtesy of the Rare Books and Special Collections, McGill University Library).

Modern view of Old Montreal from Saint Helen's Island, from approximately the same location as above. Note the contrast between the views of the cathedral towers, which dominate the nineteenth-century view, but which are difficult even to discern in the twenty-first century (they are approximately in the center of the photo). Indeed, the area between the Saint Lawrence River and the cathedral has been so built up that it's not easy even to find a vantage point from which to take a picture showing the cathedral (photo credit: V. Slonosky).

Lieutenant (Major General) C. J. Buchanan Riddell (R. A. watercolor from a portrait in possession of Captain E. W. Creak. Toronto Reference Library, J. Ross Robertson Collection 1063).

Reconstructed officer's writing desk at Fort York (photo credit: A. J. Slonosky).

First Toronto Magnetic and Meteorological Observatory (1852): Observatory (1840), near S.E. corner of Convocation Hall, Toronto Ont. by William Armstrong (1822–1914). (English artist. Toronto Reference Library, J. Ross Robertson collection JRR 1069 Cab.) Erected in 1840, it was built of logs roughcast on the outside, plastered inside with a layer of air between logs and plaster. No iron was used; nails were of copper and locks of brass. Numbers in red indicate 1) observatory, 2) anemometer house, 3) detached experimental lab, 4) shed for (magnetic) inclination circle, and 5) barracks for officers (Toronto Public Library 1967, p. 95).

Portrait of John Henry Lefroy. *Scene in the Northwest – Portrait* (c. 1845–1846) by Paul Kane, courtesy of The Thomson Collection © Art Gallery of Ontario.

Reconstructed barracks, Fort York, Toronto (photo credit: A. J. Slonosky).

Sleeping and living arrangements inside reconstructed barracks, Fort York. By the late 1830s, efforts were underway to reduce the number of soldiers and their families living in common rooms, with the new barracks built for single soldiers and larger rooms given over to single-family occupation. Still, these give an idea of the difficulty in keeping the observers concentrated on their tasks and housing their families in the conditions recreated here (photo credit: A. J. Slonosky).

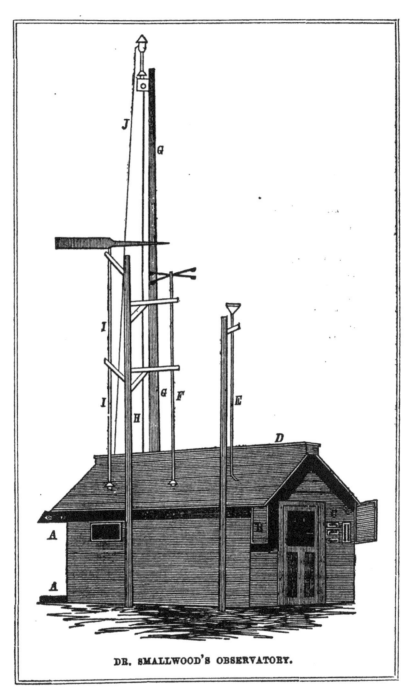

Smallwood's Saint Martin Observatory (sketch by Archibald Hall, originally published in the *Canadian Naturalist and Geologist*, Hall and Smallwood 1858).

Portrait of Dr. Charles Smallwood, 1872 (courtesy of McGill University Archives; credit: William Notman/McGill University Archives, Acc. 0000-2094 PR013186).

Photograph of Dr. Archibald Hall by William Notman (1866; McCord Museum I-20974.1).

The second observatory building, erected 1855 (Toronto Reference Library, J. Ross Robertson collection 1070). The numbers in red refer to 1) the observatory on the site of the first building, 2) weather offices, 3) self-recording magnetic instruments, and 4) the transit building. The dome was added in 1882 (water color, Toronto Reference Library, J. Ross Robertson collection 1071; Toronto Public Library 1967, p. 95).

Professor George Templeman Kingston, Director of the Toronto Magnetic and Meteorological Observatory and of the Dominion [of Canada] Meteorological Service 1855–1880 (colored photograph, water color, Toronto Reference Library, J. Ross Robertson collection 1066).

The Public Institutions of the Dominion of Canada: McGill College University, Montreal, by Eugene Haberer, Canadian Illustrated News. The observatory can be seen in the leftmost portion of the painting. The main college building is in the centre (Bibliothèque et Archives Nationales du Québec, Image Number 2723947).

Clement McLeod and students with surveying instruments (courtesy of the McGill University Archives, PR010554, photographer unknown).

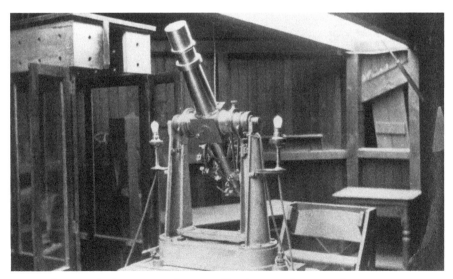

Transit telescope. Absolute time was determined by observing the transit of certain stars. The McGill Observatory was an important source of time measurement and dissemination, which was of critical importance for the determination of longitude. Payment for time signal services helped keep the observatory running (photographer unknown; photograph courtesy of McGill University Archive, PU010571).

Barometer and clock "Used for Time Signals all over McGill and the City of Montreal," McGill University, circa 1962 (photograph from the R. V. Nicholason collection of the Atmospheric and Oceanic Sciences Department of McGill University, courtesy of Prof. Frederic Fabry; McGill University Archives Acc. 0000-1101, PR010550).

Equipment used to relay time signals from the McGill Observatory across the continent through the telegraph (photograph from the R. V. Nicholason collection of the Atmospheric and Oceanic Sciences Department of McGill University, courtesy of Prof. Frederic Fabry).

The old McGill Observatory in winter, circa 1962 (photographer from McGill News; courtesy of the McGill University Archives, PR035572).

Snow gauge, circa 1962 (from the R. V. Nicholason collection of the Atmospheric and Oceanic Sciences Department of McGill University, courtesy of Prof. Frederic Fabry).

Sunshine recorder, circa 1962 (from the R. V. Nicholason collection of the Atmospheric and Oceanic Sciences Department of McGill University, courtesy of Prof. Frederic Fabry).

Current automatic weather station on McTavish Street (photo credit: V. Slonosky).

PART II

Meteorology Takes Shape

CHAPTER SEVEN

Meteorology and the Military

As the nineteenth century advanced, a Romantic, rather than a mechanistic, view of an integrated, interdependent Earth developed, largely as a result of Alexander von Humboldt's unified view of nature as a cosmic whole (see, e.g., Holmes 2008). This vision accorded well with the globally expanding European empires and decades of relative peace. The organization and apparatus of the military turned towards a global investigation of the geosciences. Considerable contributions were made by the British Empire's military officers to meteorology, climatology, and geoscience generally in nineteenth-century Canada. With the close of the Napoleonic Wars in 1815, the peacetime armies and navies of both the United Kingdom and France had the men, resources, and support structure to carry out worldwide observations in meteorology, geomagnetism, naval surveying, and cartography. In Canada, now firmly a part of the British Empire, the Royal Artillery worked with McCord and the Natural History Society (NHS) of Montreal to inaugurate bihourly observations at the Montreal garrison. The Royal Artillery, under the direction of Colonel Edward Sabine (1788–1883) in Woolwich, England, was also responsible for establishing the Toronto Magnetic and Meteorological Observatory. The Royal Engineers enthusiastically and faithfully took up the challenge of daily observations and kept weather records at most of their principal garrisons.

Humboldtian Science

The Age of Discovery and the expansion of the British and French Empires over the globe brought enormous quantities of new information from the fields of botany, zoology, geology, and climatology, which in turn sparked new interest in determining the relationship between landscape, geomorphology, the climate, and the distribution of plants and animals. The voyage of the HMS *Beagle* is perhaps the best known of these voyages, thanks to the *Beagle*'s famous passenger, Charles Darwin. It is somewhat less well known that Darwin was invited to participate in the voyage as a companion to Robert FitzRoy, the ship's captain, who was later in charge of the British Board of Trade's Meteorological Office. FitzRoy was the first to issue public weather forecasts and the daily weather maps that are now such a staple feature of our newspapers.[1] FitzRoy also devised the storm warning system still in place in ports and harbours, which considerably reduced loss of life and damage to ships caused by storms (Anderson 2005). Tragically, the strain of the scientific controversy over "predicting" the weather (FitzRoy coined the term "forecast" to distinguish them from astrological and other nonscientific "predictions") eventually cost FitzRoy his life. Today he is remembered as a great meteorologist and the founder of the world-renowned U.K. Meteorological (Met) Office (UKMO). However, it was an even earlier voyage than that of FitzRoy and Darwin aboard the *Beagle* that set the direction of earth sciences research for the early nineteenth century: the voyage of Alexander von Humboldt and Aimé Bonpland and the subsequent publication of Humboldt's discoveries during his travels to the Americas.

Humboldt's immense status as a scientific superstar in the first half of the nineteenth century makes it difficult to provide a brief sketch of his work, impact, and lasting influence. Born in Berlin in 1769, he decided early on to become a scientific explorer and was particularly interested in botany. He eventually left for South America in 1799 with Bonpland, a French naturalist with whom he spent the next four years exploring South America and dispatching letters describing his discoveries back to Europe. These letters were widely circulated and made Humboldt into a scientific celebrity while he was still toiling through the jungles of South America. Humboldt travelled with many scientific instruments and kept observations of temperature and pressure, using atmospheric pressure as a means of determining elevation

---

1. These were, unfortunately, not as accurate as hoped and proved something of a public relations disaster. See Anderson (2005).

(Helferich 2011). In the seventeenth century, Pascal established that pressure diminished with height, and a later generation of investigators such as Edmond Halley and William Derham (Derham 1698) attempted to quantify this decrease in pressure with height. Humboldt, and later McCord, inversed this idea to establish elevation by measuring pressure differences at different altitudes. In fact, using barometers to establish the heights of mountains was first proposed by the Académie Royal des Sciences as far back as the early eighteenth century (Maraldi 1709; Derham 1698). So valuable did Humboldt find this technique that he was never without his barometer during his famous voyage to South America.

Humboldt was interested in defining biogeographical regions and in mapping the distribution of plants as related to temperature. He was one of the first, therefore, to break out of the ancient linking of temperature to latitude and instead recognised that temperature varied spatially in ways not always connected to the strict parallel lines of latitude. Instead, Humboldt displayed spatial variations in temperature by the revolutionary use of isolines—lines connecting points of equal value, such as topographical maps showing lines of equal elevation—to map the spatial patterns of temperature across the globe. Humboldt's use of isolines also provided a way of condensing pages of numerical information into a visually compact form. Applied to temperature, maps of isotherms made clear the effect of such geographical features as the Gulf Stream, the Rocky Mountains, and the influence of the oceans. Determining the average temperature of a location and the best method of observing the temperature to obtain an accurate daily, monthly, and yearly mean value became a focus of attention for nineteenth-century meteorologists such as McCord.

Humboldt also mapped magnetic declination and showed that it too varied in space and needed to be measured and mapped. Filling the gaps in the maps of temperature and magnetism became the latest scientific enthusiasm after Humboldt's publication on isotherms in 1817. "Humboldtian" science came to describe a vision where, firstly, everything relating to the state of Earth should be carefully measured and, secondly, that these measurements would reveal a fundamental unity and harmony underlying these different elements (chapter 8). This theme of unity, that everything—geology, biology, plant and animal distribution, temperature, climate, magnetism, and the atmosphere—was all harmoniously related, is behind his masterwork, the description of the physical world contained in his multivolume work *Cosmos*. Humboldt's vision of science was one where the relationship of meteorology to terrestrial magnetism would emerge inductively from the analysis of worldwide measurements, just as the relationship between the distribution of plants could be related to temperature.

There can be little doubt that McCord, among countless others of that period, was inspired by Humboldt. There are numerous pages in his notebooks devoted to experiments to calculate the height of the hills of the Montreal plain using the difference in air pressure measured by barometers, a technique used extensively by Humboldt. McCord made repeated attempts during the 1830s to establish the height of Mount Royal using barometrical measurements.

Over the years McCord made many measurements, in all seasons, of the values of pressure and temperature near the summit of Mount Royal and in his study in his house near the Saint Lawrence River, making the journey between the two places "as fast as a good carriage allows," (McCord 1843, p. 106) which seems to have been about 20 minutes. Temperature readings were necessary as a part of the elevation calculations, and these can be used to estimate the difference between city and mountain temperature and thus the urban heat island effect of McCord's era. McCord's experiments show that this effect seems to have been about 2°F (1.1°C) in the 1830s, although this estimate may be slightly complicated by the roughly 150-meter difference in elevation between the two sites.

John Herschel's Instructions for Observations

John Herschel, the English son of German astronomer William Herschel, had a long and varied career and is known primarily for his work in astronomy and photography, although he also had a considerable influence on meteorology and geomagnetism (Janković 1998; Good 2006). Herschel met Humboldt in Paris as a young man, and they remained friends throughout their long scientific careers (New World Encyclopedia 2014). Between them, Herschel and Humboldt had an impact on nineteenth-century science that would be hard to overstate; both advocated for careful observations of natural phenomena, including meteorological observations, and for using these observations to discover the laws of nature.

Herschel continued the work of mapping the heavens started by his father, William Herschel, and his aunt Caroline Herschel (Holmes 2008). While on a research trip to South Africa to catalogue the stars of the Southern Hemisphere, John Herschel produced an influential outline on meteorological observations and proposed an international endeavour of worldwide, hourly meteorological observations on four days a year. Indeed, his obituary in the

London Times notes that "Herschel's residence at the Cape was productive of benefits not only to astronomy but also to meteorology" (O'Connor and Robertson 1999). His booklet of instructions to meteorological observers was widely admired and used across the globe by, among others, military officers and colonists across the far-flung British Empire. Captain Henry Bayfield, naval surveyor of the Great Lakes, Saint Lawrence River, and Gulf of Saint Lawrence, and commanding officer of William Kelly (chapter 6), forwarded a copy of Herschel's guidelines to McCord in Montreal, while the Royal Engineers used Herschel's meteorological instructions as the guidelines for their observing practices in the 1850s. Herschel initially wrote his instructions for amateur meteorologists in the South African colony of Cape Town, and as such they were particularly adapted to colonial settings. One of the biggest hurdles described by Herschel was getting instruments such as barometers, intact and still calibrated, to their destinations in colonial locations. "It should always be carried upright . . . and over all rough roads should be carried in the hand . . . If strapped horizontally under the roof of a colonial cabin or tied upright . . . there is not a chance of its escaping destruction," Herschel wrote with the voice of bitter experience. He recommended instead that the instruments be transported by hand: "strapped obliquely across the shoulder of a horseman, however, it [the barometer] travels securely and well" (Herschel 1835, p. 8).

Only exact and careful measurements of the weather, Herschel wrote, can "furnish the materials necessary for an accurate and scientific inquiry into the laws of climate, and are the only data through which (taken in conjunction with the known laws of physics) the more general relations of Meteorology [can] . . . be successfully investigated" (Herschel 1835, p. 1). Herschel hoped that regular and extensive observations would help efforts to make "reasonable and well-grounded prediction[s] [of] the irregular and seemingly capricious course of the Seasons" (p. 2), a hope that still eludes us today. Herschel wrote that keeping meteorological observations, "though not [enabling] us to predict with certainty the state of the weather at any given time and place, yet at least [allows us] to form something like a probable conjecture as to what will be the general course of the next ensuing season" (p. 2).

Although seasonal weather prediction remains an area of active research, chaos theory, introduced in the 1960s by meteorologist Edward Lorenz in the course of researching computer models for forecasting, places theoretical limits on the predictability of the weather. Given the furor that would attend FitzRoy's attempts at weather forecasting in the 1860s as "unscientific," it's interesting that in the 1830s Herschel's widely admired and circulated monograph gave forecasting as the principal reason for keeping weather records.

In advocating for simultaneous worldwide measurements, Herschel, a renowned astronomer, stated frankly that "meteorology, however, is one of the most complicated of all the physical sciences," (Herschel 1835, p. 2) a statement few today would contest. "It is only by accumulating data from the most distant quarters, and by comparing the . . . atmosphere at the same instant at different points, and at the same point at different moments, that it is possible to arrive at distinct and useful conclusions" (p. 2).

"The Barometer is the most important of all Meteorological Instruments" (Herschel 1835, p. 6), Herschel's paper begins, giving a clear statement of the view that the most important meteorological observation is not that of temperature but of pressure; it is the barometer that measures the movements of the atmosphere, and it is these atmospheric changes that control the weather. His instructions are fascinating for showing what has changed and what remains the same over the past two centuries in climatology. Herschel covers a wide range of potential sources for errors and inaccuracies in climatological time series, from the changes in the location, exposure, and time of observation that still plague many attempts to study long-term climate change, to transcription and calculation errors. "It is perhaps unnecessary to more than mention the precaution of *counting* the days . . . so as to admit of no mistake in the divisor," Herschel (p. 17) tactfully advises, but these and other errors are inevitable when compiling statistics from raw observations. Other instructions Herschel included were as old as the development of the thermometer, echoing Reaumur's caution from 1733 that not only should the thermometer be "perfectly shaded" out of any direct sunlight, but it should also be placed "where no *reflected* sunbeams from water, buildings, rocks or dry soil can reach it" (p. 9).

On the other hand, it's clear that Herschel is addressing these instructions to amateur meteorologists observing from home. While he considered it best to have only one person taking the observations, "as this may often be impracticable, the principal observer should take care to instruct one or more of his family how to do it" (Herschel 1835, p. 4). It's impossible to know now how many of these registers, like the first McCord weather diary, were the result of collaborative work of the wives, children, or even servants of the person named on the records. Another consideration with single observers or even families of observers is a set of observing times that were regular but not too burdensome. "The best hours in a scientific point of view would be those of Sunrise, Noon, Sunset and Midnight . . . but these are not the hours adapted to general habits" (p. 4). Herschel instead recommended 8:00 a.m., 2:00 p.m., and 8:00 p.m. as standard observing times. McCord followed up on Herschel's suggestions for both regular daily and hourly observations four times a year, sending copies of his weather records to Herschel in London.

Meteorology, Climate, and the Military: The Saint Helen's Island Experiment, 1839–1841

A letter from Captain Henry Bayfield gives the results of his trigonometrical survey of the heights of the Monteregian Hills, including Mount Royal, for McCord to compare with his barometrical height calculations (Bayfield 1835). Bayfield was a career sailor, who joined the navy as a child and first sailed to Canada in 1810. He spent most of his life surveying the waterways of Canada, from the Great Lakes to the Gulf of Saint Lawrence, and was a keen amateur naturalist, contributing several essays on geology to the *Transactions of the Literary and Historical Society of Quebec*. Dr. William Kelly, author of several meteorological and climatological essays on Canada and another of McCord's correspondents, was one of Bayfield's naval surveying team.

McCord enlisted in the militia at the age of nineteen and was a militia colonel by the late 1830s when he first broached the project of having hourly temperature observations recorded by the military. His goal was to determine the true, daily average temperature by taking temperature readings every hour of the day and night rather than using the average of two or three daily observations or of the minimum and maximum temperatures to estimate the mean daily temperature. Averaging the minimum and maximum temperature is the usual method to determine the mean daily temperature in practice today. Sir John Colborne was at this time the commander in chief of British North America. He enthusiastically supported McCord's project, replying that it would be of the

> greatest satisfaction to assist you in promoting the object of your Society by insuring the meteorological observations to be undertaken at St. Helen's . . . I will request the commanding officer of Artillery to communicate with you [and] make such arrangements as may be required for commencing observations immediately. (Colborne 1837)

Despite Colborne's support of the project and his promise of setting up the observations "immediately," it would be some time before the measurements could get off the ground. In November 1837, rebellion erupted in

Canada and McCord, as a senior officer in the militia, was in the thick of it. McCord saw the Rebellions as a serious threat to the settled peace and order and to the general spirit of progress and improvement that came to characterise much of the optimistic Victorian attitude. He described the Rebellions as "the unfortunate troubles which broke out in 1837 and 1838 [and] put an end for some time to the peaceful arts" (McCord 1842, p. 3).[2]

The Patriot Rebellions of 1837 and 1838 were a defining moment in the history of the Canadian colonies and led directly to political reform and a more democratic form of government, a decade before many similar revolts took place in Europe in 1848 (Saul 2010). McCord's (1838) comments, which sound half wistful and half exasperated, wondering if "Peace should ever be granted to this distracted Province" highlight the uncertainty and volatility of those years to those living through them. McCord represented the point of view of the military and governing order who struggled to contain the Rebellions and maintain the peace in opposition to patriot rebels who were trying to make their demands for responsible government heard by means of armed insurrection.[3] The wounds of civil strife and unrest would have far-reaching consequences for McCord and the development of climatology in Canada.

Once military and civil order had been reestablished in late 1839, McCord's temperature experiments could go ahead. In his 1842 report on the results of the measurements, McCord opened his remarks by noting that it was the hourly observations taken by W. S. Harris in Leith and Plymouth (England) in the 1820s that gave him the idea of applying to Colborne for the cooperation of the military guard stationed on Saint Helen's Island for a series of similar measurements to be taken in Montreal (McCord 1842, p. 3). He made a point of mentioning that he first applied to Sir John Colborne,

---

2. The French have their rallying cry of "Liberté, Egalité, et Fraternité" and the Americans have "Life, Liberty, and the Pursuit of Happiness," but Canadians are content with "Peace, Order, and Good Government", a phrase delivered in one of Alexander Spark's sermons in 1799 (Lambert, 1984, p. 9).

3. According to Young (2014), McCord's reactionary stance and support of the British Empire against the patriot rebels, his uncompromising stance towards the rebels, and his participation as a senior officer of the militia against the patriots made his position as a lawyer in Montreal untenable in the wake of the reforms and reconciliation with the patriot faction after 1839. Young postulates that McCord's stance against the patriots led directly to his being appointed a circuit court judge in 1842 to marginalise his position and get him away from the political debates and centre of power in Montreal. His continual travels around the countryside as a circuit judge made it impossible for McCord to continue his weather register.

commander of the forces in the Canadas, as early as 1836 for military cooperation in these meteorological experiments. McCord decided on a series of bihourly measurements instead of readings every hour as the changing of the guard was every 2 hours and hourly readings would have interfered with military routine. The observations started in August 1839 and were recorded on every even hour for the first year, switching over to odd hours on August 1, 1840.

The Royal Artillery was responsible for the Saint Helen's Island observations. Corporal Tweedale was appointed the superintendent of the project. Tweedale had previous experience in keeping meteorological observations in Halifax, Nova Scotia (McCord 1842, p. 4), suggesting a history of military meteorological record keeping in Canada going back at least to the 1830s, although the only original observation sheets known today are those from the Royal Engineers. These records, if they still exist, have yet to be traced, but Tweedale's experience supports Lefroy's later claim that weather records were kept in the guardrooms of many Canadian garrisons. According to Hallman, in 1844 Lefroy "introduced observing books into each of the military guard rooms across the land [at] Queenston [Niagara], Montreal, Kingston, Toronto, London, Fredericton, Halifax and [St John's] Newfoundland" (Hallman 1950, p. 156).

By the time of the Saint Helen's Island experiment, the Royal Artillery was under the command of Edward Sabine. With the considerable support of John Herschel and the Royal Society, the Royal Artillery had set up their Magnetic and Meteorological Observatory in Toronto in 1839. McCord wrote that at the time these observations were started, he "was not aware of the contemplated establishment of an observatory at Toronto, for the purpose of recording meteorological as well as magnetic phenomena," but having started the Saint Helen's Island observations decided to continue them "if only for the purpose of comparison, particularly as that comparison would be rendered additionally useful by the distance between the places of observation, (nearly 400 miles)" (McCord 1842, p. 4–5). McCord supplied all the instruments for the Saint Helen's observations. It's not clear whether additional salary expenses incurred for observers were absorbed into the general military budget. Given the budgetary difficulties detailed in the Toronto Observatory records, McCord's provision of the instruments and sponsoring of the project may have made the record keeping in Montreal relatively easier than the establishment of a new observatory in Toronto. In Montreal, the military bihourly record keeping apparently wasn't continued beyond McCord's two-year experiment from 1839 to 1841, despite

McCord's report of the "recent resolution of this Society[4] to continue these observations" (p. 6).

As a quick résumé of the Canadian climate, and for comparison purposes for readers around the globe, McCord wrote:

> The divisions of the year into quarterly periods of Spring, Summer, Autumn and Winter cannot be applied to this latitude in America . . . We have, truly speaking, no Spring, and but few weeks of autumn: in the [Spring] the transition from snow to the intense heat of summer is almost instantaneous, and from [Autumn] . . . we have only a short interval of two or three weeks, and again all is winter. (McCord 1842, p. 5)

Part of McCord's objective was to determine which hours of the day were best suited to record the temperature such that only two daily observations would be needed to "obtain the true mean-temperature of the year." He wrote, "The labour thrown away in compiling Meteorological Tables is almost incredible, from the want of system and concert in the times and manner of observation, no two observers among the mass of tables submitted to me having recorded at similar hours" (McCord 1842, p.6).

The results of the Saint Helen's observations, according to McCord, meant that observers would only have to apply a correction factor, deduced from the tables calculated from the bihourly observations, to their individual observations at whatever observing schedule suited them to be able to obtain the "true mean" to compare to other years or other locations. This appears to be an early attempt at data homogenization, an endeavour to adjust original observations for effects in the temperature data such as different observing hours, instrument types or exposures, or changes in instrument locations in order to get the "true mean" and isolate climatic changes from other non-climatic influences. These are problems whose attempted solutions are still highly debated today.

McCord found that the annual average of observations taken at 12-hour intervals came closest to the annual average from bihourly temperatures and recommended 10:00 a.m. and 10:00 p.m. as the best hours of observation to average together to get the mean daily temperature. He apparently received the first results from the Toronto Observatory as his report was in press but found confirmation of his recommendation in the Toronto data; the annual average of the 10:00 a.m. and 10:00 p.m. temperatures came closest to

---

4. The Natural History Society of Montreal.

the annual average calculated using temperatures recorded throughout the entire day. Nevertheless, private, individual observers such as Smallwood or Skakel kept to their original schedules of early morning, early afternoon, and evening observations.

The Royal Engineers

Herschel's instructions and plea for 24 hours of observations four times a year were quickly taken up not only by amateur meteorologists, such as McCord, as early as 1836,[5] but also by those for whom meteorological observations were part of their professional duties; the British military and particularly the Corps of Royal Engineers. In the same 1838 volume of *Papers of Subjects Connected with the Duties of the Corps of Royal Engineers* that contains Herschel's *Instructions*, we also find Colonel William Reid's (1791–1858) *Law of Storms*, the document that set out the details of hurricane storms and added fuel to James Espy and William Redfield's (1789–1857) storm controversy then raging in the United States. In the introduction to that volume, an appeal is made for geoscientific observations in general and meteorological and magnetic observations in particular:

> The paper on the subject of Hurricanes, by Colonel Reid, will be read with great interest, and I trust will induce many to turn their attention to atmospheric phenomena generally. The observations of greatest interest at present are those for the purpose of determining the phenomena of terrestrial magnetism. (Denison 1838, p. x)

The editorial continues:

> The Royal Society have obtained from the government a sum of money for the purchase of the necessary instruments; and they have expressed their willingness to confide these instruments, with full instructions for their use and application, to the Officers of Engineers at the different stations where it is desirable that experiments should be made. (Denison 1838, p. x)

---

5. In a letter dated November 13, 1835, Bayfield wrote to McCord: "I have received half a dozen copies of the enclosed for distribution to scientific friends and as I know that you take great interest in Meteorological Instruments & Observations perhaps a copy may interest you" (Bayfield 1835).

It seems that there was some confusion or a duplication of efforts between the Royal Artillery and the Royal Engineers, and two sets of observations were made at Quebec, though taken at different times of day, making the various registers complementary rather than redundant. The Royal Engineers were tasked with keeping meteorological observations at the major garrisons around the middle of the nineteenth century, a duty that was later taken up by the Army Medical Corps (Fleming 1990);[6] these records were later sent to the Met Office where they are conserved in the archives.

The British military under the division of the Royal Artillery and Edward Sabine were particularly interested in geomagnetism, and it's through this interest that the Toronto Observatory was established. But other branches of the military were also interested in the weather. The navy kept instrumental observations in their logbooks with increasing frequency after the 1840s, and efforts are currently underway to explore the vast amount of information contained in naval logbooks. Some of the earliest British records in Canada are from land-based officers who kept weather diaries. Captain Alex Rose had his daily record of the minimum and maximum temperature, wind direction, and weather from Quebec from 1765 to 1766 published in the *Philosophical Transactions of the Royal Society* (Rose and Murdoch 1766). There are also scattered references to other records kept in Quebec City throughout the late eighteenth century (McCord 1847; Landmann 1852). Captain Henry Bayfield and Lieutenant William Kelly kept meteorological records in the early nineteenth century as they charted the Saint Lawrence River, the Gulf of Saint Lawrence, and the Great Lakes, though these were never published. Kelly instead published a record of the observations kept by Mr. Watt at the Quebec Citadel (Kelly 1837) along with results of experiments on the water temperature of the Saint Lawrence River, the famous fogs of the Saint Lawrence, and other river-related meteorological phenomena. Captain Bonnycastle of the Royal Engineers published an article in 1827 on auroras that mentioned thermometer readings and later published a *History of Newfoundland*, which included a section on climate and meteorology (Bonnycastle 1829). Lieutenant Andrew Noble of the Royal Artillery kept a record for the periods December 1853 to April 1854 and October 1854 to April 1856, the results of which he presented to the Literary and Historical Society of Quebec (LHSQ). As we saw earlier, the Royal Engineers were also keeping records in Quebec City from 1852

---

6. James Fleming gives an account of some the earliest military weather records in the United States, which were kept by military doctors during the War of 1812.

to 1870 as well as in Saint John's, Halifax, Kingston, and New Westminster, which is now a suburb of Vancouver.

The Corps of Royal Engineers are best remembered in colonial Canada for their canal building; the Rideau Canal, overseen by Colonel By, is now a World Heritage Site and proudly mentioned on the Royal Engineer's website (Royal Engineers 2016). A lively account of a young military engineer's experiences in the Canadian colonies at the end of the eighteenth century is given in the *Adventures and Recollections of Colonel Landmann* (Landmann 1852).[7] The capital city of Canada, Ottawa, was originally called Bytown after Colonel By, who was responsible for much of its early construction as a consequence of it being the end point of the Rideau Canal. Ottawa marks the point at which the southern portion of the Great Lakes, with the Rideau Canal starting in Kingston on the eastern edge of Lake Ontario, could connect up to the Ottawa River and access the Canadian interior through the waterways. As the country expanded westward, the Royal Engineers were often the first representatives of the British Empire on the ground, and they laid out much of the initial infrastructure of newly founded, nineteenth-century settlements (Woodward 1974).

It seems likely from references made by McCord and Lefroy that meteorological logbooks had been kept in some military garrisons from the 1830s (McCord 1842; Hallman 1950), but the earliest currently known logbooks from military sources, apart from the Toronto Observatory, are those from the Royal Engineers. Many of the logbooks in the archives of the British Meteorological Office library start in 1853. Initially kept by the Corps of Royal Engineers, the observations follow Herschel's instructions to the letter. Along with pressure, temperature, wind, and weather, they also recorded shortwave solar radiation with a "blackbulb" thermometer, which is a maximum thermometer with the bulb blackened with ink and exposed to the sun; terrestrial radiation with a minimum thermometer exposed on the surface of the grass; wet-bulb temperature for humidity measurements; minimum and maximum temperatures; and clouds, rain, and snow. The only point on which they deviated from Herschel's advice was the time; they took their observations at 9:30 a.m. and 3:30 p.m. The Royal Engineers also faithfully carried out Herschel's 24 (or rather 36) hours of continual observations on the solstices and equinoxes (Meteorological Office 1851, 1862, 1864, 1866a,b, 1871a,b, 1876, 1880). Each sheet had the names of the observers and was checked before being signed by the commanding officer. In Quebec City,

---

7. With thanks to Denis Robillard for introducing me to Colonel Landmann.

Major General Ward was the person ultimately responsible for the meteorological records. William Cuthbert Ward, as a colonel in Toronto in 1839, had the unenviable task of trying to find a suitable location for and then building an observatory for another geoscience endeavour: the magnetic crusade.

References

Anderson, K., 2005: *Predicting the Weather: Victorians and the Science of Meteorology*. University of Chicago Press, 331 pp.

Bayfield, H., 1835: Letter to John Samuel McCord. McCord Family Fonds 1766–1945, Papers P001-0820. McCord Museum Archives.

Bonnycastle, R. H, 1842: *Newfoundland in 1842: a sequel to "The Canadas in 1841"* H. Colbourn, 367 pp.

Colborne, J., 1838 unpublished manuscript: Letter to John Samuel McCord. McCord Family Fonds 1766–1945, Papers P001-0820. McCord Museum Archives.

Denison, W., 1838: *Papers on Subjects Connected with the Duties of the Corps of Royal Engineers*. Vol. 2. Barker and Crane Court, 274 pp.

Derham, W., 1698: Part of a letter of Mr. William Derham, Rector of Upminster, dated Dec. 6. 1697. Giving an account of some experiments about the heighth of the mercury in the barometer, at top and bottom of the monument: And about portable barometers. *Philos. Trans. R. Soc. London*, **20**, 2–4, https://doi.org/10.1098/rstl.1698.0002.

Fleming, J. R., 1990: *Meteorology in America, 1800–1870*. Johns Hopkins University Press, 264 pp.

Good, G., 2006: A shift of view: Meteorology in John Herschel's terrestrial physics. *Intimate Universality: Local and Global Themes in the History of Weather and Climate*, J. R. Fleming, V. Jankovic, and D. Coen, Eds., Science History Publications, 35–68.

Hallman, E. S., 1950: 110 years ago. *Weather*, **5**, 155–158, https://doi.org/10.1002/ j.1477-8696.1950.tb01175.x.

Helferich, G., 2011: *Humboldt's Cosmos*. Tantor, 384 pp.

Herschel, J. F.W., 1835: *Instructions for Making and Registering Meteorological Observations in Southern Africa*. Bradbury and Evans, 17 pp.

Holmes, R., 2008: *The Age of Wonder: How the Romantic Generation Discovered the Beauty and Terror of Science*. Harper, 552 pp.

Janković, V., 1998: Ideological crests versus empirical troughs: John Herschel's and William Radcliffe Birt's research on atmospheric waves, 1843–50. *Br. J. Hist. Sci.*, **31**, 21–40, https://doi.org/10.1017/S0007087497003178.

Kelly, W., 1837: Abstract of the meteorological journal kept at Cape Diamond, Quebec, from the 1st of January, 1824, to 31st December, 1831, with some remarks on the climate of Lower Canada. *Trans. Lit. Hist. Soc. Que.*, **3**, 46–71.

Lambert, J. H., 1984: *One Man's Contribution: Alexander Spark and The Establishment of Presbyterianism in Quebec 1784–1819*. St. Andrew's Presbyterian Church, 26 pp.

Landmann, G., 1852: *Adventures and Recollections of Colonel Landmann, Late of the Corps of Royal Engineers*. Vol. 1. Colburn and Co., 347 pp.

Maraldi, G.-F., 1709: Comparaison des observations du barometre faites en differens lieux. *Mém. Acad. Roy. Sci.*, **1709**, 233–245.

McCord, J. S., 1838 unpublished manuscript: Letter to Dr. Skey. McCord Family Fonds, Papers P001-827. McCord Museum Archives.

——, 1842: Report of the meteorological observations made on the island of St. Helen's. Natural History Society of Montreal, Ed., Lovell, 18 pp.

——, 1847: Observations on meteorology. *Br. Amer. J. Med. Phys. Sci.*, **3**, 228–229.

Meteorological Office, 1851 unpublished manuscript: Climatological returns for Fort Simpson, Canada, North America (DCnn: 9FTS). ARCHIVE W02.B2-A3, Meteorological Office Archives, 1849–1851.

——, 1862 unpublished manuscript: Climatological returns for Kingston, Canada, North America (DCnn: 9KIS). ARCHIVE Z18.K1-Z17.C3, Meteorological Office Archives, 1853–1861.

——, 1864 unpublished manuscript: Climatological returns for Halifax Dockyard, Canada, North America. ARCHIVE Z18.K1-Z17.C3, Meteorological Office Archives, 1853–1863.

——, 1866a unpublished manuscript: Climatological returns for Halifax, Citadel Hill, Canada, North America. ARCHIVE Z18.K1-Z17.C3, Meteorological Office Archives, 1854–1865.

——, 1866b unpublished manuscript: Climatological returns for New Westminster—British Columbia, Canada, North America (DCnn: 9NEW). ARCHIVE Z18.K1-Z17.C3, Meteorological Office Archives, 1859–1865.

——, 1871a unpublished manuscript: Climatological returns for Newfoundland, Canada, North America (DCnn: 9NFL). ARCHIVE Z18.K1-Z17.C3, Meteorological Office Archives, 1852–1870.

——, 1871b unpublished manuscript: Climatological returns for Quebec, Canada, North America (DCnn: 9QBC). Vol. 915513-1001, Meteorological Office Archives, 1853–1870.

——, 1876 unpublished manuscript: Climatological returns for Halifax, Nova Scotia, Canada, North America (DCnn: 9HFX). ARCHIVE Z18.K1-Z17.C3, Meteorological Office Archives, 1852–1875.

——, 1880 unpublished manuscript: Climatological returns for Manitoba, St. Johns College, Canada, North America (DCnn: 9MAT). ARCHIVE Z18.K1-Z17.C3, Meteorological Office Archives, 1873–1879.

New World Encyclopedia, 2014: John Herschel. Accessed 01/12/2016, http://www.newworldencyclopedia.org/p/index.php?title=John_Herschel&oldid=978796.

O'Connor, J. J., and E. F. Robertson, 1999: The Late John Herschel. John Frederick William Herschel, MacTutor History of Mathematics Archive, accessed 01/20/2017, http://www-history.mcs.st-andrews.ac.uk/Biographies/Herschel.html.

Rose, A., and P. Murdoch, 1766: Abstract of a journal of the weather in Quebec, between the 1st of April 1765, and 30th of April 1766. By Cap. Alex. Rose, of the 52d regiment; communicated by Rev. P. Murdoch, D. D. R. R. S. *Philos. Trans. R. Soc. London*, **56**, 291–295, https://doi.org/10.1098/rstl.1766.0037.

Royal Engineers, 2016: History: A brief history of the Corps of Royal Engineers. Accessed 01/13/2016, http://www.army.mod.uk/royalengineers/26315.aspx.

Saul, J. R., 2010: *Extraordinary Canadians: Louis Hippolyte Lafontaine and Robert Baldwin*. Penguin, 253 pp.

Woodward, F. M., 1974: The influence of the Royal Engineers on the development of British Columbia. *BC Stud.: Br. Columbian Quart.*, **24**, 3–51, http://ojs.library.ubc.ca/index.php/bcstudies/article/view/817/860.

Young, B., 2014: *Patrician Families and the Making of Quebec*. McGill-Queen's University Press, 452 pp.

CHAPTER EIGHT

The Magnetic Crusade and the Founding of the Toronto Observatory

While it may seem obvious from our twenty-first-century point of view that meteorological observations are a basic and useful scientific measurement—how many of us, particularly readers of this work, obsessively consult our weather apps or start our day with a complaint about temperature?—it was not, in fact, an interest in weather, climate, or meteorology that was the impulse behind the 1839 founding of the Toronto Observatory. It was magnetism.

The expansion of trade and travel during the Age of Empire in the nineteenth century made the accurate determination of location, especially at sea, ever more urgent. Research into terrestrial magnetism, the variations and declination of Earth's magnetic field, and its influence on compasses gained in importance and urgency, not least of all for the competing European governments with their expanding global empires. With the end of the Napoleonic Wars in 1815, military officers convinced the British government to use Britain's Navy and expanding colonies to commit funding to expeditions investigating terrestrial magnetism. Alexander von Humboldt thought that the simultaneous observation of magnetic storms could help to establish fixed longitudes. It was further hoped that with a full understanding and spatial mapping of the phenomena known as magnetic declination, the angle by which magnetic north deviated from true (geographic) north at a given place, charts of declination could be used to determine longitude at sea, the most pressing scientific and navigational problem for sea-based empires in the eighteenth and early nineteenth centuries (Barrie 2014, p. 68).[1] At the time, the principles of electricity and the connection between electrical forces and magnetic phenomena were still being established. The discoveries of Charles-Augustin de Coulomb (1736–1806) in late-eighteenth-century France, Michael Faraday (1791–1867) in early-nineteenth-century England, and Joseph Henry in 1830s United States of the relationship between magnetism and electricity also spurred an interest in understanding terrestrial magnetism.

John Cawood explains how magnetic observations, though long recorded at the Paris Observatory, received new attention following the results reported by eighteenth-century solider, scientist, and explorer Louis de Bougainville during his 1766 circumnavigation of the globe and voyage of scientific discovery. A prize was offered by the Académie Royal des Science in Paris to the person who developed the best method of constructing magnetic measuring instruments that were properly aligned with the magnetic meridian and accounted for the regular diurnal (daily) variation of the magnetic flux. Dutch scientist Jean Henri van Swindon and Coulomb won this prize jointly in 1777. Coulomb was then appointed to work on magnetic observations at the Paris Observatory, as an advisor to Jean-Dominique Cassini (Cassini IV), the great-grandson of Jean-Dominique (Giovanni Domenico) Cassini and last director of the Paris Observatory before the Revolution. France was the only country with a regular program of observation in the eighteenth century, keeping systematic records at the Paris Observatory. For the French, magnetic research was part of their national scientific program pursued for both knowledge and national prestige (Cawood 1977). In Britain, geophysical research was more haphazard in the eighteenth century, with individual researchers reporting mainly to the Royal Society.

An element of competition in the age-old rivalry between Great Britain and France also played a role in the burgeoning interest in terrestrial mag-

---

1. Although chronometers had been proven reliable for determining longitude in the 1770s, they remained expensive instruments, out of reach for many navigators, as well as requiring complex adjustments for thermal expansion. Astronomical reckoning also remained complicated (Boistel 2010). Magnetic declination had been used for centuries as a means of recording location. Samuel de Champlain recorded magnetic declination along with latitude when determining location in Canada in the early 1600s (Champlain 1907).

netism, as did the relative decline of some aspects of French observational science during the revolution, when the Paris Observatory was pillaged and ransacked by mobs looking for treasure. During that time, many of the scientists and members of the academy, being either aristocratic or having close ties to the French monarchy as the source of their employment, left Paris and laid low in the countryside or fled abroad. The mapping of France, which had been ongoing since the middle of the eighteenth century under the direction of the Cassini family at the Paris Observatory, was subsumed into the New Republic's Bureau des Longitudes, to the enduring bitterness of Cassini IV.

There is a close relationship between space and time where cartography is concerned, as longitude was determined by the time difference between two locations. As Earth revolves 360° around its axis in 24 hours, local noon or local midnight proceeds around Earth at a rate of 15° of longitude every hour. Astronomy is used to determine absolute time by observing when the sun (during the day) or specific stars at night reach a certain position, usually their maximum height in the sky or zenith. The difference in time between two locations can then be used to determine differences in longitude. If the difference in local midnight between two places is 4 hours, they are 60° of longitude apart. Astronomical observatories thus had an important role to play in both mapping and timekeeping. And once some measure of stability was restored, "in Napoleonic France, science was very much in official favour" (Cawood 1977, p. 560; Sartori 2003). For much of the early nineteenth century, France's Paris Observatory and Bureau des Longitudes led the world in cartography, navigation, and magnetic research. Magnetic observations, such as the declination of the magnetised suspended needle, had been observed and reported at the Paris Observatory since the days of La Hire in the early eighteenth century (e.g., La Hire 1710, 1712, 1714).

The magnetic crusade, as it came to be called in the 1850s, was one of the most ambitious, large-scale, organised geoscientific data collection efforts of the nineteenth century, comparable to the expeditions organised in the 1760s to observe the transits of Venus in 1762 and 1768. It was enthusiastically, if perhaps a trifle hyperbolically given the eighteenth-century attempts at organised meteorological observations, described by British scientist William Whewell (1794–1866) as "by far the greatest scientific undertaking the world has ever seen" (Cawood 1979, p. 493). The original data collection effort was intended to last for only a few years, but in several cases, including that of the Toronto Observatory, permanent observatories were established from the original temporary sites. The magnetic crusade came to be so called in recognition of the organised and standardised methods of observation and data collection, such as fixed observing times and standardised instruments, in contrast to the "guerrilla warfare" (Loomis 1853) of individual observers, each with their own personal observing schedule and instruments. The principle behind the crusade seems to have been largely driven by Humboldt's vision of a united cosmos, in which gathering enough information would make geophysical phenomena such as magnetism, lightning, aurora, and weather intelligible and thereby reveal the laws describing the relationships between these phenomena, much as Newton had done for physics with his gravitational theory and laws of motion. Humboldt was convinced that meteorological and magnetic phenomena were related, citing examples of cirrus clouds forming during aurora (Humboldt 1997, p. 196).[2] "[With] the development of knowledge . . . portions [of physical science] which have long stood isolated become gradually connected," wrote Humboldt (p. 163) in 1845. Terrestrial magnetism and electricity were thought to hold a key that might be able to explain all terrestrial phenomena, from the weather and climate to volcanoes and earthquakes.

This Humboldtian view of an integrated, explicable cosmos had inspired a generation of scientists, including Edward Sabine and John Herschel (chapter 7). A few years after serving in Canada during the War of 1812, Sabine, a soldier scientist of the Royal Artillery, served as astronomer to various British polar expeditions in search of the Northwest Passage. General interest in terrestrial magnetism as well as Sabine's lifelong passion for the subject were sparked by the magnetic observations taken during these expeditions to Canada's north, which, among other discoveries, determined the location of the magnetic North Pole. Sabine was also inspired by the 1817 publication of Christopher Hansteen's (1784–1873) "The Earth's Magnetism," which revived Edmond Halley's (1656–1742) erroneous theory of a four-pole magnetic Earth, later supported by Carl Friedrich Gauss. On his return to Britain, Sabine campaigned tirelessly for official support for widespread magnetic observations throughout the British Empire. During the 1820s, experiments by Hans Christian Ørsted (1777–1851) led to the discovery of the relationship between electricity and magnetism at the Paris Observatory by André-Marie Ampère (1775–1836); Ampère favoured the (correct) two-pole magnetic theory of Earth. In 1827, Humboldt left France, no longer able to postpone his recall to his native Prussia, from where he continued

---

2. For a more complete history of terrestial magnetism, see Courtillot and Le Mouël (2007). For a current theory on how solar and terrestrial magnetism affect climate, see Courtillot et al. (2007).

to lobby for observational science. In 1829, Humboldt gave a speech in Saint Petersburg on geomagnetism as a global phenomenon and the need for coordinated observations that would "trigger a huge international collaboration" (Wulf 2015, p. 214).

Following Humboldt's Saint Petersburg speech, Sabine seized the opportunity to engineer the writing of a letter from Humboldt to the Royal Society asking for collaboration in an international observation program in 1836. In 1838, Herschel returned from his South African sojourn and lent his prestige and connections to the terrestrial magnetism project. In 1846, the endeavour was summarised in the *Transactions of the American Philosophical Society*:

> The objects of the "magnetic crusade," as it has been called, have been:
> 1. To examine, as far as possible, all of the circumstances and conditions of this force, called magnetism, in all parts of the globe; and,
> 2. To endeavour to deduce, from the results, some law or laws which shall, like the laws of gravitation, embrace all the phenomena, and, like that law, enable us to predict, not only the present condition of that force at any particular place, but its condition at any future time, and all of the intermediate revolutions which it may have undergone. (Locke 1846, p. 284)

Sabine was the main organizational force behind the British efforts at global observations, using the resources of the British Empire's military deployed in colonies across the globe to further scientific research. Once again, Canada's position as a colony in a global empire would link the development of science in Canada to a far-flung network of investigation.

In the early nineteenth century, the British needed pragmatic and political reasons to give a project serious government support, while the French outlook remained more scientific in conception. "The French achievements in terrestrial magnetism were constantly used to provoke Government action" (Cawood 1977, p. 585) to fund British voyages of exploration and systematic observations. Suzanne Zeller described this pragmatic approach to science and data collection as "inventory science" (Zeller 2009, p. 4). "Of crucial importance in justifying inventory science was utilitarianism," a utilitarianism Zeller links to the Scottish Enlightenment in particular, with its belief in the "use of statistics to improve the quality of life" (p. 5). This motive could well be ascribed to Spark's assiduous and diligent record keeping in Quebec City as well as to the other observers with Scots roots or connections, such as the Skakel brothers or Archibald Hall.

McCord's observations, and the puzzle of climate in North America discussed by Daubenay, Herschel, and others in 1838, also played a role in advancing the case for a meteorological and magnetic observatory to be established in Canada under the auspices of the British military. Daubeny's report on his visit to North America, during which he visited McCord and performed a series of experiments with him, was seized upon by Sabine to further promote the establishment of observatories throughout the British Empire (Cawood 1977, p. 552; Daubeny 1840). Herschel used his influence to combine the meteorological and magnetic studies in geophysical observatories throughout the empire. He was elected to the Geomagnetic and Meteorological Committee of the Royal Society. The British government agreed to finance four stations in the colonies: Canada, Saint Helena's, Cape Town, and Van Diemen's Land (Tasmania), with the East India Company taking responsibility for four additional observing stations in India.

The Founding Years of the Toronto Observatory: 1839–1853
The Toronto Observatory[3] was initially intended to be only a temporary building to serve as a base for scientific expeditions to northwestern Canada. One goal was to take magnetic measurements in the Hudson Bay region to determine more precisely Earth's magnetic field in the region of the magnetic North Pole, then located near Hudson Bay. The observatory was also to keep meteorological records in order to determine whether there were any connections between terrestrial magnetism and weather. Sabine originally intended to be the scientific officer on this expedition, but his persistence in advocating for an empire-wide magnetic observing network finally paid off in 1839 when he was appointed superintendent of the colonial observing posts. This position required him to remain at the central communications hub of the Royal Artillery headquarters in Woolwich, England. Lieutenant

---

3. There are several accounts of the founding of the Toronto Observatory and the transformation of the Imperial Magnetic Observatory into the Meteorological Service of Canada (MSC). MSC Director Robert Stupart wrote an early account in 1912 in the *Journal of the Royal Astronomical Society of Canada*, while many of the original letters from the 1840s were published by A. D. Thiessen in his series of articles from 1942 to 1945 in the same journal. More recently the founding of the Toronto Observatory has been described in detail in Suzanne Zeller's *Inventing Canada: Early Victorian Science and the Idea of a Transcontinental Nation* and in Morley Thomas's *The Beginnings of Canadian Meteorology*.

Charles Riddell was nominated instead to establish the Toronto Observatory in 1839.

The letter books—copies, often on blotting or carbon paper, of letters sent—of the directors of the Toronto Observatory are conserved in the archives of the MSC in Toronto. All outgoing letters from Riddell's arrival in 1839 until John Lefroy's departure in 1853 were copied into the books. They provide a fascinating account of the process of conducting science in early-nineteenth-century Canada, with an emphasis on the difficulties and setbacks entailed in setting up and maintaining a scientific operation in a frontier environment. Conducting scientific measurements and experiments in a military setting had advantges and disadvantages: funding, materials, and manpower were supplied by the army but sometimes only with considerable difficulty. The long chain of command, stretching nearly 6,000 kilometers (3,700 miles) from Toronto to London, could be both a blessing and a curse. Of neccessity, colonial officers had a certain degree of latitude and were expected to show initiative and judgement in carrying out their orders, but this could also lead to trouble when their initiative wasn't appreciated by those higher up in the chain of command. Similarly, the long wait for orders, supplies, and payment often led to difficulties during the long months of making do until supplies finally arrived. In the pages that follow, the officers and directors of the Toronto Observatory tell the story of science in a colonial setting in their own words.

Charles Riddell: Building the Observatory
Riddell sailed to Canada in 1839, intending to construct the observatory on Saint Helen's Island, where McCord had already instituted a series of bihourly temperature readings, presumably unbeknown to Herschel and Sabine. Difficulties started from the outset: arriving in Montreal in autumn of 1839, Riddell was told by Captain Henry Bayfield, the superintendent of naval surveying, who had spent more than a decade surveying the Gulf of Saint Lawrence and the Saint Lawrence River for the Royal Navy, that the underlying geology of the Saint Lawrence valley would interfere with magnetic observations.

On, October 17, 1839, Riddell wrote from Montreal that

> from information I received from Capt. Bayfield RN Commanding Naval Surveying Service in Canada . . . I am led to conclude that the Island of St. Helen's would not be an eligible situation . . . from the considerable local magnetic attraction which has been observed in its immediate vicinity.

> I am however of opinion founded on Capt. Bayfield's information that Toronto must be free from such influences, or at all events more likely to be free of local attraction than the country below it.[4]

Riddell was then obliged to hurry on to Toronto, where Bayfield thought the geology of the Niagara Escarpment would be more suitable for long-term magnetic observations, in the hopes of getting the observatory built before the onset of winter. He was too late; by the time he arrived in November 1839, winter had set in and it was too late in the season to start building the observatory. Moreover, Colonel Ward of the local detachment of Royal Engineers never received the plans and orders from the military headquarters in Montreal, and several weeks were spent in tracking down the missing orders. Ward was later promoted to Major General and was the officer in charge of the Royal Engineers' observations in Quebec City in the 1850s (Colburn 1858, p. 311).

The main military garrison in Toronto was Fort York. Fort York had originally been built during the American Revolution and was destroyed in April 1813 when the British, retreating from an invading American Army, blew up the fort's gunpowder supplies to prevent them falling into enemy hands. The American Army abandoned York later that year, and the fort was rebuilt by the British Army between 1813 and 1815.

The building of the observatory, initially planned for the autumn of 1839, did not go smoothly. Orders were lost or misplaced. The permission required for building the observatory, the type of building to be constructed, and the location of the observatory were all subject to much back-and-forth communication between the Royal Engineers in Toronto and headquarters in Monteal. These issues escalated up the chain of command to eventually involve Commander in Chief Sir John Colborne. Riddell was determined to find the best site available for the observatory, and his frequent changes of plan seemingly exasperated the commanding officers of the Royal Engineers, Colonel Oldfield in Montreal and Colonel Ward in Toronto. Several sites attached to Fort York were rejected as unsuitable by Riddell because they were too close to the military parade grounds, where both the presence of large quantites of iron and the firing of weapons during drill were likely to disturb the magnetic instruments. Elsewhere, the land given over to the military reserves consisted of "swamps in the neighbourhood [that] were worse than I had at first imagined and likely to be very unhealthy."[5]

---

4. Riddell to Deputy Adjunct General Royal Artillery, November 14, 1839.
5. Riddell to Ward, Toronto, November 14, 1840, MSC letter book.

It was perhaps an inauspicious moment for the Royal Artillery to be establishing scientific observations in Canada; the year before the Rebellions of 1837–1838 had seen the most serious challenge to the civil government by the certain elements of the populace looking to move towards more democratic and representative local government. The response to this armed insurrection was the most intensive military deployment in Canada since the War of 1812. Colonial authorities feared an alliance between the rebellious reformers and the Americans to overthrow the colonial governments in Upper Canada (Ontario) and Lower Canada (Quebec). American invasions of Canada during the American Revolution and the War of 1812 resulted in a state of constant vigilance against the threat of another potential American invasion. There was, therefore, an extensive military presence in Canada and considerable investment in the construction of defensive military works throughout the colonies, including Fort York in Toronto.

In the end, the Rebellions of 1837–1838 led to political reform and some of the earliest representative governments of the empire being formed in the Canadian colonies, a decade before 1848's "year of revolt" in Europe (Saul 2010).[6] There was no further civil unrest in Canada until the Red River Rebellions of the 1870s and 1880s, although there were times of heightened tension and fears of an invasion from the United States during the Maine–New Brunswick and Oregon boundary disputes during the 1840s and during the American Civil War of 1861–1865. In response to the 1837 Rebellions and the potential threat from the United States, a second fort, in addition to Fort York, was built near the western boundary of Toronto and completed in 1841, along with additional defences to the north of the town. In these circumstances, the demands made for building and supplying a separate observatory building must at times have seemed a frivolous distraction to the local commanders. Nevertheless, significant advances were made thanks to the determination and persistence of the officers in charge of the observatory.

The Temporary Observatory on Bathurst Street
A temporary observatory was established in some unused barracks on Bathurst Street while the debate over the site of the permanent observatory continued. Even then, it was proposed that the rooms in between the small

---

6. According to Benn (2007), the colonial government had in fact been moving in this direction already, and responsible government was granted to Nova Scotia, where there had not been any unrest, before the provinces of Upper and Lower Canada. The Rebellions may in fact have slightly retarded reform.

observatory detachment barracks and the instrument room be given over to the miliatry wives as a maternity ward or "lying-in room." It's hard to imagine who would have been most disturbed had that plan come to pass: the observers, who would have had to share space with women in labour and newborn babies, or the new mothers with the observers going past every hour to take measurements. Still, as the living conditions in the fort barracks had as many as eighteen men and their families sharing one large room, the presence of the observers may not have been any more disturbing to the new mothers than their usual living conditions.

Much of Riddell's official correspondence over the next year, as preserved in the letter book, was taken up with trying to get basic supplies for his detachment, setting up a temporary observatory, and patiently explaining that living quarters next to the instruments was indispenable. In response to the idea that the observatory be built adjoining the fort, Riddell explained why hanging outdoor thermometers next to a parade ground or firing range was not a good idea:

> I have the honor to state that I consider that position pointed out by Lt. Col. Oldfield as objectionable for the site of the Magnetic Observatory from its contiguity to the Fort, and from its being part of the drill Ground of the Troops, where during the summer months there would be frequent firing both of Artillery and Musquestry. The number of Meteorological Instruments must be placed outside the Observatory as well . . . the experiments & observations that will be required to be made in its vicinity are all strong reasons against placing the Observatory in the immediate neighbourhood of a garrison of 7 or 800 men.[7] (Riddell et al. 1839)

Complaints from the commanding officers about the observatory and Riddell's intransigence appear to have filtered through back to Sabine in Woolwich, and Riddell devoted several long letters to explaining himself and his various changes in the planned location for the observatory: first, from Montreal to Toronto, secondly, from the military compound to an isolated location outside the city and garrison, and thirdly, from a swampy location to a healthier one.

The request for separate cottages to serve as barracks for the magnetic observatory detachment was also contentious. Riddell had to explain at length and on several occasions why it was necessary for the observers to

---

7. Riddell to Ward, Toronto, November, 14, 1839, letter book, MSC archives.

be lodged close to the observatory building and the instruments, emphasizing the necessity of taking hourly observations and of being at hand to take measurements during unexpected events of short duration such as the magnetic variations during thunderstorms or aurora:

> As the barracks will be too far distant from the Observatory to serve as quarters for myself it being necessary that the Instruments should be observed during sudden storms, etc. I have the honour to request your approval of a cottage being also built adjoining it.[8]

And again, a month later, Riddell explained:

> The ordinary forms of a Garrison would alone occasion great inconvenience and delay in passing to and from the Observatory at night, especially as the Gate leading to it would not be the one used as an entrance, and to go round by the other would cause a considerable circuit over a very bad road and through the Artillery stables the gates of which are locked at night. In the event of a repetition of the last years' outbreaks[9] these considerations would deserve still more weight.[10]

The issue of living space would be a source of recurring problems throughout the twelve years that the Toronto Observatory remained a military operation.

By December 1839, negotiations were underway to site the observatory on land belonging to King's College, later the University of Toronto. It was the fourth site considered for the observatory in two months and one that proved to be acceptable to all parties. Although the magentic observatory was only expected to be in operation for the three years of the magnetic crusade, Riddell hoped that by associating the observatory with the university, its function as a scientific institution might be extended beyond the original three-year project. Riddell's successor as the observatory director, John Lefroy, eventually suceeded in establishing a permanent observatory in the 1850s. The scientific institution they founded still exists today as the Meteorological Service of Canada. In his review of the founding of the ob-

---

8. Riddell to Montreal Headquarters, Montreal, October 4, 1839, letter book, MSC archives.

9. This is a reference to the Rebellions of 1837 and 1838.

10. Riddell to Sep. Adj General Royal Artillery November 14, 1839, letter book, MSC archives; Riddell et al. 1839.

servatory, Thiessen noted that "the affililation with the University, under whose auspices much of the work has been done, has from the beginnng been a happy one, ensuring in 1853 the permanence of the establishment, and providing always the stimulus of active and productive research" (Thiessen 1940, p. 317).

The Permanent Observatory at King's College
The observatory and adjoining living space was built during the spring and summer of 1840 while observations continued in the temporary station set up in the Bathurst Street barracks. The transfer to the permanent building onto university grounds was completed by September of 1840, although difficulties in obtaining materials, furnishings, and transport from the army stores continued.

Even the college grounds didn't prove to be entirely safe from firearms, however, as they were a popular place to hold shooting matches, according to Riddell's letter of complaint:

> I . . . call . . . attention . . . to the danger and annoyance to which persons are constantly subjected from shooting matches taking place within and in the vicinity of the University Grounds . . . I may mention as an instance that yesterday afternoon 5 different discharges passed through the windows of the observatory on the ground floor.[11]

Riddell became ill in September 1840 and requested medical leave in December so as to spend the winter months in a warmer colony, but his health had improved enough in January 1841 for him to request to stay on a few months longer. He was finally sent home to the United Kingdom on sick leave on February 17, 1841, where he was posted as Sabine's second in command as general director of the Imperial Observatories. In a chatty letter to McCord in 1848, Lefroy let McCord know that "My friend Capt. Riddell is ADC to General Riddell commanding in Scotland. You have no doubt heard of his marriage to a daughter of Sir Huw Ross our Adjunct General" (Lefroy 1848).

Charles Younghusband as Caretaker Director
Riddell's post as director of the Toronto Observatory was temporarily taken up by Lieutenant C. W. Younghusband, a twenty-year-old artillery officer who had spent some time learning the observatory routine and measure-

---

11. Younghusband to Bursar of Kings College, October 5, 1846 (Riddell et al. 1839).

ment duties from Riddell as a volunteer. Lefroy described Younghusband as being, in 1837, "the youngest and smallest officer in the service, beaming with boyish spirits and quite unabashable" (Lefroy 1895, p. 19).

Younghusband had a difficult task in building and maintaining the observatory, organizing the observers, and, even more arduous, in obtaining the necessary funds from the quartermaster of the Royal Artillery, whose orders concerning the annual budget tended to arrive late from the United Kingdom. Officers, noncommisioned officers (NCOs), and ordinary soldiers, invariably listed as "gunner and driver" in military correspondence, were assigned away from their regular regiments to observatory duties. This meant that the observatory detachment was outside the normal military hierarchy and thus also outside the regular chain of provisionment. Letters requesting rations, blankets, clothing, oil and wood, and winter coats were constantly being sent to the commanding officers in Toronto, Montreal, and even the British North American Army's roving commander in chief, as the local quartermaster in Toronto had often not received any orders about supplying the observatory. The directors frequently had to pay for the day-to-day expenses of keeping the observatory warm, lit, and in reasonable shape out of their own pockets, hoping to eventually be reimbursed by the army. Other matters occupying their attention, from evidence in the letter book, included requesting back pay and funding for the functioning of the observatory[12] and tracking down a box of instruments that had remained undelivered for over a year.[13]

Lefroy and the Northwest Magnetic Survey

Meanwhile, the arrangements for the northwest magentic survey were underway in the United Kingdom during the early 1840s. Sabine, who had initally proposed the survey, was given the post of director of all imperial observatories and Riddell had been sent home on sick leave. John Henry Lefroy, who had been appointed to oversee the Saint Helena's Observatory in the South Atlantic at the same time as Riddell was sent to Canada, was now given the task of conducting the northwest survey. Given Riddell's uncertain health, Lefroy was also appointed to the position of director of the Toronto Observatory, although it was Younghusband who stayed in Toronto as interim director while Lefroy was on his surveying expedition from April 1842 to November 1844 (Thomas 1991; Zeller 2009).

---

12. Lefroy to Sabine, Toronto, October 25, 1842.
13. Lefroy to Deputy Ordinance Storekeeper, Toronto, November 16, 1842.

Lefroy arrived in Canada in the summer of 1842, spent three weeks in the United States meeting natural scientists such as Louis Agassiz, Arnold Guyot (1807–1884; later a professor at Princeton and a colleague of Henry's, working also on the Smithsonian meteorological network), Alexander Bache (1806–1867; then the head of the U.S. coastal survey), and Joseph Henry (later at the Smithsonian) as well as paying his respects to the president, General Andrew Jackson. He finally arrived in Toronto that October. He spent the winter in Toronto, where he found the work at the observatory had "fallen terribly in arrears" (Lefroy 1895, p. 63). He left on his northwest survey with the Hudson's Bay Company (HBC) canoe flotilla in April 1843, having obtained passage with the HBC fleet thanks to the connections between the Royal Society and the HBC. There was to be a nasty surprise later, however, when Sir George Simpson, chairman of the HBC, presented Sabine with an enormous bill for Lefroy's passage and keep. The details of the expenditure would be haggled over for years and cause much ill feeling between Lefroy and Sabine for years to come (Lefroy 1895). In the spring of 1843, however, Lefroy and Corporal William Henry travelled up the Ottawa River, across the Great Lakes, and spent the next eighteen months journeying into Canada's northwest, travelling as far north as Fort Good Hope, 145 kilometers northwest of Norman Wells on the Mackenzie River and only just missed crossing the Arctic Circle.[14] They returned to Montreal in November of 1844, and Lefroy arrived in Toronto some ten days later to take up his position as director of the Toronto Observatory. Humphrey Lloyd, an eminent magnetic researcher, founder of the observatory in Dublin, and one of Sabine and Herschel's "crusader" colleagues, described Lefroy's survey results as "probably the most remarkable contribution to our knowledge of magnetic disturbance" (Thomas 1991, p. 24).

The Observers and Their Families
Keeping staff at the observatory was also difficult. Things started off well, with a staff of five soldiers: Corporal James Johnston, Bombardier James Walker, Acting Bombardier Thomas Menzies, Gunner George Watson, and Gunner Joseph Graham deployed from the United Kingdom with Lieuten-

---

14. The unexpected range of their expedition considerably added to their travel expenses, leading to the surprise invoice presented to Sabine, mentioned above; this led to some recriminations and strained relations between Sabine and Lefroy. Sabine appears to have managed to quarrel with many of his colleagues and supporters (e.g., Cawood 1979).

ant Riddell in 1839 to form the observatory detachment. In September 1840, they were recommended for promotion by Riddell, who praised their "most exemplary" conduct. The NCOs were also commended for having "shown the utmost attention and ability in the execution of the duties entrusted to them."[15] The two gunners, Watson and Graham, were listed as being married with five children between them, for whom rations had to be wrested from the stores with some difficulty. In January 1842, Younghusband listed these same men as having served satisfactorily under his comand for the past eleven months, "their conduct has been most exemplary and . . . [they] continue to show the utmost attention and ability in the duties entrusted to them."[16] Acting Bombardier Thomas Maline was added to the observatory detachment that month. The next day, however, Younghusband was called upon to explain an unpleasant incident with Mrs. Watson:

> My servant Gunner and Driver G. Watson was confined to the Hospital on the 17th December . . . Asst. Surgeon Elliott . . . recommended my getting another servant his time of return being very uncertain. I Therefore . . . was under the necessity of desiring Mrs. Watson to provide herself and family with Lodgings until her husband were able to resume his duties. I had put myself to a great deal of inconvenience on her account and felt disposed to do much more had not her conduct on this occasion been so extremely violent and disgraceful to me personally, I had no alternative but to turn her at once from the Observatory, her subsequent conduct has proved that I have not too soon got rid of a very bad Woman.[17]

On January 18, 1842, George Watson was declared medically unfit for further service.

At the time, while the wives and children under the age of thirteen of active soldiers were permitted to share the barracks with their husbands, there was only enough space for 6 out of 100 soldiers, chosen by lottery, to have their wives and young children live with them in the barracks. Upon the death of a soldier, their widows and children were given 72 hours to vacate their barrack space.[18] Finding adequte living quarters for the soldiers' families was a constant preoccuptation in what was still very much a frontier

---

15. Riddell to Deputy Adjunct General Royal Artillery, September 22, 1840.
16. Younghusband to Sabine, Toronto, January 17, 1842.
17. Younghusband to Story, Captain Royal Artillery, Toronto, January 18, 1842.
18. Courtesy of the historic Fort York guided tour.

town. As can be seen from these numbers, the soliders associated with the observatory had a considerably higher percentage of families living with them in the observatory barracks. The presence of soldiers' families would be a cause for constant tensions and difficulties for the observatory staff.

On July 9, 1842, the hitherto exemplary Acting Bombardier Thomas Menzies was "removed from the detachment at the Observatory" for "repeated misconduct." John McNaught was sent to replace Menzies, but Riddell had to ask Thomas Menzies to stay long enough to train McNaught in his duties, the observatory being too short handed to function without Menzies's help.[19]

Letting Menzies go for misconduct would soon prove to have been a highly regrettable decision, however, as those sent to replace him tended to be even worse. After a week, Riddell stated, "I have made myself acquainted with [John McNaught's] capabilities . . . I consider him unfit for the duties required of him here."[20] Menzies's position was still vacant in December, and on December 6, 1842, Lefroy wrote asking for his return to the observatory: "If the conduct of Gunner Menzies has been to your satisfaction . . . I am disposed to give him another trial . . . in the hope that the lesson he has received will make his conduct more satisfactory for the future."[21] Lefroy's ideal soldier for the observatory detachment was

> a man who in addition to the quality of steadfastness, can read, write a good hand, understands simple arithmetic and is intelligent and able to master the duties he will be called upon to perform here. He must be an unmarried man, there being no accommodation for a family.[22]

Lefroy had earlier requested permission to build on additional barack space to hold the observers and their families,[23] which was not forthcoming.

In May 1844, Younghusband was still trying to get Menzies back.[24] Instead, on May 25, Younghusband reported the arrival of Acting Bombardier

---

19. Younghusband to Maricott, Commanding Royal Artillery Toronto, Toronto, August 3, 1842.
20. Younghusband to Maricott, Commanding Royal Artillery Toronto, Toronto, August 12, 1842.
21. Lefroy to Lieutenant Colonel Maclachlan, Commanding Royal Artillery, Kingston, Toronto, December 6, 1842.
22. Lefroy to Captain Story, Royal Artillery, December 11, 1842.
23. Lefroy to the Military Secretary, Toronto, October 25, 1842.
24. Younghusband to Maclachlan, Toronto, May 7, 1844.

William Grace and Gunman and Driver Cannon; Cannon, unfortunately was married, and there was no accommodation for a married man's family, especially one who was expected to take observations day or night.[25]

Finally, on May 28, 1844, the observatory got Menzies back. It wasn't until October that Younghusband found out that Menzies had married while away from the observatory; his marriage certificate was dated January 22, 1844. Where his wife lived during the intervening months remains unknown. Given the difficulties Lefroy and Younghusband had over the past two years in finding someone to replace him, Younghusband was happy to sanction Menzies's marriage, although there were already two married men with families to squeeze into the barracks:

> Gun. Menzies having made me acquainted with the circumstances and his general character being very good I have regarded him as married *with leave* ... Gun. Cannon with his family occupied one room while Gunmen Mahoney, Jones and Cole with the family of the latter occupied another.[26]

Troubles with misconduct surfaced again in 1846, and once again, Lefroy pleaded for "an intelligent, unmarried NC [noncommissioned] officer or Gunner."[27] After listing basic requirements, he adds there are no accommodations for a married man, though he would prefer a married man "of suitable trustworthiness and intelligence" over an unmarried man found wanting of those qualities.

Another difficult situation arose in March 1846. Gunner Cooper, who had acted as Lefroy's servant since September 1837, lived in the barracks with his wife, Amy Cooper, and their two children. After "repeated acts of drunkenness" and "having contracted the vice for which he is now reported, and under circumstances which give me little hopes of his reformation,"[28] Lefroy discharged Cooper, who was sent to the main military garrison in Kingston, while his wife was imprisoned. It soon turned out that Cooper had contracted debts that would be repaid out of his salary. The situation of

---

25. Younghusband to Maclachlan, Toronto, May 25, 1844.

26. Younghusband to Macklachlan, Toronto, October 10, 1844. Emphasis in the original; with only 6 of 100 men being granted living space that is, a bunk bed for their families, soldiers theoretically had to get permission to marry.

27. E.g. Lefroy to Lieutenant Colonel Dalton, July 12, 1846; Lefroy to Colonel Sabine, December 8, 1846.

28. Lefroy to Lieutenant Colonel Dalton, March 27, 1846.

his wife was even worse, having gone from jail to being confined in a lunatic asylum. According to Younghusband's "Memorandum of Amy Cooper's Case," she was insane by reason of intemperance.[29]

The Cooper children were left without their parents, though Younghusband tried to ensure Cooper's salary was forwarded to those who had taken the children in charge. The fate of the Cooper family left Lefroy with regrets for not having intervened in the situation sooner, as he recorded in his autobiography:

> I got involved in a very painful business. My eyes were opened to the infamous life led by the wife of my soldier servant, to the habitual drunkness of both of them, and to the disgraceful condition in which they lived in an old blockhouse on Spadina Avenue . . . Utlimately the woman died of drink and destitution . . . I had the bitter reflection that timely severity and greater exertion on my part might possibly have saved the whole family. I record this painful experience for the lesson it gives of the consequences of evading responsibility and indulging a good nature, which is nothing but sloth. (Lefroy 1895, p. 107)

This passage gives a startling glimpse into the life and conditons of some of those connected to the Toronto Observatory at the time. Lefroy had recorded instances during his military training in the United Kingdom in the 1830s of "death by destitution" among the soldiers' families (Lefroy 1895, p. 18). Things seem not to have been much better in Toronto a decade later, although there are also references to the work the soldiers and their wives put in to keeping the barracks clean and well ordered. Efforts were also made by the British Army to improve conditions in the barracks by steadily reducing the number of people expected to live in an open barrack room. Eventually, the barracks rooms were divided into small apartments for individual families, while many of the unmarried soldiers went to live in the new fort upon its completion in 1841 (Benn 2007, p. 24).

The above passage from Lefroy's autobiography clearly displays Lefroy's moral values and the role they played in his scientific life. Lefroy, a devout member of the Anglican Christian Church, had the motto "not slothful in business, fervent in spirit, serving the Lord" printed on the title page of his autobiography. He found in his religious faith inspiration for his dedication to science. The Victorian values of zeal, useful activity, and temperance,

---

29. Younghusband to MacKay, October 29, 1846.

which derived much of their power from Victorian religious values, were motivating factors for many of the early Canadian weather observers. Activity, including scientifc activity, stood in contrast to the Victorian vices of sloth and intemperance.

By 1850, the problem of soldiers and alcohol seemed to have reached a crisis point in Toronto. Lefroy reported with exasperation that the corporal of the local Royal Artillery detachment "cannot recommend [any man] of the Detachment for Sobriety" (Thiessen 1945, p. 275). Difficulties with finding sober, responsible, and preferably unmarried soldiers who could be trained as scientific observers continued, with another four soldiers dismissed from the observatory detachment for unsteadiness, drinking, or general attitude.

On the other hand, those who were considered by the officer directors Riddell, Younghusband, and Lefroy to be better suited as scientific observers had their work praised and were recommended for promotion. The detachment as a whole and the NCOs Corporal John Johnston, Bombardier Walker, and Assistant Bombardier Menzies in particular were praised by Riddell for their exemplary conduct, both in "the care that was taken of the instruments under their charge during their voyage and long inland journey, and since in the general performance of their duties they have shewn very great intelligence and ability" (Thiessen 1940, p. 346).

Lefroy's Decade as Director of the Observatory
On Lefroy's arrival at the observatory and Younghusband's return to England to act as Sabine's deputy in 1842, the calculations and data returns were much in arrears (Lefroy 1895, p. 63), as there was too much work for the detachment to handle. In an age before electronic computing, even the simplest of summarizing statistics, such as a mean value, could involve what we might consider today laborious and intensive calculations by hand. Performing sometimes complex corrections or adjustments to observations to account for issues such as instrument calibration added to the formidable task of "reducing" the mass of hourly recorded values to a manageable and succinctly communicated abstract. Lefroy and his NCOs made a heroic effort, and Lefroy was happy to praise the NCOs for their zealous efforts "to clear off the accumulation of work in arrear" (Thiessen 1942c, p. 460).

Menzies was singled out for praise and Lefroy sought to have his rank, reduced for the misconduct alluded to in 1842, returned to that of acting bombardier. Long after the observatory ceased to be a military operation, Menzies remained at the Toronto Observatory as a senior observer, continuing "on active duty until a few weeks before his death in 1883, having devoted half a century of his life to the service," being especially valued for his mechanical skill (Carpmael 1888, p. 209). His son, William Menzies, took over his position and was listed as an observer as late as 1917.

In November 1846, Lefroy sought a pay increase for Serjeant Johnston as senior NC officer, who was also acting as pay serjeant and serjeant major to the detachment (Thiessen 1945, p. 226).[30] Gunner Charles Jones was recommended for promotion and praised for "excellent conduct" in January 1848 (Thiessen 1945, p. 229), and James Walker was recommended for extra pay:

> I beg to bring to your favourable notice the excellent conduct and long service of Sjt. James Walker . . . [who] has performed the laborious and trying duties of an assistant at this Observatory with great intelligence and unremitting application . . . The amount of work required from the NC Officers is . . . attended with peculiar restraint and confinement, and is trying both to the constitution and to the eyesight. (Thiessen 1945, p. 271)

As a testimony to the arduous work of observing, calculating, and transcribing the magnetic and meteorological returns, Sgt. James Johnston requested his discharge in 1849, shortly after having received a good conduct medal, with a medical note reporting that "his general appearance is that of a man worn out by length of service" (Thiessen 1945, p. 272). Lefroy himself applied for medical leave in 1845, citing the strain on his eyes as the cause of a deterioration in his health. He was granted leave in the spring in 1846 and left for Great Britain on a combined leave and honeymoon in April, having recently married Emily Robinson, daughter of prominent Toronto citizen Sir John Robinson. The newly married couple spent some time with the Sabines upon their arrival in England. Mrs. Sabine was working on a translation of Humboldt's *Cosmos* with John Herschel acting as editor, while Lefroy and Colonel Sabine compared magnetic results (Lefroy 1895, p. 108).

Lefroy also kept in touch with other Canadian observers and made efforts to obtain and publish climatological data. A letter to McCord from Lefroy dated June 7, 1848, asks for McCord's cooperation in providing the publisher of the "Canadian Almanac" with tables of weather summaries for Toronto and Montreal. "I should be very glad," wrote Lefroy, "to avail myself of your very valuable and complete observations at Montreal." Lefroy proposed a list of mean monthly meteorological values to publish, including mean, highest and lowest temperature, "elasticity of vapour, humidity of air, quantity of

---

30. Lefroy to Sabine, November 24, 1846.

rain and snow, number of frosts at night during the summer months, and prevalent winds." Lefroy ended the letter to McCord with an account of the future of the Toronto Observatory: "We discontinue our hourly observations at the end of this month, which will complete six years of them ... The future observations will be on a reduced scale" (Lefroy 1848).

Sabine was "very hot upon introducing magnetic registration by means of photography at Toronto" (Lefroy 1895, p. 109), an endeavour that would cause Lefroy much difficulty in trying to make a developing solution for the photographic traces upon his return to Toronto. Indeed, trying to run the observatory and get the photographic equipment to work caused him so much stress that he again applied for medical leave in 1850, citing overwork and eyestrain.

The Transition from a Military to a Civil Institution

The Toronto Observatory, staffed and funded by the British Army, was seen as a colonial outpost of an imperial scientific establishment and as such was in touch with the latest scientific developments. Riddell and Lefroy often travelled through the United States on their way to and from Great Britain and took the opportunity to visit scientists interested in magnetism. Riddell demonstrated magnetic and meteorological instruments to Alexander Bache, who became one of the United States' leading nineteenth-century scientists (Good 1986, p. 36). Lefroy also met Bache, as well as Joseph Henry of Princeton, Guyot (later a professor at Princeton and a colleague of Henry's, working also on the Smithsonian meteorological network), and William Redfield and Agassiz, who had studied with Humboldt and became well known for his geological work on the Great Ice Ages of the past, founding the discipline of glaciology. Lefroy would put all these connections to use in his efforts to keep the work of the Toronto Observatory going as a civil institution once the magnetic crusade wound down and he and his soldiers, along with the support and funding the Toronto Observatory had received through the Royal Artillery, left Canada. The magnetic crusade and the Toronto Observatory had only been intended as a temporary experiment, a short-term effort of several years to collect enough magnetic observations to be able to use them to come up with a general theory of terrestrial magnetism through scientific induction from the data collected. But Lefroy wanted the observatory to become a permanent institution. Lefroy lobbied tirelessly to get the provincial government of Upper Canada, later Ontario, to take over the functioning of the observatory. In this he was more successful than his scientific colleagues such as McCord or Smallwood in Lower Canada (Quebec).

Lefroy called upon all his North American colleagues and allies with whom he had visited and corresponded over the past decade to send letters of support of the observatory, encouraging its continuing scientific mission as a Canadian establishment rather than one of the British government via the British Army. Letters of support came in from Lower Canada, including ones from the Natural History Society of Montreal and the Literary and Historical Society of Quebec. American professor Elias Loomis (1811–1889) at the City University of New York considered that "the meteorological observations" alone were sufficiently important "for maintaining your Observatory" as well as providing a central reporting location for all British North American weather observers (Thiessen 1945, p. 314). Loomis proposed bringing the matter before the 1850 meeting of the American Association for the Advancement of Science (AAAS) to obtain a resolution in support of the continuation of the Toronto Observatory. Joseph Henry, writing on behalf of the Smithsonian, assured Lefroy that the Toronto Observatory had the support of the Smithsonian Institution, and the matter would be brought up before scientific and cultural groups such as the American Philosophical Society, the American Academy, and the American Association for the Advancement of Science (p. 400). Professor W. C. Bond (1789–1859), first director of the Harvard Observatory, for his part, felt that "the loss of the Toronto Observatory would, I fear, prove fatal to the development of an extended system of observation . . . throughout the United States" (p. 402–403).

Historians Suzanne Zeller and Morley Thomas described Lefroy's promotion of science in Toronto, how he made the observatory one of the points of pride in the burgeoning city and promoted science as one of the landmarks of civilisation in the rapidly growing, and still largely agrarian, colony of Upper Canada (Zeller 2009; Thomas 1991). Lefroy was one of the founders of the Canadian Institute, which published a journal of its transactions. He not only established a scientific and intellectual milieu while in Toronto but ensured it would continue when he left it behind upon being recalled to the United Kingdom.

References

Barrie, D., 2014: *Sextant: A Voyage Guided by the Stars and the Men who Mapped the World's Oceans*. William Colllins, 348 pp.

Benn, C , 2007: *Fort York: A Short History and Guide*. City of Toronto Culture, 34 pp.

Boistel, G., 2010: Training seafarers in astronomy: Methods, naval schools, and naval observatories in eighteenth- and nineteenth-century France. *The Heavens on Earth*, D. Aubin, C. Bigg and H. O. Sibym, Eds., Duke University Press, 148–173.

Carpmael, C., 1888: *Sessional Papers of the Second Session of the Sixth Parliament of the Dominion of Canada*. Vol. 11. Government of Canada, 855 pp.

Cawood, J., 1977: Terrestrial magnetism and the development of international collaboration in the early nineteenth century. *Ann. Sci.*, **34**, 551–587, https://dx.doi.org/10.1080/00033797700200321.

Cawood, J., 1979: The magnetic crusade: Science and politics in early Victorian Britain. *Isis*, **70**, 492–518, http://www.jstor.org/stable/230719.

Champlain, S. D., 1907: *Voyages of Samuel de Champlain, 1604–1618*. W. L. Grant, Ed., C. Scribner's Sons, 377 pp.

Colburn, 1858: *Colburn's United Service Magazine and Naval and Military Journal*. Vol. 88, part III (Sept–Dec 1858) Hurst and Blackett, 634 pp.

Courtillot, V., and J.-L. Le Mouël, 2007: The study of Earth's magnetism (1269–1950): A foundation by Peregrinus and subsequent development of geomagnetism and paleomagnetism. *Rev. Geophys.*, **45**, RG3008, https://doi.org/10.1029/2006RG000198.

Courtillot, V., Y. Gallet, J.-L. Le Mouël, F. Fluteau, and A. Genevey, 2007: Are there connections between the earth's magnetic field and climate? *Earth Planet. Sci. Lett.*, **252**, 328–339, https://doi.org/10.1016/j.epsl.2006.10.032.

Daubeny, C., 1840: Letter to J. McCord. McCord Family Fonds, Papers P001-0820. McCord Museum Archives.

Humboldt, A. V., 1997: *Cosmos: A Sketch of the Physical Description of the Universe*. Vol. 1. E. C. Otté, Trans., Johns Hopkins University Press, 375 pp.

La Hire, P. D., 1710: Observations de la quantité d'eau qui est tombée à l'Observatoire pendant l'année 1709, avec l'état du Thermomètre et du Baromètre. *Mém. Acad. Roy. Sci.*, **1710**, 139–143.

——, 1712: Observations sur la Pluye, sur le thermomètre & sur le baromètre, à L'Observatoire Royal, pendant l'année 1711. *Mém. Acad. Roy. Sci.*, **1712**, 1–5.

——, 1714: Observations sur la Pluye, sur le thermomètre & sur le baromètre, à l'Observatoire Royal, pendant l'année 1714. *Mém. Acad. Roy. Sci.*, **1714**, 1–3.

Lefroy, J. H., 1848 unpublished manuscript: Letter to Mr. Justice McCord. McCord Family Fonds, Papers P001-820. McCord Museum Archives.

——, 1895 unpublished: *Autobiography of General Sir John Henry Lefroy*. C. Lefroy, Ed., Pardon and Sons, 342 pp.

Locke, J., 1846: Observations Made in the Years 1838, '39, '40, '41, '42, and '43, to determine the magnetical dip and the intensity of magnetical force, in several parts of the United States. *Trans. Amer. Philos. Soc.*, **9**, 283–328, https://doi.org/10.2307/1005361.

Loomis, E., 1853. Physics. *New York Quarterly* p. 265.

McCord, J.S., 1838: Meteorological register for 1838 kept at Montreal, Lower Canada. *Amer. J. Sci. Arts*, **36**, 180–183.

———, 1841: Meteorological Summary of the Weather at Montreal, Province of Canada. *Amer. J. Sci. Arts*, **41**, 330–331.

———, 1842: Report of the meteorological observations made on the island of St. Helen's. Natural History Society of Montreal, Lovell, 18 pp.

———, 1843 unpublished manuscript: Notebook on climate and meteorology of North America. McCord Family Fonds, Papers P001-828. McCord Museum Archives.

———, 1847: Observations on meteorology. *Br. Amer. J. Med. Phys. Sci.*, **3**, 228–229.

McKenzie, R., 1976: *Admiral Bayfield, Pioneer Nautical Surveyor*. Fisheries and Marine Service, 13 pp.

Riddell, C., C. W. Younghusband, and J. H. Lefroy, 1839 unpublished manuscript: *Observatory Letters 1839–1847*. IS408, Vol. 1839A. Meteorological Service of Canada.

Sartori, É., 2003: *L'Empire des Sciences: Napoléon et ses savants*. Ellipses, 640 pp.

Saul, J. R., 2010: *Louis Hippolyte Lafontaine and Robert Baldwin*. Extraordinary Canadians Series, Penguin, 253 pp.

Thiessen, A. D., 1940: The founding of the Toronto Magnetic Observatory and the Canadian Meteorological Service: Introduction Part I–IV (with plates XIII–XVI). *J. Roy. Astron. Soc. Can.*, **34**, 308–348.

———, 1942c: Her Majesty's Magnetical and Meteorological Observatory, Toronto: Part IX. *J. Roy. Astron. Soc. Can.*, **36**, 469–472.

———, 1945: Her Majesty's Magnetical and Meteorological Observatory, Toronto: *J. Roy. Astron. Soc. Can.*, **39**, 221–230, 267–278, 311–319, 355–369, 394–408.

Thomas, M. K., 1991: *The Beginnings of Canadian Meteorology*. ECW Press, 308 pp.

Toronto Public Library, 1967: *Landmarks of Canada: A Guide to the J. Ross Robertson Canadian Historical Collection in the Toronto Public Library*. Vol. 1. Toronto Public Library, 383 pp.

Wulf, A., 2015: *The Invention of Nature: Alexander von Humboldt's New World*. Knopf, 473 pp.

Zeller, S., 2009: *Inventing Canada: Early Victorian Science and the Idea of a Transcontinental Nation*. McGill-Queen's University Press, 372 pp.

CHAPTER NINE

Medical Meteorology

By the mid-nineteenth century, interest in climate change seems to have largely petered out. Other urgent weather-related problems had arisen from the 1830s onward that needed solving: storms and epidemics. Devastating scourges of cholera and typhus were sweeping the globe in an era of increased transportation between the far-flung colonies of worldwide empires. It's been speculated that the 1830s cholera pandemic might have been triggered by the disturbed and difficult conditions in Southeast Asia following the disastrous eruption of Mount Tambora in 1815. Today the pandemic is known to be caused by bacteria and related to unsanitary water sources; in the nineteenth century, the origin of these epidemics was still sought in climatic conditions. Dr. Charles Smallwood, founder of the McGill Observatory, thought the cure for cholera might lie in the disinfectant properties of atmospheric ozone.

Science itself was becoming more and more the domain of professionals. Over the course of the nineteenth century, weather record keeping and analysis shifted away from the amateur "lay scientists" such as Alexander Spark or Alexander Skakel, towards more formal professions, notably the medical profession and the military or, as in the case of the Army Medical Department, both.

Mid-Nineteenth-Century Developments in Meteorology: Storms and Storm Warnings

By the time of the 1857 American Association for the Advancement of Science (AAAS) meeting in Montreal, the first time this meeting was held outside of the United States, controversy as to the nature of storms had been raging for more than 25 years. The eastern United States, where much of the population lived in the early nineteenth century, was subject to post tropical hurricanes (with their tight bands of rotating winds moving in an arc from a northwestern to a northeastern direction up the Atlantic coast), the travelling high and low pressure systems that sweep across the continent from west to east, and the small-scale convective storms of summer. It took decades of often acrimonious debate before there was a classification of these different types of storms. The storm controversy is described in detail by James Fleming (1990) and Charles Moore (2015), with a briefer account outlined by Vladimir Janković (1998).

Two "amateur" meteorologists, William Redfield of the United States and Lieutenant Colonel William Reid of the U.K. Royal Engineers, noticed the predominantly circular pattern of the debris left behind by rotating storms such as hurricanes and proposed "whirlwind" or a kinetic "law of storms," with the main cause of the storms arising from the rotation of Earth. James Espy, who as the Director of the Meteorological Committee of the American Philosophical Society could be considered as a more experienced, "professional" meteorologist, held that storms were thermally driven, with the latent heat of condensation as the main source of energy driving storms. These two competing theories became a bitter dispute, with both Redfield and Espy travelling to Europe to seek support from scientists in France and the United Kingdom. British scientists such as John Herschel, John Dalton, and John Daniell favoured the rotational kinetic theory of Redfield and Reid. French scientists, exploring the newly developed theories of heat described by Joseph Fourier (1768–1830), tended to favour Espy. Eventually, the main mechanisms proposed were found to each have a role in the formation of storms. The heat released by condensing water vapour during convection, the rotating character of the winds caused by the turning of Earth and the deflection of air as it moves from regions of high pressure to areas of "rarefied" or low pressure, and the collision between warm and cold air masses all have their role to play in understanding weather systems.

While the storm controversy raged in North America, in the British Empire, dependent on shipping for transport and communication, storm warnings were becoming of considerable practical and scientific interest. Storms were dangerous not only on the open ocean but also, in Canada, on the Great

Lakes and the Saint Lawrence River, the main corridors of transportation and settlement in the colony. Indeed, to this day half of all Canadians live along the Saint Lawrence River or in the Great Lakes Basin. Admiral Robert FitzRoy, the former captain of the HMS *Beagle*, at whose invitation Charles Darwin embarked on his famous voyage to South America, had been the first to organise oceanographic and meteorological information as a systematic government-sponsored exercise in 1854. Unlike the magnetic crusade, which operated under the auspices of the army's Royal Artillery, FitzRoy's office was not within his home institution of the navy but was instead under the authority of the Board of Trade, as the vast amount of meteorological information collected by the ships' logs was of considerable importance in reducing shipping losses and delays. Institutional differences between the civilian Board of Trade and the Royal Navy made FitzRoy's task difficult, and he lacked the institutional and scientific support given a decade earlier to the magnetic crusade.

FitzRoy's extensive experience at sea (he had been in the navy since 1819) and his keen interest in preventing shipwrecks by applying knowledge of wind and weather patterns, prompted him to go one step further than mere analysis of winds and currents. He instituted a system of storm warnings communicated visually by combinations of flags and cones hoisted at English ports in 1861. His system is still used today. FitzRoy was emboldened by the success of his storm warnings at ports in reducing losses of ships, lives, and goods at sea to send synoptic maps and publish forecasts, or probabilities of upcoming weather, in the daily press. This proved to be too much like prophesying or fortune-telling for some members of the British scientific establishment, who considered FitzRoy's forecasting to be unscientific. FitzRoy's forecasts were based largely on his decades of personal experience and years of synthesizing data at the Board of Trade and were not formalised as a set of rules or equations. His forecasts were also a step too far for government civil servants, who baulked at the expense of collecting weather reports by telegraph and collating the information. In 1865, the morning after he had met with his American counterpart Lieutenant Maury, FitzRoy committed suicide while, the inquest found, "in an unsound state of mind" (Anderson 2005, p. 121). FitzRoy's tragic story has been told in two recent accounts: Katherine Anderson's *Predicting the Weather* (2005) and Peter Moore's *The Weather Experiment* (2015).

FitzRoy's storms warnings, which had been cancelled for being "unscientific," were reinstated under pressure from the public in 1868 by Robert Henry Scott (1833–1916), FitzRoy's successor and head of the Royal Society's

Meteorological Committee. Scott billed *his* forecasts as "actual *facts*, not *prophecies*" (Anderson 2005, p. 128), though from Anderson's description it would be difficult to spot the difference between Scott's storm warnings and FitzRoy's. The interest in storm warnings would finally prompt the Canadian government, after decades of appeals, to institute organised observations and centralization of meteorological information under the Department of Marine and Fisheries in 1874.

Weather and Health
The potential connection between weather and health continued to be an area of concern throughout the period since Gaultier's time. Before Louis Pasteur's germ theory of disease was developed in the 1860s, it was not known how diseases spread, especially those caused by unsanitary water, such as cholera, or by insects, such as malaria. As these seemed to be seasonal, it had been thought since ancient times that prevailing weather influenced prevailing disease and illnesses; even the word "malaria" comes from "bad air." These preoccupations can be seen in the weather records, letters, and diaries described in previous chapters. Gaultier, of course, as the doctor in charge of the health of both the French Army and Navy stationed in New France as well as the health of the colonists, kept records of prevailing illnesses alongside his weather registers. McCord also noted illnesses such as cholera and typhus in his weather diary, acknowledging the fear upcoming epidemics spread in the community; with reports of cholera in Europe, residents knew it was only a matter of time before it came to Canada. In the letters discussing the site of the magnetic observatory to be built in Toronto, Riddell rejected several sites situated in swampy ground as being too unhealthy.

Gaultier was slightly ahead of his time in keeping weather records. In 1778, France established the Société Royale de Médecine (the Royal Society of Medicine) under the patronage of Louis XVI, with the goal of establishing and maintaining correspondence on matters both medical and meteorological throughout the kingdom of France (Kington 1980). Louis Cotte composed the *Traité de Météorologie* in 1774, in which "detailed instructions were issued on instrumental operation and exposure and observational procedure, with the request that standardised observations of pressure, temperature, wind, humidity, rainfall, evaporation, state of sky and significant weather should be made three times a day at specified times" (Kington 1980, p. 12). By the time of the French Revolution, the network included over seventy observers. This included observers in the colonies (McClellan and Regourd 2000,

p. 34). Meteorological observations remained connected with colonial and military medical officials in both the French and British Empires throughout the nineteenth century, though the connection was stronger in the French Empire. In the British Army, meteorological observations were kept by the Royal Artillery and Royal Engineers from the 1830s until the 1860s, when the Army Medical Department took over responsibility for weather observations (Napier Shaw and Austin 1932, p. 167). In France, medical climatology remained of interest well into the twentieth century, as can be seen by the massive three volume publication of the *Traité de Climatologie Biologique et Médicale* (*Treatise on Medical and Biological Climatology*) in 1934 (Piery et al. 1934).

In the United States, the army surgeons during the War of 1812 were convinced that the weather and climate were responsible for the loss of lives because of diseases such as dysentery and typhus (Fleming 1990, p. 13). The army's surgeon general instituted regular meteorological observations as a part of the military medical staff's duties, and, after the war, meteorological observations became a routine part of the Army Medical Department (p. 15).

Meteorology in Montreal: The Medical Community

As the nineteenth century wore on, there were fewer clergymen or professionals in Montreal keeping records and publishing results. Interest in meteorology had shifted to medical professionals such as Drs. Charles Smallwood, Archibald Hall, and William Sutherland, whose weather diaries are conserved in the McGill University Archives. Hall and Smallwood were both obstetricians. Perhaps the high infant mortality, especially in the summer months (Olson and Thornton 2011), spurred their interest in the connection between weather, climate, and disease; or perhaps the recurrence of infectious diseases such cholera and typhus throughout the century brought about a renewal of interest in the ancient weather and health relationship, as exemplified earlier by Gaultier. Smallwood was particularly interested in the effect of ozone as a disinfectant and published research into the connection between atmospheric electricity, the production of ozone, and the prevalence of cholera.

Charles Smallwood

Charles Smallwood was the most prominent meteorological figure in Montreal in the second half of the nineteenth century. Born in 1812 in Birmingham, England, he arrived in Canada with a medical degree in 1833. Much of the biographical information on Smallwood comes from notes made by Nancy Bignell in preparation for an article commemorating the centenary of the McGill Observatory in 1963, just before it was demolished to make way for a new, modernist humanities building (Bignell 1962). His licence to practise medicine in Canada East is dated July 16, 1834. According to an autobiographical sketch, Smallwood "acquired one of the largest medical practises in the country embracing a circle of some 40 miles" in 1834 (Smallwood 1861). In an article written in 1858, Smallwood referred to twelve years of observations, which implies his observatory in Saint Martin was functional by 1846 at the latest. Marshall's biographical sketch of Smallwood suggests that Smallwood set up his medical practice, residence and observatory in Saint Martin by 1841 (Marshall and Bignell 1969, p. 483). The earliest logbooks of Smallwood's meteorological observations in the McGill Observatory Archive collection start in 1849, with the observations made in a scattered and irregular fashion, with no consistent observation schedule.

All his life, Smallwood combined his medical profession with his meteorological hobby, though what he considered to be a hobby encompassed as many instruments and observations as most professional observatories. Although engaged as a professor of meteorology to McGill University, the appointment was an honorary one, and his income derived from his medical practice. He was a governor of the College of Physicians and Surgeons of Lower Canada from 1851 to 1865 and affiliated with Bishop's University in Lennoxville, some 150 kilometers southeast of Montreal. In the spring of 1871 Bishop's created a second Faculty of Medicine located in Montreal, and appointed Smallwood the professor of midwifery and diseases of women and children; on March 10, 1871, the medical faculty elected him dean of Bishop's Faculty of Medicine. At the end of May, however, he received word that the government of Canada would lend financial support to his meteorological observatory, which was housed at McGill since late 1863. He resigned his post of dean of medicine at Bishop's the next day. The Faculty of Medicine of Bishop's met on June 12 to try to persuade Smallwood to change his mind and stay on as dean, but Smallwood finally had to choose between medicine and meteorology; meteorology won.

Smallwood had started publishing the results of his meteorological observations in the *British American Journal of Medical and Physical Sciences* (later simply the *British American Journal*) in 1849 and in the *Canadian Journal* in 1853. He started writing articles summarizing his observations, under the title "Contributions to Meteorology," in 1852 in the *American Journal of Science* and in the *Canadian Naturalist and Geologist*, which was the major scientific publication of the Montreal Natural History Society (NHS), in 1857. As well as straightforward summaries of the weather, he presented papers and published research on ozone, crystal formation of snowflakes, aurora borealis, and photographs of solar eclipses. His scientific work and, presumably, increased participation in the NHS led to the awarding of a legum doctor (LLD; doctor of laws) and his appointment to an honorary—unpaid—professorship of meteorology at McGill University in 1856. This was the high point of nineteenth-century meteorology in Montreal; after Smallwood's death in 1873, the professorship in meteorology lapsed for eighty-eight years (McGill Stormy Weather Group 1968, p. 7).

Smallwood's home observatory was considered a marvel of ingenuity. He constructed it entirely on his own and set it up to be as self-recording as possible. Dr. Archibald Hall wrote a "sketch" of Smallwood's observatory that was read to the Canadian Institute on February 20, 1858, and was subsequently published in the *Journal of the Canadian Institute* and reprinted in the *Canadian Naturalist and Geologist* in 1858.

Smallwood had rigged up an impressive number of instruments to monitor the atmosphere on the outside of the observatory and made extensive use of electricity to automate his readings. The outside instruments included a snow gauge on the lawn, a direction dial to gauge the direction of cloud movement, a lantern raised on a staff to collect atmospheric electricity and conduct the electric current via a copper wire down to an insulated conductor inside the building, a wind vane that rotated a dial inside the observatory building to show wind direction, an anemometer, and a rain gauge, which collected the rain and funnelled it down into the building, along with marking devices attached to a strip of paper moved by a clock at a constant rate that marked the time any rainfall started and ended. The thermometer and wet-bulb thermometer were mounted on an outside wall and screened on both sides to prevent any direct or indirect sunlight from falling on the thermometers. "On the left [are] the scales with which experiments are conducted throughout the winter to ascertain the proportional evaporation of ice" (Hall and Smallwood 1858, p. 352).

Inside the building, a clock was attached to a wheel which moved

> slips of paper along little railways, on which the anemometer by dots registers the velocity of the wind; the rain gauge, the commencement and end of showers; and the wind vane, the continually shifting currents of the wind. This is effected [*sic*] by a pencil, kept applied by a spring to a piece of paper. (Hall and Smallwood 1858, p. 353)

Smallwood also used photography as an auto recording device and an electrical apparatus to measure "the force of the electric fluid." His electrical system was well grounded to avoid "the fate of the unfortunate Richman,"[1] (Hall and Smallwood 1858, p. 353) electrocuted in Saint Petersburg while measuring atmospheric electricity during a thunderstorm. "The whole of this apparatus, even to the electrometers, is the result of [Dr. Smallwood's] ingenuity and mechanical skill," an admiring Hall wrote (p. 353). Three barometers, a standard Newman, a Negretti [made by Enrico Angelo Ludovico Negretti (1818–1879) and Joseph Warren Zambra (1822–1887)], and one of Smallwood's own construction, measured atmospheric pressure, with the average of the three readings recorded in the register. Smallwood also had a parabolic mirror with a self-registering spirit thermometer at its focus to measure terrestrial radiation as well as a transit telescope to determine the time: "not the most perfect of its kind, but simply adequate for all its uses" (p. 353).

Smallwood's observatory had telegraph cables connecting it, through the Montreal telegraph network, to the "principal places in the United States" (Hall and Smallwood 1858, p. 354). Meteorological observations were taken "at 6 and 7am, 2, 9 and 10pm" (p. 355); the 7:00 a.m., 2:00 p.m., and 9:00 p.m. observation times were the standards set by Joseph Henry at the Smithsonian Institute (Fleming 1990, p. 82) and were added to Smallwood's original schedule of 6:00 a.m., 2:00 p.m., and 10:00 p.m. Smallwood may have telegraphed his observations directly to Henry at the Smithsonian as well as sending in the monthly returns.

Smallwood was a part of the NHS network as well as other social institutions such as churches or public institutions. He, like McCord, was careful to compare his observations with others. When Smallwood recorded an exceptionally high pressure reading of 30.876 inches of mercury (1045.6 hPa) at 4:00 p.m. on December 3, 1859, he took care to note, in his published report, that McCord at Temple Grove had measured 30.865 inches (1045.2 hPa), while Hall had a daily average pressure reading of 30.744 inches (1042.1 hPa); meanwhile, in Toronto the value for that day was 30.392 (1029.2 hPa; Smallwood 1860a, p. 308–309). On the other hand, McCord expressed personal doubts of Smallwood's records in 1859; on June 18, a printed table in a newspaper clipping listed Smallwood as recording a maximum temperature of 98.2°F. "Is this possible; my therm[ometer] only gave 85.5 at the same time," noted McCord in his diary entry for that day (McCord 1859).

---

1. Georg Wilhelm Richmann (1711–1753).

Research on Atmospheric Ozone: A Natural Disinfectant?

Smallwood also kept measurements of atmospheric ozone, the most direct link between his profession as a medical doctor and his interest in meteorology. Ozone, now known to be an unstable form of oxygen with three oxygen atoms instead of the usual two atoms found in atmospheric oxygen gas, was first described by Christian Schönbein (1799–1868) in 1839. He gave it the name ozone, from the Greek word for odour, after he recognised the distinctive acrid smell of ozone while decomposing water using an electrical current; Schönbein recognised it as the same smell produced by lightning. The chemical structure of ozone as three oxygen molecules wasn't determined until 1865, but it was recognised as a powerful oxidant with bleaching and thus disinfectant properties. It was this disinfectant property that interested Smallwood. Was atmospheric ozone related to epidemic diseases? Did it disinfect the atmosphere and were lower levels, or even the complete absence, of ozone in the atmosphere the reason why epidemic diseases such as typhus and cholera broke out when they did?

Smallwood measured ozone concentrations using paper or calico coated with a solution of starch and potassium iodide, cut into strips. These strips were placed 5 feet above the ground and sheltered from the sun and rain. In the presence of ozone, the coated strips turned light blue, blue, and then blue black; the concentration was determined using a colour scale. Smallwood also measured the ozone at a height of 80 feet as a comparison between ozone in the free air and ozone at ground level. He was interested in the levels of ozone during the potato blight, placing strips of coated paper next to diseased plants, and noted that the levels of ozone were particularly low during the cholera epidemic of 1854.

Smallwood at the 1857 AAAS Meeting in Montreal

Smallwood presented the results of his ozone research to the American Association for the Advancement of Science, which held its annual meeting for 1857 in Montreal. He had been recently appointed an honorary member of the Montreal NHS. He was thus part of the Montreal delegation that attended the 1856 AAAS meeting in Albany and successfully persuaded the American scientists to cross the border for their next meeting. Smallwood extended a warm, even fulsome welcome to his colleagues from the United States:

> Until the present time, these Annual Meetings have been confined to the United States alone, separated by an imaginary boundary, which has now

been removed, for here we meet as one family . . . the gentle breeze that wafts the red cross banner of St. George and merry England alike unfurls the Stars and Stripes . . . Long may these two flags entwine in peace . . . and may that masterpiece of scientific genius, the electric cable, which is at this moment being laid beneath the Atlantic Sea . . . be that peaceful band, that will cement more firmly the destinies of the two great nations of the earth. (Smallwood 1858b, p. 190)

It's apparent from this speech that Smallwood, like McCord before him, considered himself to be British and Canada to be a part of the United Kingdom and British Empire. Addressing the opening of the meeting, the vice president of the association, Alexis Caswell, also made the following point:

We of the United States are here assembled on British soil, little thinking that we have passed beyond the protection of American law, little thinking, amid the generous hospitalities of this enterprising commercial capital of a noble province of the British Empire, that we are alien to the British Constitution. The American Eagle has left us, but I assure the gentlemen who have invited us here, that we feel in no danger of being harmed by the British Lion. (Smallwood 1858c, p. 163)

Here was an example of the republic of science making a conscious effort to transcend national boundaries.

These speeches from the meeting underscore at the same time the connections between North American scientists in the two countries and the alienation arising between the scientists of two very different traditions who considered themselves to be of distinctive, sometimes even distrustful, nationalities. The intermittent and endemic wars of the eighteenth and early nineteenth centuries between the French and the British, and later between the British and Americans in North America, left a long legacy of mistrust. The War of 1812 was forty years past, however, and just eight years later, partly as a result of fresh invasion fears stemming from the American Civil War, Canada would become its own country within the British Empire as the Dominion of Canada.

Smallwood's reputation as an avid instrumentalist and experimentalist is evident here, as he describes the construction of his "ozonometer" and his general conclusions. These were that "the presence of ozone in the atmosphere is accompanied by a low reading of the barometer, which generally continues while the ozonic period lasts; this period is accompanied or terminated almost invariably by precipitation" (Smallwood 1858b, p. 194). Temperature was not a factor, Smallwood having detected ozone at all temperatures from −20° to 80°F, but found that:

> it would appear that a moist state of the atmosphere was necessary for its production . . . for when the difference between the dry and wet bulb thermometer is little . . . ozone in considerable quantity is invariably present; but when the difference between the two thermometers is considerable, no ozone is appreciable by the ozonometer . . . In reviewing these observations, there is no condition of the atmosphere . . . that indicates the presence of ozone, except the presence of vapour or humidity. (Smallwood 1858b, p. 194)

He found that ozone was only produced in the presence of water, and days with low humidity produced nearly no ozone; Smallwood concluded that water vapour was necessary for the presence of ozone at low levels.

Smallwood found that, after over 6,000 observations of both ozone and the electrical state of the atmosphere, Schönbein's assertion that a "high electrical state of the atmosphere was always present when ozone was developed . . . was not sustained" (Smallwood 1858b, p. 196), though presumably a source of energy was necessary to split apart the molecules of molecular oxygen (O_2) or water (H_2O) to form ozone (O_3); this energy was often lightning. Smallwood's observations seemed to him to indicate that water vapour in the atmosphere was necessary for the formation of ozone, which would account for the presence of ozone near the sea. This notion of disinfecting ozone being related to sea air would have certainly helped popularise the idea that sea air was beneficial to health and continued the tradition of seaside hospital towns.

Considering the effect of ozone on health, and whether it could be considered a general disinfectant to prevent epidemics, Smallwood continued cautiously:

> When largely diffused in the atmosphere it causes, like chlorine, very unpleasant sensations . . . it kills small animals very quickly . . . and it has the power, in a large degree, of destroying *miasma* . . . The small amount of ozone in 1854, which was the year of the last visitation of cholera, would tend to favour the opinion that there was a deficiency of ozone in the atmosphere during the prevalence of that epidemic. A deficiency was observed in almost every month of that year, although the number of days on which

rain or snow fell was almost equal with the other years. (Smallwood 1858b, p. 192, 195)

After considering the possible effect of ozone on the recent potato blight, Smallwood concluded his 1857 address on ozone to the AAAS with a slightly sceptical call for more study:

> It cannot be doubted that an agent so active as ozone, if really present, must exert a great influence on the health of individuals . . . I have, as you will perceive, offered no theoretical deductions. If, as our continental brethren assert, it does possess such powerful and wonderful properties, it must be evident that [we] should take up the subject, in a way that we may arrive at some conclusions. (Smallwood 1858b, p. 197)

In a second presentation, this time to the Montreal NHS, Smallwood expounded on the claims of European scientists regarding the properties of ozone and

> proclaimed it to be the instrument [that] provides for the production of the grand phenomena of nature, that its action can explain the formation of all meteors, as well as the fluctuations and diurnal changes in pressure. (Smallwood 1859, p. 169)

This seems quite a large claim. Smallwood himself was more cautious and limited himself to describing the results of his investigations with little theorizing. He placed his ozone-detecting strips "among plants, over drains, in the sick-chamber" (Smallwood 1859, p. 344) in his efforts to determine what effect ozone had, if any, on health. His attempts to link ozone to health, however, "have up to the present time given no decided results" (p. 409). Smallwood also tried to test the influence of light on ozone production by exposing his test papers to light under different coloured and polarised lenses. His conclusions were that green light prevented the formation of ozone the most, whereas direct light affected it the least.

We now know that ozone is an allotrope of oxygen, which had been suspected by Schönbein and Smallwood, but its exact molecular composition of three oxygen atoms was not confirmed until 1865. It's usually formed high up in the free atmosphere under strong sunlight, the energy from the sun giving the required energy to dissociate water molecules and provide a free oxygen molecule to combine with atmospheric oxygen.

Smallwood and Joseph Henry at the Smithsonian

The AAAS meetings of 1856 and 1857 may have been the occasions when Smallwood made the personal acquaintance of Joseph Henry, the Smithsonian Institute's first secretary. The correspondence, Smallwood's half of which is conserved in the Smithsonian Archives, starts off in formal terms in 1852, with Smallwood sending a copy of his observations to Henry, following this up with a request for blank forms in 1853. Both Smallwood and Archibald Hall participated in the Smithsonian's meteorological observation program, sending their observations to Henry's network centre in Washington. By 1858, however, the correspondence had warmed from a salutation of simply "Sir" to "my dear Professor Henry." In his request for a rain gauge, Smallwood also passed on Mrs. Smallwood's "kindest remembrance to you and the Misses Henry" (Smallwood 1858a). In a following letter, along with his thanks for the rain gauge, Smallwood expressed his regrets that he was unable to attend the 1858 AAAS meeting in Baltimore:

> as much for the sake of the meeting as of seeing Washington & the Institution over which you preside. I hope to present at Springfield & to meet you there & I also trust that you will take a trip back with me & pass a few days with me in the *Canadian backwoods* with any part of your family that may join you ... My family desire me to remember them to the Misses Henry & yourself [emphasis in the original]. (Smallwood 1860b)

Smallwood continued to send his observations to Henry at the Smithsonian and later to the Signal Service at the U.S. War Office.

Dr. Archibald Hall

Archibald Hall was the son of merchant Jacob Hall and Rebecca Ferguson. He was born in Montreal in 1812 and baptised in the Anglican parish of Christ Church by Jacob Mountain, John Bethune's sometimes patron, antagonist, and superior as bishop of Montreal. Hall's early education, like McCord's, was at Alexander Skakel's Royal Grammar School, where he "in early life exhibited a strong love for the study of nature" (Canada Medical Journal 1868, p. 429). At sixteen, Hall was apprenticed to Dr. Robertson, one of the most respected medical practitioners in Montreal at the time, and became a student at McGill's Faculty of Medicine. As an advanced student, Hall was placed in charge of the fever sheds during the 1832 cholera epidemic, where he worked tirelessly. According to the *Canada Medical Journal*, "it was customary in those days, as it is still with all who can afford it, for medical students to repair to the mother country for the purpose of completing their studies at some of the centres of learning abroad" (p. 430). Hall went to study in Edinburgh, which was considered to have the best medical faculty in the British Empire, where he studied from the autumn semester of 1832 to the summer of 1834, receiving both his licence from the Royal College of Surgeons and his medical doctorate from the University of Edinburgh, his dissertation having been on the "respiratory function of plants" (p. 430).

Hall returned to Canada around 1835, when he received his Canadian medical licence and was elected to the Literary and Historical Society of Quebec (LHSQ). On his return to Montreal he was appointed a lecturer in the Medical Faculty of McGill College and elected physician at the Montreal General Hospital. He held the position of chair of the Department of Chemistry from 1842 to 1849, becoming professor of the Department of Midwifery in 1854, a position he held until his death in 1868.

As well as his demanding career as a physician and medical lecturer, Hall continued to pursue his interests in natural philosophy. He received the Gold Medal of the NHS of Montreal in 1839 for his dissertation on the "Mammals and Plants of the District of Montreal" (Canada Medical Journal 1868, p. 431). He started two scientific journals. The first was the *British American Journal of Medical and Physical Sciences*, which ran from 1845 to 1852, when it "ceased publication from want of pecuniary support" (p. 432). Undaunted, Hall next edited the *British American Journal* from 1850 to 1853, when it also folded from a lack of financial support. As Hall was interested in the influence of weather and climate on health, he often published monthly tables of daily meteorological observations. These observations were taken either from his friend and mentor Alexander Skakel or, after Alexander's death, from William Skakel's continuation of the Skakel observations from the house on St. James Street, or from Charles Smallwood. In some cases, the printed record in Hall's periodicals is the only source of weather data for these years. Hall had earlier sent abstracts of Skakel's observations to the *Edinburgh Philosophical Review* in 1836. In the 1860s, Hall kept his own meteorological records and was also an enthusiastic participant in Joseph Henry's volunteer network of weather observations at the Smithsonian.

In May 1838, Hall married Agness Burgess of Edinburgh; Alexander Skakel was one of the witnesses. Despite all of Hall's skill as an obstetrician and training in the diseases of women and children, their newborn son died two days after his birth in December 1843; Agness died four days after their son. Mother and child were buried together on December 22, 1843.

Dr. William Sutherland

William Sutherland (also sometimes spelled Sunderland) was a native Montrealer who graduated from the Faculty of Medicine at McGill in 1836; his doctoral dissertation was on asphyxia. He spent some years on what was then considered to be "out west" on the Niagara frontier, returning to Montreal in the early 1840s. He and some like-minded colleagues established the Montreal School of Practical Medicine and Surgery, where lectures were given twice a day, once in English and once in French, to disseminate medical knowledge to students in both languages. Sutherland also established a self-supporting dispensary where indigent patients could be treated by students for free; this was discontinued as too many people took advantage of the free medical service (Canada Medical and Surgical Journal 1875, p. 378). He was appointed chair of the Department of Chemistry at the McGill Medical Faculty in 1849, where he took over the position left vacant by Hall. His weather observation record in the McGill University Archives runs from 1844 to 1848. It seems that with his appointment as professor of chemistry at McGill, he no longer had time to pursue his meteorological observations. He died in 1875, his lungs damaged from a lifetime of breathing in noxious chemical substances (p. 474).

The *Canada Medical and Surgical Journal* states that "his punctuality and attention to the business of his profession enabled him . . . to enjoy that quiet and solace of opulence" (Canada Medical and Surgical Journal 1875, p. 380). This underscores the fact that much nineteenth-century science, meteorology included, was largely undertaken by men with private means of funding their observations, especially for the purchase of instruments and equipment. It wasn't until 1874 that Montreal had, in Clement Henry McLeod (1851–1917), someone who could be considered a professional meteorologist, and even then McLeod was primarily a professor of civil engineering with an interest in the determination of time and longitude. Charles Smallwood, professor of meteorology at McGill, received no salary payment and presumably derived most of his income from his medical practice.

It was only towards the last third of the nineteenth century that weather record keeping and analysis became partially subsidised by the national government, though even after the establishment of the MSC the vast majority of observers were volunteers.

References

Anderson, K., 2005: *Predicting the Weather: Victorians and the Science of Meteorology*. University of Chicago Press, 331 pp.

Bignell, N., 1962: Official time signal: 100 years. *McGill News* (summer), 16–22.

Canada Medical and Surgical Journal, 1875: The late William Sutherland M.D. *Can. Med. Surg. J.*, **3**, 377–380.

Canada Medical Journal, 1868: The late Archibald Hall, M.D., L.R.C.S.E. *Can. Med. J.*, **4**, 429–432.

Canada Medical Record, 1875: The late William Sutherland M.D. *Can. Med. Rec.*, **3**, 473–474.

Cotte, L., 1774: *Traité de Météorologie*. Imprimerie Royale, 634 pp.

Fleming, J. R., 1990: *Meteorology in America, 1800–1870*. Johns Hopkins University Press, 264 pp.

Hall, A., and C. Smallwood, 1858: The observatory at St. Martin, Isle Jesus, Canada East. *Can. Nat. Geol.*, **3**, 352–363.

Janković, V., 1998: Ideological crests versus empirical troughs: John Herschel's and William Radcliffe Birt's research on atmospheric waves, 1843–50. *Br. J. Hist. Sci.*, **31**, 21–40, https://doi.org/10.1017/S0007087497003178.

Kington, J., 1980: Daily weather mapping from 1781: A detailed synoptic examination of weather and climate during the French Revolution. *Climatic Change*, **3**, 7–36, https://doi.org/10.1007/BF02423166.

Marshall, J. S., and N. Bignell, 1969: Dr. Smallwood's weather observatory at St. Martin's. *Nat. Can.*, **96**, 483–490.

McClellan, J. E., III, and F. Regourd, 2000: The colonial machine: French science and colonization in the ancien regime. *Osiris*, **15**, 31–50, http://www.jstor.org/stable/301939.

McCord, J. S., 1859 unpublished manuscript: Diary, January 1 1859–December 31 1859. McCord Family Fonds, Papers P001.B1, Item 0410. McCord Museum Archives.

McGill Stormy Weather Group, 1968: Three McGill weather observatories. McGill Stormy Weather Group, 19 pp.

Moore, P., 2015: *The Weather Experiment: The Pioneers Who Sought to See the Future*. Farrar, Straus and Giroux, 395 pp.

Napier Shaw, W., and E. Austin, 1932: *Manual of Meteorology: Meteorology in History*. Vol. 1. Cambridge University Press, 343 pp.

Olson, S., and P. Thornton, 2011: *Peopling the North American City: Montreal, 1840–1900*. McGill-Queen's University Press, 524 pp.

Piery, M., M. Milhaud, and R. van der Elst, 1934: Traité de Climatologie Biologique et Médicale. Librairies de l'Académie de Médécine. Vol.1. Masson et Cie, 904 pp.

Smallwood, C., 1858a: unpublished manuscript: Letter to Prof. Henry, June 10th 1858. Smithsonian Institution Archives, Record Unit 60, Meteorological Project, 1849–1875 (data from 1820) Records.

——, 1858b: On ozone. *Proc. 11th Meeting of the American Association for the Advancement of Science, Montreal, Canada*. American Association for the Advancement of Science, 190–196.

——, 1859: On ozone. *Can. Nat. Geol.*, **4**, 169–172, 343–345, 408–410.

——, 1860a: Contributions to meteorology reduced from observations taken at St. Martin, Isle Jesus, C.E. *Can. J.*, **27**, 308–312.

——, 1860b unpublished manuscript: Letter to Joseph Henry. Meteorological Project, 1849–1875, Record Unit 60. Smithsonian Institution Archives.

——, 1861 unpublished manuscript: Letter. Office of the Principal and Vice-Chancellor, Dawson Correspondence, RG 2, Container 10, ACC 927/460, Item 359. McGill University Archives.

CHAPTER TEN

The Establishment of the Meteorological Service of Canada

The increasing professionalization of science over the course of the nineteenth century meant there was less of a place in the structure of scientific activity for the interested amateur or lay scientist. Recording the weather was one of the exceptions to this as, by their nature, weather systems span hundreds and even thousands of kilometers and need a diffuse but widespread network of observations to be fully captured. Volunteer weather observers collect data for governmental weather agencies in North America even today, and anyone with an instrumental weather station can plug into an online network with data-sharing platforms.

The international horizons nineteenth-century Canadian meteorologists looked to were in two main directions. The first, towards Britain, was not considered "international" but was often described as "home," even by those who had been born in Canada. The second, the United States, was considered as not only foreign but periodically hostile. Nevertheless, important links were established between scientists in British North America and the United States. John Samuel McCord, who had grown up during the War of 1812, showed no hesitation in communicating with American scientists and even became a member of U.S. scientific societies such as the Albany Institute. Charles Smallwood and Archibald Hall in Montreal contributed regularly to Joseph Henry's Smithsonian Institute Meteorological project, with Smallwood visiting Henry in the United States (Smallwood 1858). John Henry Lefroy, Charles Riddell, and Charles Younghusband all visited scientists in the United States, and American scientists went to the Toronto Observatory to be trained in using the latest magnetic and meteorological instruments (Good 1986, p. 36). As described in the previous chapter, the American Association for the Advancement of Science (AAAS), with Smallwood as one of the principal instigators, held their 1857 meeting in Montreal, and Smallwood, Hall, Lefroy, and Kingston were all AAAS members.

Yet there were tensions also between Canadian scientists, the British scientists operating in Canada, and the Americans. Some were matters of scientific reputation and national honour, but others reflected ongoing tensions between the United States and Canada, as a representative of the British Empire. After the War of 1812, tensions were raised again during the Rebellions of 1837–1838 in Canada, which the Americans were perceived as aiding and abetting. Hardly had those tensions died down when the Oregon boundary dispute of 1846 once again raised the prospect of war between Canada and the United States. Lefroy indicated the willingness of the soldier scientists of the Toronto Observatory, recruited from the Royal Artillery, to fight if the dispute came to open warfare (Good 1986, p. 42).

The breaking point, after which Canadian scientists seem to turn more towards Britain than the United States, came after the U.S. Civil War. Relations between the United States, reeling from the destruction of war, and Britain, which though officially neutral was perceived as siding with the Confederates, were poor. Canadians regarded the huge armies amassed by their neighbours with trepidation fearing that, once united after the end of the Civil War, they would turn their attention, once again, towards an invasion of Canada. This concern was an impetus towards the confederation of some of the colonies of British North America into the Dominion of Canada in 1867. Paradoxically, links to Great Britain seemed to be strengthened in the last quarter of the nineteenth century after confederation, with the British Association for the Advancement of Science holding their 1884 meeting in Montreal.

Networks and Standards

The more observations are taken, the more necessary standards become to be able to analyse, compile, and compare data from difference places and observers. Two mid-nineteenth-century meteorological networks recorded standardised observations in Canada. The British Army with the Royal Engineers, whose observing practices were based on John Herschel's 1835 *Instructions for Meteorological Observers*, incorporated weather observations as a part of the regular professional military duties at bases across the country. Records kept by the Royal Engineers at Saint John's, Newfoundland; Halifax, Nova Scotia; Quebec City, Quebec; Kingston, Ontario; Saint John's College (near Winnipeg), Saskatchewan; and New Westminster (near Vancouver), British Columbia, are kept in the archives at the U.K. Met Office. Records from Fort Simpson, Northwest Territories; Fort York on Hudson Bay, Manitoba; and Okak, Killineck, and Nain on the Labrador coast are also part of this collection, though not associated with the British Army (Meteorological Office 1851, 1862, 1864, 1866a,b, 1871a,b, 1876, 1880). Across the British Empire, responsibility for weather observations was later taken over by the Army Medical Department.

The U.S.-based Smithsonian Institution Meteorological Project, run by Joseph Henry, was based on volunteers (Fleming 1990, p. 170). Many of the Canadian contributors to the Smithsonian network were professionals, including doctors such as Archibald Hall, Charles Smallwood, and Dr. William Craigie in Hamilton, Ontario. Other Canadian contributors to the Smithsonian include John Delany and his sons in Saint John's, Newfoundland; Henry Poole, at Albion Mines (currently Stellarton), Nova Scotia; Reverend Professor Hemsley and Professor Hoe at King's College, Windsor, Nova Scotia; Professor Stuart and Professor C. Fred Hartt at Acadia College, Wolfville, Nova Scotia; Lawrence Clarke Jr. at Fort Rae in the Northwest Territories; John MacKenzie at Moose Factory [a Hudson's Bay Company (HBC) post]; and Donald Gunn at the Red River settlement (Manitoba), to name some of the over twenty observers sending reports to Joseph Henry at the Smithsonian across the territories known as British North America (Smithsonian Institution 1857, 1869; Fleming 1990). The Smithsonian weather network broke down during the U.S. Civil War, putting an end to the international cooperation between volunteer weather observers in Canada and United States, and the Smithsonian ceased to act as a focal point for the collection and analysis of the voluntary observations made by Canadians. John Henry Lefroy, George Kingston, Charles Smallwood, and others had been lobbying for a Canadian national meteorological service since the withdrawal of the Royal Artillery from the Meteorological and Magnetic Observatory of Toronto in 1853. At this point, the Toronto Observatory came under the authority of the provincial government and was administered by the University of Toronto. George Kingston, originally appointed professor of natural philosophy at the university, found, upon his arrival, that his position had been claimed instead by his brother-in-law John Cherriman. Kingston was instead appointed to Cherriman's position of professor of meteorology and named director of the Toronto Observatory, a position he held until 1880.

After confederation in 1867 gave Canadians direct control over their internal affairs, the Meteorological Service of Canada (MSC) was established. Kingston, as director of the Toronto Observatory, was also appointed as the director of the MSC. The Toronto Magnetic and Meteorological Observatory thus became the forerunner of the MSC. Although originally only funded to provide storm warnings along the crucial shipping routes that stretched from the Great Lakes to the Atlantic Ocean, under Kingston's tenure the MSC developed into a national institution, growing as the country expanded rapidly westward and northward. In 1874, not even ten out of nearly ninety stations reporting to the MSC were west of the Great Lakes; by 1890, over a third of the more than 300 stations were in the West.

Kingston and the Establishment of the Meteorological Service of Canada

The Toronto Observatory, the first civilian observatory maintained with public funding in Canada, became the central meteorological office in Canada in 1871 with Kingston as its head while also continuing as the director of the Toronto Observatory. The Toronto Observatory became the central collecting office of official meteorological observations in Canada, although most observers remained unpaid volunteers or continued within their established positions, with Smallwood of the McGill Observatory in Montreal or Naval Officer Captain Edward Ashe of the Quebec Citadel Observatory being but two such examples. In 1875, Kingston described the Canadian Meteorological Office as "an offspring of the Toronto Observatory, with the work of the MSC carried on within the walls of the Toronto Observatory" (Kingston 1880).[1]

The director's letter book, consisting of books made of blotting paper containing the traces of the original letters written by the directors of the MSC, gives an account of the subjects addressed by Kingston in his national and international correspondence. Only a portion of the immense collection is presented here, but they provide a portrait of the early years in the organization of the national meteorological service.

---

1. Letter to Scott, January 27, 1875.

Early International Cooperation

Kingston was in correspondence on behalf of Canada with the members of the International Meteorological Organization (IMO) in Vienna, including Christophe Buys-Ballot (1817–1890), to discuss setting and conforming to international standards for meteorological observations. A letter from Kingston to Buys-Ballot sums up his position, and thus that of the MSC, in a letter dated 1874:

> Your letter dated Utrecht Dec 31 asked me to send to the permanent Committee for International Meteorology, a report on the changes that have been carried out in Canada in conformity with the decisions of the Vienna Conference. I regret that I have not seen an account of the transactions at Vienna, and am therefore unable to state whether it will be practicable or not to bring the Canadian System into perfect conformity with those decisions. I assure you, however, that I shall gladly do what is possible to bring about such conformity . . . I am willing to comply with the recommendations in the circular except that which is related to the use of the centigrade and metric scales. It would be impossible to abolish the use of our present instruments, but I am willing to present the mean values in the centigrade and metric scales.
>
> As regards the establishment of observatories in remote places, I have to inform you that in addition to the stations in older settlements, I have stations at Fort Garry, and in British Columbia, and hope to establish others during the coming summer near the Arctic Circle on the Mackenzie River." (Kingston 1880)

The Vienna Conference on International Meteorology was the second attempt to harmonise meteorological observations across national boundaries. The first meeting was held in Brussels in 1853 and had ended inconclusively, with no general agreement on international standards of measurement. The 1873 conference, as Kingston's letter shows, didn't hold out much more promise. Kingston was not particularly cooperative, insisting on keeping imperial units; Canada would finally change to metric units in 1975.

In his letter to Buys-Ballot, Kingston listed the instruments used at the Toronto Observatory, which included thermometers by Fastré of Paris, a barometer by Newman, and Robinson's anemometer.

Kingston also had extensive correspondence with Robert H. Scott of the Royal Meteorological Society in the United Kingdom, who was Robert Fitz-Roy's successor at the U.K. Met Office (Anderson 2005, p. 230).[2] Kingston

---

2. Kingston, G.T., (1874): Letter to Scott, June 2, 1874, and June 26, 1874.

regularly obtained instruments from Scott to send into the field in Canada. Once the MSC was established in 1871, after years, or even decades, of campaigning for public support of meteorological observations, Kingston finally had the budget he needed to spend on upgrading the instruments of the volunteer observers and for expanding the network of observations. In 1874, Kingston recruited the newly appointed bishop of Athabasca William Bompas as an observer, having ordered the necessary instruments for him through Scott at the Met Office and arranged for their delivery to Bompas (Kingston 1880).[3] Kingston ordered instruments for sixteen stations in the northwest in 1875, with six class I stations, including Athabaska (northern Saskatchewan), York Factory (on the western shore of Hudson Bay), Swan River, and three stations of the North-West Mounted Police, the forerunners of the Royal Canadian Mounted Police. Five of the stations were under the supervision of Bompas, bishop of Athabaska: Fort McPherson, (now in the Northwest Territories), Rampart House (now in Yukon Territory), Fort Simpson, (now in the Northwest Territories), Fort Revolution (probably a misprint of Fort Resolution, now in the Northwest Territories), Great Slave Lake (now in the Northwest Territories), and Fort Chippewyan (now in Alberta). Another two, Fort Stanley at English River (now in Saskatchewan) and the Devon, Cumberland Mission at The Pas (now in Manitoba), were kept by clergymen (Meteorological Service of Canada 1875, p. 11). Kingston also sent instruments to Mr. J. Bunn at Edmonton. Many of these locations were fur-trading posts or missionaries or both (see, for example, Meteorological Service of Canada 1875, p. xiii).

In 1879, Kingston ordered instruments for the use of Joseph Mason at Moose Factory, a HBC post on the southwestern shore of James Bay in northern Ontario (Kingston 1880).[4] Kingston was also happy to obtain observations from other groups, such as the Moravian missionaries operating in northern Labrador (Kingston 1880).[5] Directing the shipping of instruments from London to their final destination in northern Canada from his office in Toronto could be a complicated procedure, and Kingston relied considerably on Scott to oversee the dispatch of instruments to their intended observers:

Nov 20 1879 RH Scott. Dear Sir, Will you be kind enough to forward to Mr. Jos Mason of Moose Factory, Hudson's Bay, by the ship leaving London

---

3. Kingston, G.T., (1874): Letter to Scott, June 2, 1874, and June 26, 1874.
4. Kingston, G.T., (1874): Letter to Scott, May 9, 1879.
5. Kingston, G.T., (1874): Letter to Scott, November 20, 1879.

in June, 5 minimum thermometers the same as sent last year, and a small prismatic compass. An invoice of the goods should be sent to Mr. Mason and another to this office with the corrections to the thermometers. Please forward the box to the Hudson's Bay Co. in Lime Street City, asking them to give it in the charge of the captain. I should also be glad if you would pay the freight and charge it on the invoice sent to us. (Kingston 1880)

In the short five years between 1873 and 1878, Kingston commissioned more than 700 instruments, including a documented 210 minimum thermometers, 147 maximum thermometers, and 68 barometers from Scott. The expenditure listed in a sample of his letters to Scott alone amounts to nearly £1,720 (nearly $10,000, at a rough estimate, or CAD $300,000 in today's terms; DeGreef 2017).[6] As is still often the case in government spending, any surplus left in the allocated budget at the end of the fiscal year was confiscated and often even eliminated from the next year's budget rather than being carried forward. This not unnaturally tended to lead to a scramble at the end of the fiscal year and pleas for instruments, or at least their invoices, to be sent before the deadline of June 30, while also considering transportation issues and the opening of the shipping season:

Feb 17, 1876 RH Scott: I shall be glad to obtain as large a portion as possible of the above named articles, in time to admit of my paying for them not later than June, as any funds on hand on 30 June (the end of our fiscal year) are confiscated. There is no need to wait for the opening of navigation, as thermometers travel safely through Portland [presumably Maine]. (Kingston 1880)

On a visit to Scott during the course of a return to the United Kingdom, Kingston found much to his liking at the Met Office and sent an instrumental wish list to Scott on March 15, 1877, which included a small barometer that he had seen in Kew and an inquiry as to the cost and shipping estimate for a Pentagraph "similar to the one in use in your office" (Kingston 1880).

In the copies of letters sent by Kingston or his deputy Charles Carmichael, some concerns stand out. One was having the instruments manufactured in Europe, chiefly by the London instrument-makers, packed and transported in such a way as to arrive intact. Another recurring concern was the calibration and testing of the thermometers down to sufficiently low temperatures

---

6. The actual total must be even higher.

to capture the Canadian cold. Both these concerns date back to the time of Gaultier, although as Kingston was ordering large numbers of instruments to set up a Canadian network, instrument breakage was less of a concern for him than it had been for the individual meteorologists of earlier times. Kingston's concerns in the 1880s about the shipping of instruments are reminiscent of Herschel's instructions issued some forty years earlier on the transportation of delicate instruments over rough terrain (Herschel 1835, p. 8).

One incident in particular incensed Kingston: in 1874, specially calibrated Kew barometers sent for from London were duly received—in pieces:

> October 1 1874
> RH Scott
> Dear Sir,
> You will sympathize with my vexation when you know that of eight Kew Barometer which arrived here on the 28th in *SS Circassion*, <u>seven</u> were utterly smashed—my belief is that the cases must have fallen and one from the top of a pile of boxes . . . Others I understand are just landed & I tremble for their fate, & still more for the self-recording instruments.
>
> I am so much indebted to your good offices that I do not like to appear to find fault, but certainly the barometers were not packed as those which come from New York, nor as those which I send about this continent none of which have been broken.
>
> I propose to send some of the broken barometers to Adie for repair [emphasis in the original]. (Kingston 1880)

Evidently the European carriers and instrument-makers were not as used to the poor conditions of transport in North America as were Kingston and the New York instrument-makers. The barometers were sent back to Patrick Adie, the renowned instrument-maker in London.

Kingston received another disappointment a week later:

> Oct 8 1874
> RH Scott
> Dear Sir,
> I this day received by *Scandinavian* eight more barometers of which four were utterly smashed & another injured, I fear very seriously. I do not know that any procuring proceeding would ensure the barometers from breakage when exposed to the treatment inflicted on the route adopted, but they certainly were <u>not</u> properly packed . . . Every one [is] disclaiming responsibility, & the

Express Co. saying that they cannot be responsible for such delicate articles as barometers [emphasis in the original]. (Kingston 1880)

By the following spring, Kingston still had no news of what had become of the smashed barometers and referred caustically to the shortcomings of the shipping and overland transportation companies:

15 April 1875
Mr. P Adie, 15 Pall Mall, London, England
Sir, On the 4th November 1874, I wrote informing you that I had sent you the shells of eight (8) marine barometers the tubes of which had been smashed by the joint effort of the Allan Line & the Canadian Express Co. (Kingston 1880)

The smashed Kew barometers were finally repaired and shipped by the autumn of 1875. The difficulties were not completely resolved, however, as more broken instruments arrived in 1878.

The recurring difficulty of getting thermometers tested and calibrated to sufficiently low temperatures to record the Canadian cold became a greater concern as Kingston dispatched more instruments to a rapidly expanding network of observers to the northwest of Canada. Kingston repeatedly specified the ordinary thermometers he ordered from the United Kingdom were to be tested and graduated to the freezing point of mercury (−38°F; −38.8°C) and the minimum thermometers were to be graduated to as low as −70°F (−57°C; Kingston 1880).

He was not hesitant in criticizing the design of the instruments, particularly when elaborate designs impeded the practical functioning of the thermometers at the low temperatures reached in Canada. Kingston often included sketches in his letters to try to convey his meaning. Although he appreciated the quality of Negretti and Zambra's thermometers, he decidedly did not appreciate their attempts to add ornamentation to the design of the instruments, calling them "unbusinesslike" in appearance and stating "the clumsy mass of wood so near to the bulk is certain to impair its visibility . . . I do not understand why he does not mount . . . them with a light porcelain scale, instead of the ugly block of wood" (Kingston 1880). But when thermometers with porcelain scales were sent out the following year, these too were met with exasperated criticisms on the apparent inability to manufacture thermometers with a clear line of sight to enable accurate reading down to the coldest temperatures.

The graduation of thermometers at low temperatures was an eternal preoccupation for Canadian observers, as we saw with Jean-François Gaultier in the eighteenth century (chapter 2) and John Samuel McCord in the 1830s (chapter 5). Reliable thermometers were even more critical for accurate observations in the areas near Hudson Bay. Kingston requested Scott to contact British instrument-makers on his behalf, and to ensure that "all thermometers sent to Canada should be tested at as many places between 32 and −37.9 [°F] as possible" (Kingston 1880). There can be little doubt that the constant urging of Kingston, like McCord and Gaultier before him, to the European instrument-makers to supply thermometers capable of recording extremely low temperatures, was one of the key factors in the ongoing development of meteorological instruments.

Kingston was also aware of, though he seemed to be only marginally involved in, the ongoing efforts to develop international standards in meteorological observations. In 1876, meteorologists were trying to address, at an international scale, the difficulties involved in reconciling observations taken at a variety of different observing schedules, as faced by McCord forty years earlier. Establishing a rigorous observing schedule was not easy when most weather observers were volunteers with most of their time taken up by their regular occupations. Kingston found 9:00 p.m. preferable to midnight, for example, as "people are generally at home by that time and not usually in bed, whereas at noon few businessmen are at home and at midnight most are in bed" (Kingston 1880). As Kingston wrote to Scott in the letter quoted below, it was not clear or indeed obvious how snow or rain that fell overnight should be registered: should overnight precipitation to be entered on the day it was measured or the day it probably fell?

> 10 February 1876
> RH Scott
> Dear Sir, As I am about to print a new edition of Instructions to Observers and wish to make it harmonize with the recommendations of the Permanent Committee of the Vienna Conference . . . I would desire specifically of be informed on the following points:
>
> On p 62, it is said "Rain measured at first morning observation is to be put down in the register of preceding day." Do you consider that this is intended for observers who are able to measure the rain at midnight, so that the entry may compare the rain included in the meteorological day. The *apparent violation of truth* [emphasis added] in this rule, in the case where rain, measured Tuesday at 7am & known to have fallen say at 5am, is nevertheless entered as the rain of Monday, is got over with us, by filling up the column showing time of beginning and ending of rain, which of course, entered for the day in

which the rain actually fell—do you consider this to be in accordance with the resolutions. (Kingston 1880)

Similar concerns were raised with respect to cloud observations. These difficulties relating to the time of observation, and the definition of the meteorological day, are with us still, as reflected by current debates as to the proper adjustment of historical records to account for differences in long-term means caused by variations in the time of day at which the temperatures were recorded (the time of observation bias).

Kingston's efforts developed the Canadian meteorological network into a national institution, and observations from Saint John's to Vancouver were collated and published in yearly reports as early as 1874. Some of the names listed in the early years of the MSC annual reports are those listed earlier from the Smithsonian network, such as John Delaney from Saint John's. As his correspondence with the HBC and the bishop of Athabaska shows, Kingston was actively engaged in extending the network as far northward and westward as he could. This was not always simple, as there were often no experienced weather recorders in those places to the north and west from which observations were most needed. Thus, "it becomes necessary either to send an observer to the station, or to procure the services of some person on the spot, whose *premises* are suitable, and to instruct him as we best can" (Meteorological Service of Canada 1875, p. 2). Additional voluntary observers were recruited in 1875, including nine clergymen, the superintendent of Guelph Agricultural College, and the headmasters of the High Schools of Belleville and Peterborough in Ontario. In 1876, nuns from convents around Quebec who had reported rainfall to the Toronto central office stopped their observations, a loss that was much lamented, as "good quality data was now hard to get" (Meteorological Service of Canada 1877, p. x). On the other hand, a number of priests set up new observing stations in Quebec, including in the Lac Saint-Jean area. In 1875, Kingston expressed his "strong sense of the great obligations . . . to the fidelity and skill which characterise . . . the numerous observers in correspondence with this office . . . The names of the ordinary *stations* [75 stations] shews [*sic*] how much the Service is indebted to unpaid voluntary labour" (Meteorological Service of Canada 1875, p. viii). Of the 129 observers listed in the 1875 report, more than 89 were unpaid.

Indeed, the thousands of letters Kingston wrote to observers across the country and his efforts to obtain and distribute instruments to the observers in his network are a testament to Kingston's determination to encourage and support all weather observers. But a new class of professional meteorologists and dedicated observatories, often if not always supported by the government, was also developing. Record keeping was centralised at the MSC headquarters in Toronto, not scattered throughout local and national archives as personal documents. The story of the development of national and international meteorology and the impact this had on climatology from the turn of the twentieth century to World War II is left for the next instalment in the history of meteorology. In the next chapter, we'll return to McGill University to see how this transition from an individual, independent, and largely voluntary observatory to a subordinate station in a national network was experienced in Montreal.

References

Anderson, K., 2005: *Predicting the Weather: Victorians and the Science of Meteorology*. University of Chicago Press, 331 pp.

DeGreef, D., 2017: Measuring worth. Accessed 06/03/2015, http://www.measuringworth.com.

Fleming, J. R., 1990: *Meteorology in America, 1800–1870*. Johns Hopkins University Press, 264 pp.

Good, G. A., 1986: Between Two Empires: The Toronto Observatory and American Science before Confederation. *Scientia Canadensis* 10 (1): 34–52.

Herschel, J. F.W., 1835: *Instructions for Making and Registering Meteorological Observations in Southern Africa*. Bradbury and Evans, 17 pp.

Kingston, G. T., 1870 unpublished manuscript: Letter book 1855–1870. Meteorological Service of Canada Archives: MSC Correspondence and Data IS408, Vol. 1855C, Meteorological Service of Canada.

——, 1872 unpublished manuscript: Letter book of G. T. Kingston 9 May 1870–9 Dec 1872. Meteorological Service of Canada Archives: MSC Correspondence and Data, IS408 Vol. 1870B, Meteorological Service of Canada.

——, 1876 unpublished manuscript: Letter book of G. T. Kingston to Meteorological Stations, 1873–1876. Meteorological Service of Canada Archives: MSC Correspondence and Data IS408, Vol. 1873A, Meteorological Service of Canada.

——, 1880 unpublished manuscript: Letter book of G. T. Kingston to beyond Canada, 1873–1880. Meteorological Service of Canada Archives: MSC Correspondence and Data, IS408 Vol. 1873B, Meteorological Service of Canada.

Meteorological Office, 1851 unpublished manuscript: Climatological returns for Fort Simpson, Canada, North America (DCnn: 9FTS). Vol. ARCHIVE W02.B2-A3, Meteorological Office Archives, 1849–1851.

——, 1862 unpublished manuscript: Climatological returns for Kingston, Canada, North America (DCnn: 9KIS). Vol. ARCHIVE Z18.K1-Z17.C3, Meteorological Office Archives, 1853–1861.

———, 1864 unpublished manuscript: Climatological returns for Halifax Dockyard, Canada, North America. Vol. ARCHIVE Z18.K1-Z17.C3, Meteorological Office Archives, 1853–1863.

———, 1866a unpublished manuscript: Climatological returns for Halifax, Citadel Hill, Canada, North America. Vol. ARCHIVE Z18.K1-Z17.C3, Meteorological Office Archives, 1854–1865.

———, 1866b unpublished manuscript: Climatological returns for New Westminster—British Columbia, Canada, North America (DCnn: 9NEW). Vol. ARCHIVE Z18.K1-Z17.C3, Meteorological Office Archives, 1859–1865.

———, 1871a unpublished manuscript: Climatological returns for Newfoundland, Canada, North America (DCnn: 9NFL). Vol. ARCHIVE Z18.K1-Z17.C3, Meteorological Office Archives,1852–1870.

———, 1871b unpublished manuscript: Climatological returns for Quebec, Canada, North America (DCnn: 9QBC). Vol. 915513-1001, Meteorological Office Archives, 1853–1870.

———, 1876 unpublished manuscript: Climatological returns for Halifax, Nova Scotia, Canada, North America (DCnn: 9HFX). Vol. ARCHIVE Z18.K1-Z17.C3, Meteorological Office Archives, 1852–1875.

———, 1880 unpublished manuscript: Climatological returns for Manitoba, St. Johns College, Canada, North America (DCnn: 9MAT). Vol. ARCHIVE Z18.K1-Z17.C3, Meteorological Office Archives, 1873–1879.

Meteorological Service of Canada, 1875: Reports on the meteorological, magnetic and other observatories of the Dominion of Canada. Government of Canada Rep., 317 pp.

———, 1876: Reports on the meteorological, magnetic and other observatories of the Dominion of Canada. Government of Canada Rep., 528 pp.

———, 1877: Reports on the meteorological, magnetic and other observatories of the Dominion of Canada. Government of Canada Rep.

Smallwood, C., 1858 unpublished manuscript: Letter to Prof. Henry, June 10th 1858. Smithsonian Institution Archives, Record Unit 60, Meteorological Project, 1849–1875 (data from 1820) Records.

Smithsonian Institution, 1857: *Smithsonian Meteorological Observations for the Year 1855*. Smithsonian Institution, 114 pp.

———, 1869: *Meteorological Stations and Observers of the Smithsonian Institution in North America and Adjacent Islands from 1849 up to the End of the Year 1868*. Smithsonian Institution, 42 pp.

Toronto Public Library, 1967: *Landmarks of Canada: A Guide to the J. Ross Robertson Canadian Historical Collection in the Toronto Public Library*. Vol. 1. Toronto Public Library, 383 pp.

CHAPTER ELEVEN

The McGill Observatory and the Professionalization of Meteorology

Keeping Time

"A proper time-keeper," observed meteorologist William Napier Shaw, is "a primary requirement" of any observatory (Napier Shaw and Austin 1932, p. 177). When Smallwood moved his observatory lock, stock, and barrel from his home in Saint Martin to McGill University, his transit telescope, used to measure time by noting the passage of certain stars, would eventually prove to be of as much practical use to the residents of Montreal, and of a source of financial revenue for the McGill Observatory, as his weather observations. The president of the Grand Trunk Railway wanted to set up an astronomical observatory in Montreal for accurate timekeeping, which was of critical importance for railway schedules, and it was to fill this need that Smallwood offered to move his instruments to McGill in 1862 (Bignell 1962, p. 17). The stone tower of the observatory building was completed at a cost of £2,000, and Smallwood reestablished his observatory at McGill in 1863. He continued in the hope that the government of Canada would eventually subsidise the meteorological observations as well and create a national network of observatories. He had a long wait. It wasn't until the 1870s that the Meteorological Service of Canada (MSC) was created and placed under the auspices of the Department of the Marine and Fisheries. As with FitzRoy's effort two decades earlier, the aim of the new service was to reduce shipping losses because of storms, either at sea or on the Great Lakes and the Saint Lawrence River. The centre of the new government service was the observatory in Toronto.

One of the most important civic functions of the observatory at McGill was thus the time signal. As the nation depended heavily on shipping for transportation, and the only reliable way to determine position, particularly longitude, was with timekeeping instruments (chronometers), accurate timekeeping was of vital importance. The astronomical function of the observatory was more concerned with the accurate determination of time by stellar observation than with the exploration of the universe, and as Montreal was one of the principal ports of North America, Smallwood's astronomical timekeeping was relied upon by many. The observatory was connected to the main telegraph office in Montreal; at the observatory's signal, a time ball dropped in the harbour, giving the captains of the ships at the harbour a chance to synchronise their chronometers. McGill's observatory was also connected by telegraph to Ottawa, where the time furnished by McGill at noon would be signalled in Ottawa by the firing of a cannon.

Despite his position as the professor of meteorology and now director of McGill's observatory, Smallwood reestablished his medical practice at 32 Beaver Hall Hill in Montreal, located on the road between was then the commercial centre near the port on the Saint Lawrence River and McGill University, higher up on the lower slopes of Mount Royal. He continued to be actively involved in the teaching of medicine until his appointment as professor of meteorology in 1871.

Smallwood had had a long struggle to obtain official support for his meteorological observations; in 1856, the Montreal Natural History Society (NHS) had petitioned for a subsidy to publish Smallwood's meteorological oeuvre, but this request was denied and it was eventually the Smithsonian Institute in the United States that commissioned the publication of his observations. Smallwood applied again in 1859 to Canada's Executive Council for a subsidy to keep his observatory running; it was denied, but it appears he was finally successful in receiving some funds in 1860, as we learn that in 1861 that his grant of £250 would be renewed. In 1862, Smallwood applied for permission from the colonial government to move his instruments to McGill. This permission was granted, and in October 1862 Smallwood started observations at the McGill Observatory. A letter to Joseph Henry in the Smithsonian Institute Archives refers to "our worthy friend Smallwood [who] has received a government allocation of $1000 for magnetic purposes and is in correspondence at present with Col. Sabine in London about the necessary apparatus" (Hall 1860). That the subsidy was finally awarded for magnetic, rather than meteorological observations, is rather ironic given that it was due to the apparent unsuitability of the geology of the Saint Lawrence valley for magnetic observations that Riddell travelled to Lake Ontario to establish the magnetic observatory in Toronto in 1839. Magnetic field observations were added to the McGill Observatory's repertoire in the 1860s. The instruments were set up in the observatory's basement (Hall 1860).

Expansion into Meteorology

Smallwood continued to send brief annual reports to the Department of Finance throughout the 1860s. Finally, in late spring of 1871, after over fifteen years of lobbying for government support, and just after being elected dean of Bishop's Medical Faculty, comprehensive government support for Smallwood's McGill Observatory was obtained from the Canadian government, jointly with the Signal Office of the U.S. War Department! Whether the Canadian government would have acted without the spur of a foreign government subsidizing essential timekeeping and meteorological observations for storm warnings remains an open question. Smallwood doesn't appear to have hesitated, however, in his choice of medicine or meteorology; he handed in his resignation to Bishop's the following day. Unfortunately, he only had a few years to enjoy the results of his long years of work; he died in November 1873 after a short illness.

Smallwood's legacy includes the establishment the McGill Observatory, a crucially important Canadian scientific institute for its role in timekeeping and the establishment of longitudes as well as for its meteorological work. McGill's weather records are among the oldest in Canada and, as such, are a valuable scientific and historical resource. The observatory also played a vital role in collecting and conserving some of Canada's oldest weather journals, those of Spark and the first McCord diary (chapter 5).

Smallwood also established the science of meteorology in Montreal and his publications on weather and climate in both Canadian and international journals did much for Montreal's reputation as a scientific centre. He was an active member of the NHS of Montreal, serving as its president in 1865–1866. He played an important role in bringing together scientists from British North America and Canada during the American Association for the Advancement of Science (AAAS) meeting of 1858. He was an international member of several meteorological organizations, including the Société Météorologique de France, l'Observatoire Physique Central of

Saint Petersburg, l'Académie Royale des Sciences, Lettres et Beaux Arts of Belgium, the National Institute of the United States, and the Academy of Natural Sciences of Philadelphia as well as the AAAS. In 1861, Smallwood summed up his achievements as

> established the Meteorological & Electrical Observatory. Discovered the effect of atmospheric electricity on the formation of snow crystals. Instituted investigations on ozone. The object of the whole of these observations has always been directed to practical utility as having a bearing in medical sciences & contributes to the health of mankind. (Smallwood 1861)

The Problem of Smallwood

The relationship between Smallwood and G. T. Kingston seems to have been rather cool, at least on Kingston's part. Although Smallwood referred to Kingston of the Toronto Observatory in print as "my friend Prof. Kingston," whose results were "perfectly reliable" (Smallwood 1869, p. 390), Smallwood's sentiments were not reciprocated by Kingston. In private Kingston wrote to John W. Dawson, the principal of McGill who was trying to sort out what was to be done with the observatory after Smallwood's death in 1873, that "[Smallwood's] violations of the first two rules of arithmetic was so frequent as to prove him to be utterly disqualified for scientific work" (Kingston, 1874). It is indeed difficult at times to reconcile Smallwood's ledgers at McGill University with the forms that were sent to Toronto, where the recorded daily minimum temperatures are sometimes higher than temperatures recorded during the regular hourly observations. Kingston put these down to "arithmetic blunders" and complained of the "habitual inaccuracies" of Smallwood's work, to the point where he was "ashamed of Canada" when forwarding them to Washington (Kingston 1874).

The forms Smallwood sent to Kingston in Toronto show evidence of carelessness, and it seems likely that he did not consider the task of forwarding the records to Kingston in Toronto a high priority, although the records at McGill were well maintained. Reconciling the differences between the McGill records and the forms Smallwood sent to Toronto is an arduous task. The observatory Smallwood founded at McGill was a rival to the Toronto Observatory, but under Smallwood's tenure it didn't receive the same level of government support. It's likely that Smallwood saw himself as at least Kingston's equal, and it's probable that he didn't take as much care as he should in compiling his abstracts to send to Kingston. It's also not clear why Kingston was sending Smallwood's reports to Washington, as Smallwood was himself doing so, originally to Joseph Henry at the Smithsonian and later to the Signal Service. Networks of communication and authority between Kingston in Toronto, Smallwood in Montreal, and international data exchanges appear to be conflicted and tangled.

Further, although Joseph Henry and Smallwood seemed to have had a cordial relationship in the 1850s, upon Smallwood's death Henry wrote to Dawson that "though not intimately acquainted with the events of his life or with his social qualities we held him in high esteem as a zealous labourer in the cause of science whose loss will be felt by all who are interested in the subject of meteorology" (Henry 1873). Henry also felt that Smallwood "must have left a large collection of valuable meteorological records" and urged Dawson to make "some effort . . . to induce your government to make an appropriation for [the publication of] this desirable work" (Henry 1873). It wasn't until after Smallwood's death that Kingston was able to train the recently graduated civil engineer Clement McLeod to his satisfaction and bring the McGill Observatory fully under his control as a subordinate station of the MSC.

The Observatory after Smallwood

After Smallwood's death, Dawson, the principal of McGill, assumed administrative responsibility for the observatory. Kingston, in another letter dated December 30, 1873[1], explained that, from his point of view

> the relations between myself & Dr. Smallwood were two-fold.
>
> 1. The observations taken by him in virtue of his connection with McGill College have been regarded by me as voluntary contributions, and as far as these were concerned Dr. Smallwood was one of the numerous observers unpaid from the meteorological grant from whom I have received reports.
>
> 2. When a grant was first made 2 1/2 years ago for the extensions of the meteorological system which I had inaugurated, the well known industry of Dr. Smallwood suggested him as a fit recipient of a portion of that grant; and he then became officially connected with the meteorological organization of which I am the superintendent and one of the observers who to the number of twelve have reported daily by telegraph to Toronto. (Kingston 1877)

---

1. There are two separate letters from Kingston to Dawson dated December 30, 1873, one in Dawson's correspondence held at McGill University, and another in Kingston's letterbook held at the Meteorological Service of Canada.

After Smallwood's death, Kingston seems to have taken a much greater interest in the McGill Observatory, corresponding more frequently with Dawson and offering to have the equipment and instruments "put in order." It was arranged for McLeod, a recent graduate in civil engineering, to go to Toronto to receive training at Kingston's office and to take as many instruments as could safely travel with him for recalibration. Kingston seems to have had a fairly comprehensive inventory of the McGill instruments and issued detailed instructions to Dawson as to which instruments McLeod should bring with him, how they should be prepared for transport, and how he was to transport them. There was some issue as to which of the instruments had been Smallwood's personal purchase, and thus part of his estate, and which belonged to the MSC as public property. Despite Kingston's misgivings as to the site of the McGill Observatory, which continued to be entertained by the central office in Toronto into the twentieth century, pressure from McGill ensured that the McGill Observatory remained in the national network.

McLeod was sent to Toronto in December 1873 for training under Professor Kingston and resumed observations at McGill in January 1874, with reports sent by telegraph to Kingston in Toronto. McLeod didn't take up the professorship in meteorology, however, instead continuing in the Faculty of Engineering. Living quarters were added to the stone observatory building to accommodate McLeod's growing family and ensure the continuity of the observations.

Kingston's comments regarding the site of the McGill Observatory seem to have centred on the observatory being on the southeastern slope of Mount Royal, where it was too sheltered from prevailing winds. Dawson appeared to have wanted to move it to the top of the mountain, a proposal that Kingston quickly tried to damp down as diplomatically as he could:

June 3rd 1874
Prof. Dawson
Dear Sir, I am in receipt of yours of 1 June. There is no need at present to enter in the larger question of an observatory on the top of the mountain. Whatever may be done a year or two hence, when meteorology is more valued than now, I do not think that the Government would sanction the purchase on its behalf of any land for an observatory, and still less the expense of building one. What we need now is not an observatory but only permission to put up a post for the instruments . . .

On behalf of the Government I would much like to bear the cost of the post, and of putting up the apparatus. (Kingston 1880)

To overcome the sheltered position of the McGill Observatory on the mountain's slope, an anemometer, an instrument to measure wind speed, was installed on the ridge of Mount Royal, with 2 miles of circuitry to connect the instrument to the observatory: "telegraph people manage to get posts put up everywhere, and the anemometer post may be regarded as the terminus of a telegraph line" (Kingston 1880).

The continuing importance of observing star transits for accurate timekeeping, along with the related determination of longitude, became one of the major scientific preoccupations, as well as a source of revenue, of McGill's observatory under McLeod's tenure. The Harbour Commission of Montreal paid the observatory $250 per year[2] for the time signal transmitted daily from the observatory to the harbour, where mariners in the port set their watches, to be used for the determination of longitude when at sea. Indeed, "the whole of the Ocean Shipping of the St-Lawrence depend[ed] for the rating of Chronometers"[3] on the McGill time signal. The longitudinal work "besides being of scientific value is also of great Commercial importance" for the accurate determination of time and space in sea and train travel. There was, moreover, "a constantly increasingly demand, on the part of the public, for information relative climatology. The preparation of tables incidental to such demands entails much labour."[4]

By 1891, McGill was the base station for Canada and was given use of the telegraph lines free of charge for an experiment to connect McGill to Waterford, Ireland, and determine the longitude more exactly by the exchange of time signals. This endeavour was promoted by the Governor General Frederick Arthur Stanley, sixteenth Earl of Derby,[5] who, through his friendship with William Christie, the Astronomer Royal, obtained the funding for McLeod to carry out the work (Bignell 1962, p. 18). This experiment resulted in McGill having the most accurate determination of longitude, and thus

---

2. For example, see Brakenridge (1893). It's not clear if the time signal was required year-round or only during the navigation season.
3. McGill University Archives (undated).
4. McGill University Archives (undated).
5. Stanley is probably best remembered today as the founder of the Stanley Cup in hockey.

time, in North America. This was of especial importance at the time, as the U.S. Coast Guard and the U.S. Geodetic Survey (geodesy is the measurement of the shape of Earth) as well as the Canadian Dominion Land Survey were all moving westward across the continent, making the accurate determination of longitude ever more vital.

Transition to Professional Meteorology

The pace of everything grew faster in the second half of the nineteenth century, with instantaneous communication by telegraph, rapid travel by steam trains and steamships, and, in the beginning of the twentieth century, wireless communication, and the advent of automobiles and airplanes. Weather observations needed to be made and transmitted more quickly too, as with the invention of the telegraph, it was at last possible for weather data to travel faster than the weather. The age of the amateur, of homemade instruments, and local societies was diminishing as industrialization with factories and mass production, and ever more rapid communication, proceeded apace. With weather, once storm warnings became foreseeable, priority was placed on stations that were attached to a telegraph. In exceptional cases, the homes of individuals such as Smallwood could be set up to a telegraph relay, but most telegraph connections were attached to official observatories or other professional places of work. With the foundation of MSC, stations were classed into chief stations, telegraph stations, first order stations, second order stations, and so on. To qualify as a chief station, observations had to be made at least every 3 hours. For observations to be truly synoptic, or to provide "snapshots" of the atmosphere at specific times, all observations had to be made at the same universal time, which translated into odd times locally: McGill, for example, made their observations at 47 minutes past the hour and also had to allow for enough time for the observations to be transmitted by telegraph to a central observation collecting point such as Toronto or Washington. These requirements were too exacting for most volunteer observers, who still had to earn their living or tend to other duties.

Despite being seemingly left behind in the new, rushed world of telegraph reports, storm warnings, and forecasting, interested volunteer observers kept on recording, several times a day, temperature, wind, weather, and sometimes pressure. One of the things that make the personal weather diaries of such important value, compared to tables of numbers in the professional weather logs, lies in the way the observers recorded the personal, agricultural, and social impact of weather and climate on their everyday lives. We turn in the next chapter to look at decades', even centuries', worth of these amateur weather diaries to see what they can tell us today about the climate of the past in the Saint Lawrence valley.

References

Bignell, N., 1962: Official time signal: 100 years. *McGill News* (summer), 16–22.

Brakenridge, J. W., 1893: *Letter to McLeod, July 7, 1893*. McGill University Archives, R.G. 4, C. 75 item F10468.

Hall, A., unpublished manuscript, 1860: Letter to Prof. Joseph Henry. Meteorological Project, 1849–1875, Record Unit 60. Smithsonian Institution Archives.

Henry, J., 1873: unpublished manuscript: Letter to Prof. J. W. Dawson, Dec. 31st[sic] 1873. R.G.2 Office of the Principal.M.G. 1022:Dawson correspondance, c10. McGill University Archives.

Kingston, G. T., 1870 unpublished manuscript: Letter book 1855–1870. Meteorological Service of Canada Archives: MSC Correspondence and Data IS408, Vol. 1855C, Meteorolgical Serivce of Canada.

———, 1872 unpublished manuscript: Letter book of G. T. Kingston 9 May 1870–9 Dec 1872. Meteorological Service of Canada Archives: MSC Correspondence and Data, IS408 Vol. 1870B, Meteorological Service of Canada.

———, 1873 unpublished manuscript: Letter to Prof. J. W. Dawson, 30 Dec. 1873. R.G.2 Office of the Principal. Dawson correspondance, Accession number 927/469 c10, ref. 31, McGill University Archives.

———, 1877 unpublished manuscript: Letter book of G. T. Kingston to Meteorological Stations 1873–1876. Meteorological Service of Canada Archives: MSC Correspondence and Data IS408, Vol. 1873A, Meteorological Service of Canada.

———, 1880 unpublished manuscript: Letter book of G. T. Kingston to beyond Canada 1873–1880. Meteorological Service of Canada Archives: MSC Correspondence and Data, IS408 Vol. 1873B, Meteorological Service of Canada.

McGill Stormy Weather Group, 1968: Three McGill Weather Observatories. McGill Stormy Weather Group, 19 pp.

McGill University Archives (undated): *Extract of minutes regarding the observatory*. R.G. 2, C. 205. item F19275.

Meteorological Service of Canada, 1875: Reports on the meteorological, magnetic and other observatories of the Dominion of Canada. Government of Canada Rep., 317 pp.

Napier Shaw, W., and E. Austin, 1932: *Manual of Meteorology: Meteorology in History*. Vol. 1. Cambridge University Press, 343 pp.

Smallwood, C., 1861 unpublished manuscript: *Letter to Henry J. Morgan*. R.G.2 Office of the Principal. Dawson correspondance, Accession number 927/469 c10, ref. 35b, McGill University Archives

Smallwood, C., 1869: On the heavy snowstorms of the winter 1868–69 in the province of Quebec, Dominion of Canada. *Proc. Meteor. Soc.*, Vol. 4. J. Glaisher, Ed., 387–391.

CHAPTER TWELVE

What Do Three Centuries of Observations Tell Us?

For over a century and a half, from the mid-eighteenth century and through two empires, people were systematically recording the weather and climate of the Saint Lawrence valley. At the McGill Observatory in Montreal and the Quebec Citadel in Quebec City, observations continued under the Dominion of Canada and are today recorded every hour at the cities' airports under the auspices of the Meteorological Service of Canada (MSC). Are we any closer to the goal of Gaultier, McCord, Kelly, and others of determining whether there has, in fact, been any observable change in climate? Are any such changes in one direction, or do they swing back and forth over time? Can we see changes in the seasons, the length of winter, the heat of summer, the precarious growing season? Do we experience the climate in the same way as the inhabitant of the colonial French Empire, the officers of the British Army, or the lawyers and doctors of the Victorian era? Although we now have access to more instrumental measurements than McCord had, thanks to the heroic task of typing the historical weather observations contained in McCord's notes and other weather registers in digital format, performed by the citizen scientists of the Canadian Historical Climate Data Rescue Project,[1]

---

1. See the acknowledgments in the front of this book for the names of all these dedicated volunteers.

it's still interesting to, in McCord's words, "compare the seasons, the plants, the opening of navigation, first arrival [of ships], last sailings, flowers, as then recorded with the seasons of the present" (McCord 1836).

This chapter is not an in-depth analysis of the temperature, pressure, and other instrumental readings, which require detailed statistical analysis. Instead, it's a look at the world our observers lived in around the Saint Lawrence valley region, how they perceived the effect climate had on their world, and whether we relate to our world in similar ways today.

Cities and the River: Then and Now
Before going deeper into this discussion, a few caveats are in order, mainly concerning the enormous spread of our two main cities, Montreal and Quebec City, and the effect of industrialization on the local environment. The area of urbanization around present-day Quebec City and Montreal dwarfs that of the nineteenth-century cities, both vertically and horizontally, while the eighteenth-century cities would be considered little more than villages by today's standards. This has an effect not only on the temperature measurements, but also on the local winds and precipitation. The influence of the city on temperature in the twenty-first century can perhaps be seen most clearly when looking at the dates of the last frost in spring as calculated from the temperature observations at the Dorval (Montreal International) Airport on the southwest tip of Montreal Island as compared to the last frost of spring reported by the agricultural communities surrounding Montreal in southwestern Quebec. For the years 2004–2015, the median last frost date on the island of Montreal was April 25 but was twenty days later, May 16, for the surrounding countryside.[2] The median dates for the last recorded snowfall, on the other hand, are very similar: April 16 for Montreal Airport and April 15 for the surrounding country. Similarly, the first frost of autumn at the Montreal Airport, inferred from the first below freezing minimum temperature, occurs nearly thirty days later than the first frost reported in the surrounding regions: September 17 for the outlying regions compared to October 21 for the city. On the other hand, the first agricultural killing frost

---

2. For example, see Financière agricole du Québec (2004a) and other reports for frost dates in agricultural regions, and Environment Canada (1961) http://climate.weather.gc.ca/climate_data/hourly_data_e.html and other hourly and daily data for temperature records in the urban areas.

of season tends to occur closer to the date of the first recorded frost in the city, with a median date of October 18.³

Many changes have also taken place over the past century and a half to the environment of the Saint Lawrence River itself. The most notable, for their impact on the freezing of the river near Montreal, are the active intervention of the Coast Guard with ice breakers to prevent the formation of ice floes upstream of Montreal (Benoit Turcotte, personal communication, Nov 15, 2017), the construction of bridges and their ice-breaking support structures, the construction of the Saint Lawrence Seaway, and the industrialization of the Great Lakes and Saint Lawrence system, which leads to waste water and heat being added to the river system. These changes all combine to prevent the river from freezing as extensively as it used to. Though the Saint Lawrence Seaway still freezes over and closes to shipping every winter, these physical and environmental changes have added to man-made influences on the freezing and thawing of the river ice, and it's no longer possible to compare the shipping season dates of today with those of the nineteenth century. Nevertheless, with these caveats borne in mind, and attempting as much as possible to compare like observations with like, we can follow in the footsteps Jean-François Gaultier, John Samuel McCord, William Kelly, and others over the centuries. With their documents and today's records, we can also try to determine what changes, if any, have come to the climate of the Saint Lawrence valley over the past three centuries from a consideration of natural indicators, such as the first autumn frost, the flowering dates of fruit plants, and wheat harvest dates.

Some analyses of the temperature readings and estimates of the daily maximum and minimum temperatures taken from the weather journals have been published in scientific journals (Slonosky 2014, 2015), along with the technical details concerning the types of instrument used, accuracy, placement, and times of observations. The calculations used to bring together such a disparate set of observations into a single series are detailed in these papers, and some of the results of that work will be discussed here.

---

3. In the analysis presented in this chapter, modern dates for frosts, snowfall, or other weather events in the agricultural regions, as well as agricultural events such as the flowering of fruit trees or harvest dates, are taken from the biweekly reports of the *Financière agricole du Québec: L'état des cultures au Québec*, unless otherwise indicated. See www.stat.gouv.qc.ca/statistiques/agriculture/etat-cultures/ and www.fadq.qc.ca/en/salle-de-presse/bulletins-dinformation/etat-des-cultures/.

Figure 12.1. Temperature reconstructions taken from journals from the Saint Lawrence valley regions, adjusted to centre on Montreal. Average minimum temperature for December, January, and February are shown here.

The main points that can be seen from the instrumental observations of Jean-François Gaultier, Alexander Spark, Louis-Edmond Glackmeyer, John Samuel McCord, Robert Cleghorn, John Bethune, Alexander and William Skakel, Charles Smallwood, and others were that while the coldest temperatures, especially the winter minimum temperatures, gradually warmed over the course of the twentieth century from a low point in the 1880s, there was little change in the warmer temperatures.

In Figs. 12.1 and 12.2, the temperature observations from the weather observers scattered around the Saint Lawrence valley discussed in this book have been compiled together to form a single series, centred on Montreal.[4] Figure 12.1 shows the average minimum winter temperature for the months of December, January, and February. A clear increase over the twentieth century can be seen, showing that the winters are not as cold as they were, especially compared to the 1810s, 1870s, 1880s, and the turn of the twentieth century, all of which stand out as being particularly cold. Except for the 1810s, winters before the 1880s were not as cold as those of the late nineteenth century, and severe winters were rare after the late 1930s.

The average summer maximum temperatures for June, July and August,

---

4. For technical details on the calibration of source material and regression analysis to compile the various locations into one composite series, see Slonosky (2014, 2015). The Toronto observations have not yet been transcribed, compiled, and analysed in a similar way.

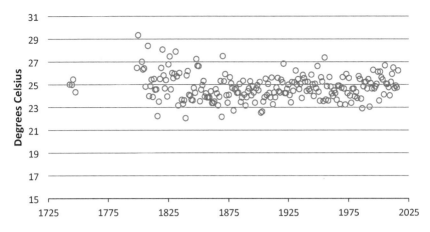

Figure 12.2. As in Fig. 12.1 but for average maximum temperatures for June, July, and August.

shown in Fig. 12.2, show very high summer average summer temperatures in the early 1800s, a sharp fall in 1809, and the severe cold of the famous "year without a summer" in 1816. The first half of the nineteenth century shows enormous variability in the summers, with cold summers such as the late 1810s followed by the warm 1820s. Similarly, the early 1830s were warm, while the late 1830s were disastrously cool. Again, 1848 and 1850 were warm, but the late 1850s and early 1860s saw another series of very cold summers; after the very warm 1870s, summers were generally cool until the 1930s. There are later periods of occasional cool summers, such as the mid-1950s and the late 1970s to early 1980s, but unlike the winter minimum temperatures, there is little overall trend in summer temperatures since in mid-nineteenth century. It's not getting warmer, but it *is* getting less cold.[5]

Interestingly, a look at the timing of events important for society and agriculture, such as the date of the last spring freeze, shows a slightly different picture, with harsher conditions towards the middle of the nineteenth century.

Transport and Communication Changes

As well as instrumental measurements of temperature, atmospheric pressure, and records of wind, rain and snow, many of the weather diarists also recorded other signs of the state of the weather, climate, and surrounding environment in their journals, such as the first snowfall of autumn, the freez-

---

5. This was the conclusion of Bill Hogg, chief of Environment and Climate Change Canada's Climate Monitoring and Data Interpretation Division, around the year 2000.

ing of the rivers in winter, or the first blossoms and leaves on trees in spring. Some recorded more events than others. This could be because some existing records, such as those of McCord, Glackmeyer, or William Sutherland are their original journals with daily entries, while others, like the Skakel extracts, are likely copies with only the information of importance to the copyist being retained. A few of the diaries are particularly interesting for their comments on the influence that weather events had on the daily life of the inhabitants. The state of the winter roads, the ice roads over the frozen rivers (especially the Saint Lawrence), and shipping conditions were all carefully noted by Gaultier, McCord, Bethune, Sutherland, and Glackmeyer.

The two communities of Quebec City and Montreal were cut off from contact by sea during the winter months when the Saint Lawrence River was frozen over. As outposts first of the French Empire and later of the British Empire, communications, orders, supplies, and personnel arriving by ship from Europe were of vital importance. Years following poor harvests or long winters led to shortages and higher prices for staples towards the end of winter. High prices for firewood and hay were recorded at the end of April 1799 by Spark in Quebec City and for the island of Montreal by Cleghorn in February 1832: "Road full of Cahots [potholes]. Markets dear, Hay and firewood in particular" (Cleghorn 1833). Warm winters could also play havoc with a society adapted to cold weather transport, as this newspaper extract from 1828 shows:

> Quebec Gazette: MONTREAL, Feby 18, 1828
> For the last three days we have had a succession of remarkably mild weather, which has almost deprived us of the little snow we previously had. Our communication with the country, by means of roads, is now extremely limited, so that our markets for provisions and fuel are very indifferently supplied, and at a great advance in price. This state of things is highly favourable to the industrious and provident farmers in our immediate vicinity, on whom we are now almost totally dependent for our supplies. (C. Mock 2015, personal communication; see also discussion in Meyer [2014])

It's possible to trace changes in the speed of communication over the centuries and to see how crucially important the links to Europe were for the diarists over time. Gaultier's annual reports were sent to his correspondents at the Royal Academy of Paris on the last sailing ship to leave for Europe every November. Spark recorded tidings of battles and the subsequent celebrations of British victories months after the events during the Napoleonic

Wars. In the 1840s, Sunderland recorded political events near and far across the British Empire. He noted the passing of the corn laws by the British Parliament, which removed the privileged access Canadian wheat farmers had to the British market, and the Anglo-Sikh war in India in December of 1845, tidings of which only reached Montreal in the spring of 1846 with the arrival of the first steamships. Even news within North America travelled slowly; it took about three weeks for rumours of the Great Fire at Saint John's, Newfoundland, of June 9, 1846, to reach Montreal and another week for them to be confirmed (Sutherland 1848).

The arrival of the telegraph changed everything. The date of the laying of the transatlantic cable was noted by several of the diarists, including Smallwood, who referred to the cables being laid on the Atlantic seabed during the 1857 meeting of the American Association for the Advancement of Science (AAAS; Smallwood 1858a, p. 190). Glackmeyer cut out a newspaper clipping headed "The Atlantic Cable: Messages between Her Majesty Queen Victoria and the President of the United States," which gave in full the messages exchanged between these two heads of state on August 16, 1858. The telegraph brought nearly real-time communication to Canada and revolutionised not only communication, but also meteorology with the rapid exchange of weather observations. For the first time, information about the weather travelled faster than the weather systems themselves. This opened up the possibility of storm warnings, vitally important at a time when most long-distance and even much of the short-distance transportation was by sea.[6]

The Saint Lawrence River: Canada's Highway

Montreal is an island city in the middle of the Saint Lawrence River. There was an intense interest, even anxiety, recorded in the weather journals concerning the annual freeze-up and breakup of the river's ice. These were two periods every year when the city was cut off from the mainland shore, with travel and communication restricted at the beginning of winter and the end of winter. With the winter freeze-up, the Saint Lawrence turned into an ice road; with the spring melting, it became an aquatic highway.

---

6. The role of the telegraph in the development of meteorology, and the controversies around storms, storm warning, and forecasting are described in James Fleming's *Meteorology in America* (1990), Katherine Anderson's *Predicting the Weather* (2005), and more recently in Peter Moore's *The Weather Experiment* (2015).

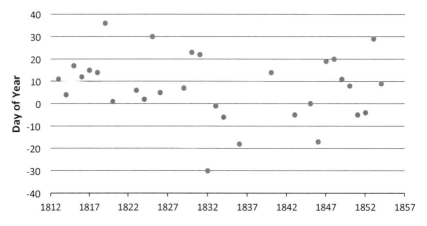

Figure 12.3. Day of the year when the first crossing of the Saint Lawrence on foot was recorded in the weather diaries. Negative values show a day in December of the prior year.

Freeze Up and First Crossing

In the beginning of the nineteenth century, the first crossing over the frozen Saint Lawrence River from the island of Montreal towards the communities of Longueil, Saint-Lambert, or La Prairie on the south shore of the Saint Lawrence tended to be around the middle of January. The ice over the river generally held firm and lasted until sometime in April.

The earliest freeze-up and crossing dates in December are recorded in the 1830s; by the middle of the decade, the first freeze-up had moved forward to the second half of December, when McCord (1835) recorded "crossing on ice to Longueuil" on December 14, 1835. The years 1836 and 1846 stand out as two years of exceptionally early freeze-up (Fig. 12.3). By the 1850s, the freeze-up was once again pushed towards January. In 1852, ferry boats went into their winter quarters on December 22, but the ice roads across the frozen river weren't passable until January 29 of 1853 (Bethune 1869).

Ice Melt and Break-up of Ice Roads

In 1815, McCord recorded that the ice was "bad," that is, starting to melt, on April 4; three days later, two men drowned attempting the dangerous crossing. It wasn't until two weeks later, on April 17, 1815, that the first boats were in operation (McCord and McCord 1826). In 1838, the ice road to La Prairie was cut off on March 24, but the first steamboats weren't in operation until April 18 (Bethune 1869). For three weeks or more, at the end of winter when supplies were running low, Montreal was cut off from travel and communication with the rest of the continent. If something similar were to happen today, it would be a humanitarian disaster for the nearly two million inhabitants of the island. During the 1998 ice storm, when more than 100 millimeters of freezing rain fell between January 4 and January 10, electricity power transmission lines came down across southern Quebec while transmission towers crumpled under the weight of the ice. Only one power line remained live across the Saint Lawrence River to the city of Montreal, while the thick ice on the roads and bridges made travel dangerous. Over one million people were left without power for a period ranging from ten days to over three weeks, while the temperature fell to −18°C in the following days. Even today, the island city of Montreal remains occasionally vulnerable to being isolated by the weather.

Winter isolation had been lessened somewhat by the building of the Champlain and Saint Lawrence Railroad in 1836; the railway from La Prairie to Saint John's on the Richelieu linked the Montreal region to the Lake Champlain steamers via a locomotive designed by Robert Stevenson. This connection brought Montreal within easier reach of New York, whose harbour was open year-round, though even here the operation of the Lake Champlain ferry ceased over the winter months, and travellers made their way overland through the snow with sleighs. With the building of the Victoria Bridge in 1859, the island city of Montreal finally had a permanent link to the rest of the continent.

The Victoria Bridge was initially a railway bridge only, still leaving all foot and horse traffic to cross the river by ferry or on the ice road, depending on the season. Even in the early twentieth century, walking across the frozen river was common: my grandmother, Eugenia Milanowska, recollected walking across the river from the harbour of Montreal to Saint Helen's Island as a child, before the construction in 1930 of Montreal's second fixed link to the south shore of the river, the Jacques-Cartier Bridge. Although the river rarely, if ever, freezes solid across Montreal's harbour to its south shore anymore, the Saint Lawrence Seaway still does and provides a winter recreational sporting venue for ice fishermen, skiers, and skaters.

Winter Roads: Snow and Mud

In addition to the ice roads, travel switched back and forth from *carioles* (sleighs) to *caleches* (carriages), depending on the extent and timing of snowfalls and snow thaws that affected the conditions of the roads, usually but not always at the beginning and end of the winter season. Midwinter thaws almost always caused travel difficulties and frustration. In Montreal in 1847, Sunderland recorded being obliged to switch back and forth between his sleigh and carriage throughout the entire winter, with "almost all the snow gone, wheels in general use today" on December 30, "wheel carriages in use" once more on January 17, 1848, "wheel carriages again in use" on January 25, but enough snow on February 2 for "sleighing fairs." "Sleighing [was] hard in town" because of a lack of snow on February 22 (Sutherland 1848). As for crossing the river, it wasn't until January 19, 1848, that there was "crossing before the town," which lasted for about two months before "rain all night" on March 27 made the "ice to La Prairie unsafe." Interestingly, wary residents had stopped crossing over the ice four days earlier: "no sleighs crossing on ice yet firm" (Sutherland 1848).

Thawing snow left the roads mired in mud and, as Sutherland's diary shows, transport had to be switched over repeatedly from wheeled carriages to sleighs on runners. From Glackmeyer's diary, we see that in Quebec City in 1852, the sleighs were out on November 14 following a snowfall, put away in favour of wheeled carriages from November 18 to 22, taken out again after a second snowfall for November 23 and 24, put way in favour of wheels again when the snow melted, and so on until December 12, switching over six times in less than a month (Glackmeyer 1859). The following year was even worse; Glackmeyer was driven to comment on December 6, 1853, that he had "left off using wheeled vehicles, I hope for the remainder of the winter." His hope wasn't realised, for on December 18 he commented again "sufficient snow for runners, till now we have been obliged to use runners & wheels alternately." In 1859, Glackmeyer switched to his sleigh on November 5 following an early snowstorm but "had to [change] to summer vehicles" when the snow thawed on November 9. A second snowfall made "roads almost impassable, [with] carrioles [sleighs] again" on November 11 (Glackmeyer 1859).

The Shipping Season

Closely related to the melting of the Saint Lawrence River's ice is the opening of the shipping season in Montreal. Most of the diarists made notes on the date of the arrival of the first ferry, the first ship from Lake Champlain, and the first ship from Europe, whether in Montreal or Quebec City. Some dates of the first vessel into port at Quebec City from the Atlantic Ocean from 1760 onward are given in Fig. 12.4. The general downward trend over the nineteenth century may be partly due to technological changes in ships, especially the predominance of faster steamships over sail.

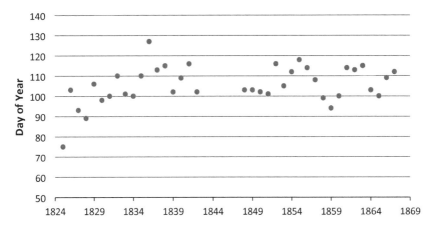

Figure 12.4. Day of the year of the arrival of the first oceangoing ship at the port of Quebec in spring, indicating the Saint Lawrence River was navigable and open to shipping for the season (MMA Collection P001, Item 410).

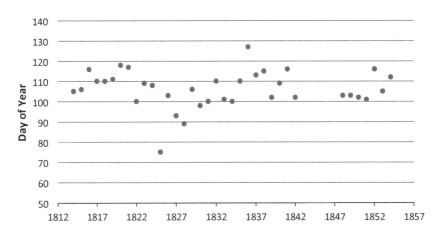

Figure 12.5. Day of the year on which the first shipping activity, usually steamboats or ferries from La Prairie or Boucherville on the southern shore of the Saint Lawrence across from Montreal, was recorded. The dates are compiled from a variety of sources (McCord 1836; Hough 1872).

Starting with McCord in 1813, and ending in 1869 with some supplemental information from Hough, the date of the opening of the shipping season is available for much, but not all, of the mid-nineteenth century. The earliest opening date of shipping on the Saint Lawrence in Montreal for which we have information from the weather diaries was in 1825, when the first river steamers were recorded on March 17, although regular ferry service between

Montreal and La Prairie wasn't established until April 6, 1825. The latest arrival of the steamships and the opening of the river for local traffic was on May 7 in 1835 (Fig. 12.5). A plot of the opening dates of the shipping season at Montreal shows some intriguing evidence of a cyclical tendency, with an alternation between early ice melt and late ice breakup on an approximately ten-year frequency, but the observations from the diaries are too sporadic to pick out any pattern with confidence.

Agriculture in Colonial Canada

Quebec City and Montreal are not that far apart, speaking from a synoptic weather point of view. Both are usually affected by the same large-scale weather systems and storm tracks sweeping from west to east across the continent and funnelling up the Saint Lawrence valley to the Atlantic Ocean, or the post tropical storms churning their way up the Atlantic seaboard from the Gulf of Mexico. Still, Quebec City is located some 250 kilometers downstream from Montreal. It's somewhat farther north, somewhat closer to the maritime climate of the North Atlantic, and somewhat higher in elevation as the Laurentian Mountains start to close in on the Saint Lawrence. These differences are enough to make the Quebec City region a more marginal area for agriculture than the Montreal plain, especially for the wheat that was so highly prized to make the staple bread on which most of the population depended, particularly in the earlier years of the colony. Peas and oats were other staples. Peas in particular provided daily nourishment and were especially important in times of scarce wheat harvests, but they seem to have been considered of only secondary importance by the diarists, Gaultier in particular (Wien 1990, p. 540).

Allan Greer gives a succinct overview of the agricultural calendar of Lower Canada (Greer 1985; Wien 1990). Once the snow had melted and the soil was dry enough, ploughing began, followed by sowing. As Gaultier made clear, whenever the autumn was mild, with late frosts and no snow cover, as much ploughing as possible would be done in the autumn, so that sowing could take place as early as possible the following. Preparing the kitchen garden also took place in early spring. Though, according to Greer, the kitchen garden was the domain of women, it's interesting to see that Gaultier, McCord, Bethune, and Glackmeyer all noted the dates of their garden vegetable sowing. In Bethune's case, the dates of hothouse sowing are also sometimes noted. After the spring sowing, the men would erect temporary summer fences around the grain fields. With long winters during which animals have to be fed indoors from stocks of hay gathered over the short summer, the hay harvest from midsummer to the end of the season was, and still is today, of prime importance in Quebec. It was customary to take the fences down after harvest and let the animals roam freely to forage over the fields in the late autumn before the first snowfall; the taking down of the fences was another important date in the agricultural calendar. The years in which two hay harvests were made were noted by Gaultier; today, three harvests are often made. In some recent years wet conditions, such as those seen in 2004 and from 2006 to 2009, or flooding, such as that of the Richelieu River in 2011, led to some harvests being left incomplete. In some southern regions of Quebec, four hay harvests were brought in over the summers of 2010 and 2011 (Financière agricole du Québec 2004b, 2006a, 2007, 2009a, 2010, 2011a,b, 2012).

From Gaultier's evidence, there was a brief period when wheat was sown in both spring and autumn, producing two crops a year: "I have to say that the autumn wheat was magnificent and they surpassed those which were sown in spring" (Gaultier 1755; see also Wien 1990).[7] Though this practice was strongly recommended by both Gaultier and Indendant Gilles Hocquart, who Gaultier praises highly, it doesn't seem to have lasted (Greer 1985, p. 29). The all-important wheat harvest usually started at the end of August or in early September, though in exceptionally warm years, such as 1746, harvesting started in early August. If the snow and ground frost held off, work began on ploughing the fields, in preparation for the following spring, and on repairing the ditches, as described by Gaultier in 1744 (Duhamel Dumonceau 1746, p. 89).

Drainage ditches were of critical importance to prevent flooding of the grain fields, increasingly so as the forests were pushed back, which led to increasing runoff to the Saint Lawrence River (Coates 2000, p. 114). The lands allocated to farmers (habitants) under the seigneurial system of Lower Canada, from the first founding of New France until the 1850s, gave each habitant a long, narrow strip of land with a small frontage on the shore of the Saint Lawrence and other rivers such as the Richelieu, where the habitant's house was built. Each individual property extended deep inland into forested areas, which needed to be cleared to create agricultural fields. These fields were thus directly in the line of runoff water going from the interior forests

---

7. The original text reads: "il faut remarquer que les bleds d'automne étoient magnifiques et qu'ils surpassent de beaucoup bonté es en quantité ceux qui avoient été semé le printemps."

down to the Saint Lawrence. The fields had to be kept from flooding by the unceasing labour of maintaining the ditches.

Herds and flocks were kept small because of the restrictions imposed by the long winter and by the necessity of having sufficient stock to feed the animals over the long winter season. Some slaughtering was done in late fall and advantage would be taken of below-freezing temperatures to freeze meat. Midwinter thaws could be disastrous if frozen meat thawed and then froze again, becoming inedible as described by Gaultier in 1754 (Gaultier 1755). After the outdoor fieldwork came the threshing, though this could be carried on inside barns over the winter and was often done in January or February (Greer 1985, p. 32; Meyer 2014, p. 34).

Spring Weather

Some of the diarists were more assiduous than others in recording the influence of the weather and climate on their surroundings. Gaultier, even more of a botanist than a climatologist, kept the most complete records of the natural environment's response to the weather, carefully noting the arrival of migratory birds, the blooming of fruit trees, and the sowing and harvesting dates of a variety of crops. Cleghorn presumably also kept careful records for his plant nursery, but only general comments at the end of each month are to be found in his weather records. Thomas McCord, Glackmeyer, and Bethune also kept sporadic botanical notes.

Some indicators of the warmth of a particular season are given by specific weather events more commonly recorded in the weather diaries, such as the last snowfall of the season or the first autumn frost, as well as by the recorded temperatures. The first thunderstorm also gives an indication of when the ground was sufficiently warmed up after the spring snowmelt to allow warm rising air to develop into a convective thunderstorm. Here, the dates of the latest spring frost and the earliest autumn frost have been compiled from two categories: those provided by a written description, such as "hoar frost" or "*gelée*" and those calculated using estimated minimum temperatures from the thermometer readings. The latest value for spring (earliest for autumn) is the one indicated in the following figures. Not all descriptions of frost necessarily coincide with the first record of below-freezing temperatures; strong radiative cooling at night can produce ground frost while the air temperature remains above freezing.

Unlike the personal weather records kept in the eighteenth and nineteenth centuries, the hourly observations from the city airports of today don't record "frost" as a description of the weather. Consequently, the dates of the last frost for Montreal and Quebec City in the modern period are determined solely from the temperature records. This may not be a completely reliable comparison as, in general, the countryside cools more rapidly than the heat-retaining cities. The frost dates calculated from the city temperature records, particularly in the twentieth century, are therefore biased to be somewhat earlier in spring or later in autumn than the historical records or the dates culled from the agricultural reports. This makes the season appear warmer than the agricultural records would indicate. With these caveats in mind, these data, while not a perfect record, provide an approximate tool with which to compare climatic indicators across the centuries. Documents from the Quebec City diarists don't overlap in time and so are each presented as individual datasets. Since there is considerable overlap in the records of the various Montreal diarists, they are presented together under the label of Montreal historical diarists.[8]

The nineteenth-century weather journals report later dates for the last spring frosts than either Gaultier's records or the twentieth-century hourly synoptic weather records from 1953 to 2015. Some years in the nineteenth century had frosts persisting later into spring than any since the mid-twentieth century, with 1816, 1842, 1859, and 1864 being notable examples (Fig. 12.6). The variation between the warmth of Quebec City in Gaultier's time in the mid-eighteenth century and the cold period of Glackmeyer's record from the mid-1840s to 1859 is striking. It should be noted that Glackmeyer's temperature records, as seen from his own notes and from statistical analysis, are skewed towards warm values because of his thermometer at times being improperly sited and not completely shaded. His written descriptions of frost are therefore more reliable than the dates calculated from the thermometer values. Gaultier's Quebec City records show earlier dates for the last spring frost than the modern record, implying warmer springs. Spark's are at the same level as the modern distribution, while Glackmeyer's values are well below, suggesting much colder springs in the middle of the nineteenth century than either before or after (Fig. 12.6).

In Fig. 12.7, there is a clear pattern in Spark's values, with warmer springs until 1808, after which the last spring frosts occur later in the spring season,

---

8. All figure axes for the spring indicators have been reversed, with the earliest dates towards the top of the graph so that, in all cases, values near the top of the graph (i.e., early dates in spring or late dates in fall) indicate warmer conditions, while values towards the bottom of the graph indicate cool conditions.

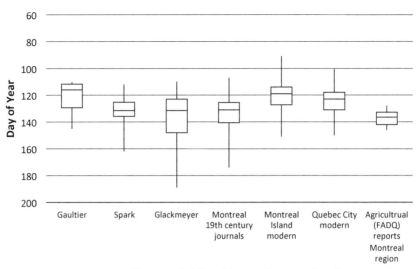

Figure 12.6. Comparisons of last recorded day of the year for spring frost for Montreal and Quebec City; boxes show the values that fall between the first and third quantiles, and lines show the earliest and latest extreme dates. Values for the nineteenth-century Montreal journals are taken from J. S. McCord (McCord and McCord 1826; McCord 1836, 1843), Alexander and William Skakel (Skakel 1852, 1869), Cleghorn (1833), Bethune (1869), Sutherland (1848), and Smallwood (1855, 1862). Modern values for Montreal and Quebec City are taken from the hourly observations at the Environment and Climate Change Canada website (climate.weather.gc.ca/) and reports from the Financiere agricole du Quebec (fadq.qc.ca/accueil/) for the Montreal region. Not enough data were found for Quebec City. The y axis has been reversed so earlier dates of the last spring frost, associated with warmer conditions, are towards the top of the graph.

culminating in the famous year without a summer of 1816 with a frost in June, although 1811, 1815, and 1817 all also recorded frosts after May 20. Although 1816, with its late frost and snow on June 8 went down in memory as the year without a summer, 1844 and 1859 also appear to be contenders for that title in Quebec City (see also chapter 13; Glackmeyer 1859). Other late spring frost dates from the Montreal records are June 16, 1862, and June 23, 1868. The latest spring frosts recorded since 1953 are May 31, 1961, and May 30, 2004. Other late frost dates in recent times are May 26, 2008 (Financière agricole du Québec 2008), and May 25, 2009 (Environment Canada 1961; Financière agricole du Québec 2008; 2009b). The earliest dates of the last spring frost for Quebec City all hover around the third week of April, indicating that the earliest possible date to start agricultural activity would be towards the end of April. Four of the six years for which we have Gaultier's recorded temperatures show the last frost day before the end of April, while

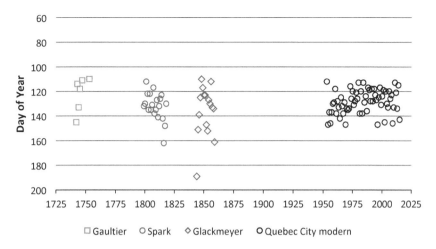

Figure 12.7. Day of the year of the last spring frost recorded for each available year as described in Fig. 12.6 for the Quebec City region. As before, the *y* axis has been reversed so that the earlier dates (warmer springs) are towards the top of the graph.

Figure 12.8. Day of the year of the last spring frost recorded for each available year as described in Fig. 12.6 for Montreal. As before, the *y* axis has been reversed. The gap between 1873 and 1953 represents weather information that has not yet been typed into digital format. A McGill University project, Data Rescue: Archives and Weather (DRAW) citsci.geog.mcgill.ca, is currently organizing a citizen science transcription effort to digitise the McGill Observatory weather registers from 1874 to 1953.

Spark recorded a cluster of early springs between 1801 and 1807; 1807 is one of the warmest summers, if not the warmest summer, on record, along with 1808 (see chapter 13).

Early springs are more common in Montreal than in Quebec City, as expected with Montreal's warmer climate, though the decreasing trend towards cooler springs and later dates of the last spring frost can be seen throughout the nineteenth century (Fig. 12.8). The modern record is interesting for the trend towards later spring frost dates towards the middle of the century and again since 2005, although the more recent last frost dates in the twenty-first century are surely due to the more sensitive reporting of frosts in the countryside in the agricultural reports, compared to the estimates from the thermometer readings at Montreal's city airport. Even today, it's considered foolhardy to start recreational gardening before the third weekend of May, which is widely considered to be the start of the summer season.

McCord described the typical sudden transition from winter to summer in the Saint Lawrence valley in 1838:

> The spring or more properly that very short period the "avant courier" of summer for spring properly called we have none, sets in at Montreal about fifteen days or three weeks earlier than at Quebec. The ice in the rivers breaks up and the snow disappears usually between the first and fifteenth of April and in like manner the falls of snow in autumn and the closing of the rivers by ice occur at about a like period after the winter has set in at Quebec. —with the difference, and a rather higher mean temperature of our summer months, Dr. Kelly's observations on Quebec will be perfectly applicable to Montreal... The transitions here are as great as at Quebec and equally sudden. (McCord 1838; see also McCord 1842b, p. 5)

Gaultier also described the sudden transition from winter to summer, with the trees and other vegetation bursting into leaf and unfurling practically as he watched, *à vue d'Oeil* (Gaultier 1748). McCord noted in 1841 that although the last snowfall and opening of the river usually took place towards the middle of April, wintry conditions tended to linger well into May: May 20, 1841, was the "first spring day." The next day summer appeared, and May 21, 1841, marked the "sudden transition from winter to summer" (McCord 1842).

A comparison of the date of the last spring snow is more consistent across the centuries, with more overlap between the records (Fig. 12.9). The last spring snow tends to occur a little earlier than the last spring frost and usually marks the beginning of the agricultural season. Here, the dates of the last recorded snow from the historical Quebec City journals are slightly earlier in the season than those from the Quebec City Airport (Fig. 12.10).

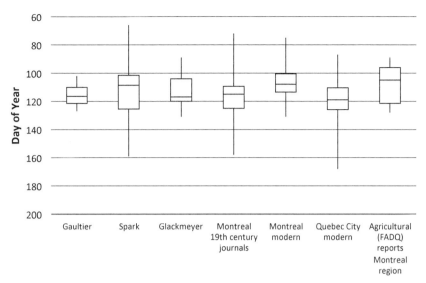

Figure 12.9. Comparisons of last recorded spring snowfall for the Quebec City and Montreal regions; the boxes show the values which fall between the first and third quantiles, while the lines show the earliest and latest extreme days. The y axis has been reversed.

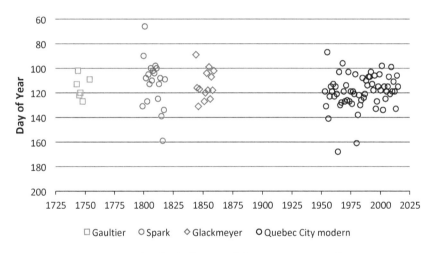

Figure 12.10. As in Fig. 12.7, but for the last recorded spring snow in Quebec City.

The opposite situation occurs in Montreal, though the dates all cluster around the second half of April (Fig. 12.11). Early springs in 1800, 1801, and 1844 all saw the last snowfall before the end of March, which happened only six times in Quebec City during the 1953–2015 period. The earliest dates for the last snowfall are March 7, 1801; March 13, 1819; and March 30, 1844

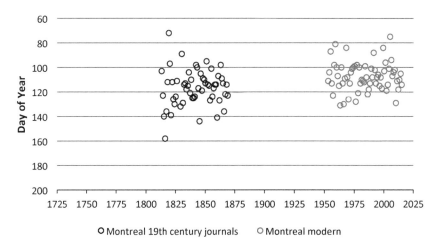

Figure 12.11. As in Fig. 12.8, but for the last spring snowfall in Montreal.

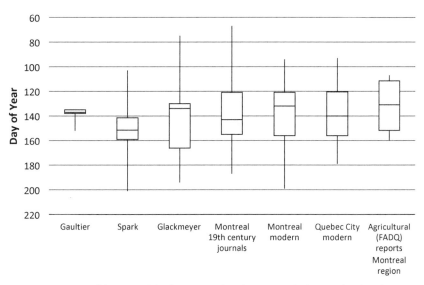

Figure 12.12. Day of the year of the first spring thunderstorm, which can often be taken to show the onset of warm and humid summer conditions when the ground has warmed enough to initiate convective activity. The *y* axis has been reversed.

(though as noted above, that didn't stop July 1844 from having episodes of frost, the latest on record). The latest spring snowfall date of the nineteenth century is the famous episode of snow on June 7, 1816, though both 1964 and 1980 recorded later snowfall dates in Quebec City than 1816, on June 16 and June 9, respectively.

The date of the first recorded thunderstorm seems more consistent over time, with most being recorded during the month of May (Fig. 12.12; to look only at thunderstorms caused by local convective heating, the handful of storms with thunder recorded before March are excluded here), although there is considerable spread over the spring and early summer. The date of the first recorded thunderstorms varies considerably in all records. The dates largely overlap among the various periods of the different records; the variability in the date of the first recorded thunderstorm is too high to be able to discern any trends or changes over time.

Spring Agricultural Indicators

Other information from the weather diaries, like dates of ripening fruit and harvest, can be difficult to use to interpret long-term climate change, given the sporadic nature of the recorded comments both in the historical records and in modern times. Dates for various agricultural markers for the 2003–2014 period have been collected and inferred from the biweekly reports published during the growing season on agricultural conditions in the province of Quebec. These reports have frequent references to the weather and its impact on various agricultural sectors. The dates given can only be approximated to within, at best, two or three days. Spring or early summer indicators from the botanical world, for which it's possible to compare dates across time, include the strawberry harvest, apple blossom dates, and sowing dates. Although other sources exist for some of these indicators, or at least the general quality of harvests, the interest here is in trying to see to what extent the meteorological records can be corroborated by other evidence in the weather journals themselves, not to provide a comprehensive analysis of agricultural returns for the past three centuries.

Gaultier's comments regarding eating strawberries in early or mid-June is startlingly early for the region; today strawberry season in Montreal tends to be in full swing in mid- to late June and early July in Quebec City. Indeed, compared to dates since 2003, the strawberries in Gaultier's time were ripe over ten days earlier than today and nearly a month earlier than the two dates in the 1840s noted by Glackmeyer a century later (see Table 12.1). Apple blossom dates should be interpreted with even more caution as there are only a handful of dates here for both the modern and historical periods and none for the modern-day Quebec City region. The nineteenth-century apple blossoms are also ten days to two weeks behind the modern records (Table 12.2). Both Quebec City in the 1740s and Montreal in the early 1800s had

Table 12.1. Strawberry maturity dates.

Strawberries ripe	Gaultier (six years between 1742 and 1754)	Glackmeyer (1844 and 1846)	Montreal region (agricultural reports 2003–2014)	Quebec City region (agricultural reports 2003–2014)
Median date	Jun 17	Jul 11	Jun 16	Jun 29

Table 12.2. Apple blossom dates.

Apple blossom	Gaultier (three dates 1743–1754)	Montreal journals (five dates between 1815 and 1850)	Agricultural reports for Montreal (only six dates between 2003 and 2014)	Sainte-Anne-de-Bellevue* (eleven dates 2004–2015; Belho, 2015)
	May 31	May 30	May 20	May 16

*M. Bleho, Macdonald Campus Horticulture Center, 2015, personal communication.

Table 12.3. Spring sowing dates for garden produce, peas, and cereals.

Spring sowing	Gaultier	Glackmeyer	McCord (1815–1821)	Bethune	Montreal (2003–2014)	Quebec City
Garden	Apr 23	May 24		Apr 22		
Peas		May 10	May 4	Apr 18		
Cereal	May 25	May 6	May 21/June 9 (1815)	Apr 29	May 7	May 10

flowering dates nearly ten days later than in the twenty-first century to date. On the other hand, in 2006 the agricultural reports note that the apple blossom date of the apple-growing region south of Montreal (the Montérégie), May 23, was ten days earlier than normal, and there are later reports of the early blossoming apple flowers being later damaged by late frosts (Financière agricole du Québec 2006b, 2008, 2009b). This suggests that June 2 was considered the normal flowering date before 2006 and is more in line with the few dates available in the historical journals. The sowing of cereals such as wheat, corn, and barley usually took place as soon as the snow had melted and the ground could be tilled and worked. Again, the weather journals don't have complete records, but, from the dates available, the sowing of grains began earlier in the 1840s (most of the notes on agriculture from Bethune and Glackmeyer were made in the 1840s) and somewhat later in the 1740s and in the 1810s (Table 12.3).

Autumn Weather

At the other end of the summer season we have equivalent events marking the end of summer and the onset of winter: the first autumn frost, the first snowfall, and the first agriculturally defined "killing frost," when it gets sufficiently cold to freeze most plants beyond recovery. Up until the early nineteenth century, the onset of the first autumn frosts occurred at about the same time as today. In the mid-nineteenth century we see an abrupt change towards the earlier onset of winter, with early first autumn frost dates noted by Glackmeyer and the diarists in Montreal (Fig. 12.13).

Seasons with late frost dates stand out easily: 1754, 1800 (Quebec City; Fig. 12.14), 1835, 1870, and 1873 (Montreal; Fig. 12.15) and in more recent times 1968 (Quebec City) and 1995, 1998, and 2005 (Montreal). In some cases, but not all, these late frost dates coincide with exceptionally warm summers. The warmest summer on record from the temperature values is 1808, with 1870, 1998, and 2005 also in the ten warmest summer half years (March 15 to October 15) in the Saint Lawrence valley. Late frosts in autumn were particularly beneficial in that farm animals could forage outside for longer periods of time and weren't dependent on being fed hay indoors. Autumns when the animals were still foraging outside into late November were noted by the Quebec City diarists in the years 1745, 1747, and 1852. At the other ex-

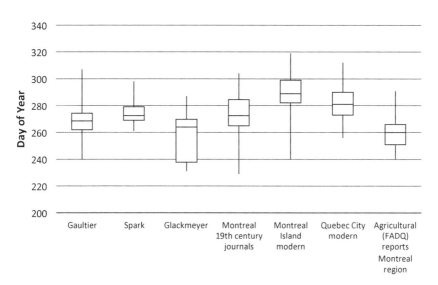

Figure 12.13. Comparisons of the first recorded day of the year of autumn frost for the Quebec City and Montreal regions; the boxes show the values that fall between the first and third quantiles, while the lines show the earliest and latest extreme dates.

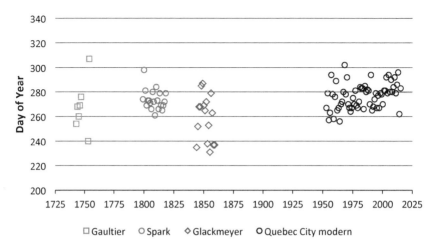

Figure 12.14. Day of the year of the first autumn frost recorded for each available year from the weather diaries and modern observations for the Quebec City region.

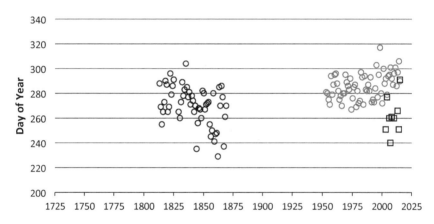

Figure 12.15. Day of the year of the first autumn recorded for each available year from the weather diaries and modern observations for the Montreal region.

treme, in 1859 it was as early as the end of October when Glackmeyer noted that the "cattle have to be housed and fed" (Glackmeyer 1859).

Autumn Agricultural Indicators

Both the records from Montreal and Quebec City point to a mid-nineteenth-century cold period, with a series of early autumn frosts from the mid-1840s to the 1860s but particularly concentrated in the second half of the 1850s. Killing frost dates are harder to come by; Gaultier recorded a median date of October 9 over five years between 1742 and 1754, with November 22, 1754, being by far the latest date ever recorded for a killing frost in either the Quebec City or the Montreal regions. In comparison, the median date for the first killing frost in the Montreal region since 2004 is October 18, with the earliest date recorded since 2004 being September 18 in 2014 and the latest being October 28 in 2011. In the central Quebec region, the median killing frost date is September 18, the earliest occurred on September 5, 2013, and the latest killing frost was on October 5, 2004.

Either the first persisting snow or the first hard frost could put an end to the agricultural season and outdoor work in the fields as well as the period during which the farm animals could forage outside (Fig. 12.16). Gaultier wrote that it was generally considered better for the first hard frost to occur before the first snowfall, as that way agricultural pests such as insects, their eggs, and the roots of the weeds would be killed, whereas if an insulating blanket of snow covered the ground before it froze, their chances of overwinter survival were better (Gaultier 1755). If there had been time enough for autumn ploughing before the first hard frost or snow, the furrows kept their shape better if the ground was frozen. On the other hand, Gaultier wrote, if the ground froze hard before the first snowfall, the following spring

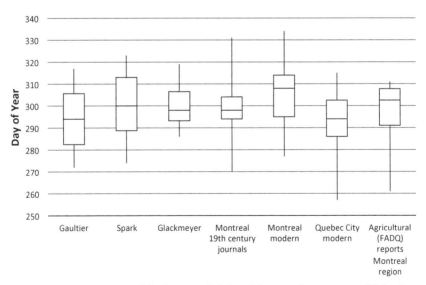

Figure 12.16. Comparisons of the first recorded day of the year of autumn snowfall for the Quebec City and Montreal regions; the boxes show the values that fall between the first and third quantiles, while the lines show the earliest and latest extreme dates.

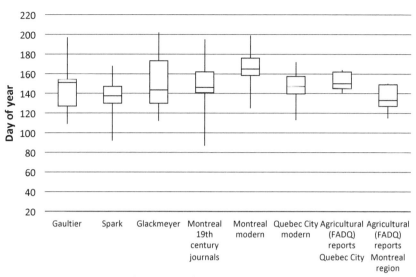

Figure 12.17. Comparisons of the length of the frost-free season for the Quebec City and Montreal regions; the boxes show the values that fall between the first and third quantiles, while the lines show the earliest and latest extreme days of the year.

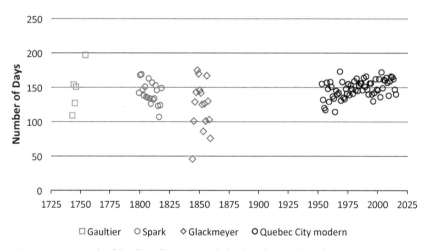

Figure 12.18. Length of the frost-free season, defined as the number of days between the last recorded frost in spring and the first recorded frost in autumn, for each available year from the weather diaries and modern observations for the Quebec City region.

it took longer for the frozen ground to thaw sufficiently to be worked. The frost-free season, or the number of consecutive frost-free days, is calculated here as the number of days between the last frost of spring and the first frost of autumn. This gives an idea of the length of the growing season from the various sources, though as not all frosts are necessarily killing frosts, the

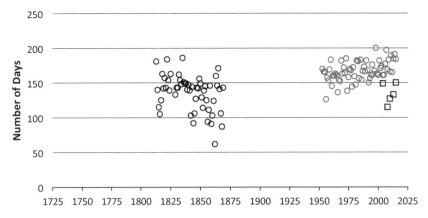

Figure 12.19. Length of the frost-free season, defined as the number of days between the last recorded frost in spring and the first recorded frost in autumn, for each available year from the weather diaries and modern observations for Montreal.

actual growing season for some years could be longer than shown here (Figs. 12.17, 12.18 and 12.19).

Wheat and Hay Crops

Gaultier noted that hot, dry summers were to be preferred over rainy ones in Quebec, at least as far as wheat is concerned (Gaultier 1745):[9]

> The great heats are always more to be wished for [in manuscript: "in Canada" is added in] than rain, which proves the proverb that never has a dry year been sterile. In fact, since I've been in Canada, I've constantly observed that the dry years are the abundant ones and the most fertile in useful and necessary crops. (Gaultier 1748)[10]

---

9. The original text reads as follows: "Les années où il y a de grandes chaleurs sont celles qui sont ordinairement les plus abondantes, es ou la récolte est très bonne" (Gaultier 1745). LeRoy Ladurie (2004, p. 410) made a similar comment for France and eastern Europe.

10. "Les grandes chaleurs sont toujours plus à souhaiter [in manuscript: en Canada] que la pluye, ce qui prouve le proverbe que jamais année Seche n'a été stérile. En effet depuis que je suis au Canada j'ay constament observé que les années seches sont les plus abondantes es les plus fertile en production utiles es necesaires" (Gaultier 1748).

Gaultier's preoccupation with the wheat harvest shows the reliance here on wheat as the principal crop rather than crops more adapted to cool weather like oats or peas. Peas, fortunately, were a staple fallback that were seldom mentioned as being problematic.

It's clear that the agricultural window was shorter in the mid-nineteenth century, with short growing seasons from the 1840s to the 1860s. Gaultier's period—the 1740s and 1754—is notably variable, with good years alternating with mediocre ones, though no shorter than the Quebec City region today (Table 12.4). None of the years for which we have records in the 1740s have frosts as late as in the 1840s and 1850s, although the record is not complete for the 1742–1754 period. Of the ten summers for which Gaultier recorded a description, eight are described as "very warm," with droughts recorded for four of the summers (1743, 1746, 1748, and 1754); 1744 was warm in June and July but rainy in August and September; and only 1753 is described as *mauvaise temps*, cold and rainy. While the median date for the first hay harvest tends towards the later end of the modern range, the median date of the wheat harvest from Gaultier's diaries is several days earlier than the earliest date reported in the twenty-first-century reports for the Quebec City region.

Towards the end of the French regime in Canada, wheat imports into the colony increased, partly as the escalating war brought more and more soldiers who needed to be fed and partly because of the destruction of war itself; British major-general James Wolfe in particular led a scorched earth campaign in which his army set fire to the crops. Poor harvests related to bad weather, as recorded by Gaultier, also increased the demand for imported grain. In 1753, Gaultier described eloquently the dire situation in the Quebec City region caused by the poor harvest that had been ruined by rain and frost. According to his account, however, Montreal had a bumper harvest in 1753. Why grain couldn't be transported between regions remains unclear. The next year, 1754, the situation was reversed; it was so hot and dry that the wheat crop in Montreal failed because of the drought, while Quebec City had an abundant harvest (see chapter 13).

The harvest failed several times in the later part of the eigthteenth century. However, 1769, 1783, 1789, and 1796 were years of grain shortage and high wheat prices (Greer 1985, p. 65). From the weather diarists, we can see also that in 1799 the "spring [was] extremely backward, & weather wet, cold & unpleasant. Hay sold in market at 20 dollars per hundred," according to Spark (1819).[11] In July 1816, McCord noted that there were "no mill[s] turning

---

11. Hay sells for about $3–$5 per 100 pounds today.

Table 12.4. Dates of hay and wheat harvests.

	Gaultier (dates)	Agricultural reports Montreal region (dates)	Agricultural reports Quebec City region (dates)
Hay (first harvest)	Jul 15–Jul 24	Jun 4–Jul 15	Jun 14–Aug 3
Wheat	Aug 4–Sep 7	Aug 5–Sep 15	Aug 23–Oct 12

for want of Wheat" (McCord and McCord 1826), while on August 22, 1847, Glackmeyer recorded the appearance in Quebec of the potato blight that had so devastated the Irish. Despite these years of shortage, Lower Canada (present-day Quebec) had been an exporter of wheat in the late eighteenth century up until sometime before the 1830s. In the 1830s, according to Greer (1993), systematic wheat crop failures occurred because of insect infestations that migrated up the Richelieu River from New York.

The first decade of the nineteenth century also had longer growing seasons, with a decline towards shorter seasons in the 1810s. The four years 1815–1818 were cool with short growing seasons in Montreal, which resulted in the wheat shortage of 1816 noted by McCord, but longer summer seasons, well within the modern range, returned in the 1820s. The 1830s brought considerable variability before the growing season length dropped precipitously in the 1840s, also with considerable year-to-year variation. A tendency towards shorter growing seasons in the middle of the century is seen in Glackmeyer's record. It is this variation from one growing season to the next, and the uncertainty as to which crops would fare best, that would create difficulty in adapting to the changeable and increasingly unreliable climate, rather than the colder conditions in themselves.

It is generally accepted that there was an ongoing agricultural crisis in Lower Canada at the beginning of the nineteenth century (Coates 2000; Greer 1985, 1993). Insofar as this crisis might be weather related, rather than caused by other factors such as exhausted soils, insect infestations, or by an increasing population expanding into marginal lands, it would not appear to be due to cold conditions per se but perhaps rather to the increased variability and swinging back and forth between warm and cold conditions—the relatively cool 1810s and the warmer 1820s, for example—that created uncertainty. Although the growing season in the late 1830s was not as short as it would become in the 1840s and 1850s, the number of consecutive frost-free days diminished after the long summer of 1835. The summers of the late 1830s in particular were cool, and there was hunger and hardship, particularly in the northeastern regions, although, as Greer points out, the regions near Montreal where the Rebellions were focused were not as severely affected.

There is not, based on this evidence, a simple story of a colder past and a warmer present. Rather, there is considerable variability across the past three centuries, with both warm and cold years, as well as early and late seasons, occurring in all periods. One thing that seems to stand out is a cold period in the mid-nineteenth century; the period during which Glackmeyer and Bethune were principally recording. It might be that rather than considering the twentieth century as extraordinarily warm in the Saint Lawrence valley, it was the nineteenth century, especially the second half from the late 1850s, and particularly the 1880s to the turn of the twentieth century, that was particularly cold. It would seem from an agricultural point of view that the mid-nineteenth century was the last stand of the Little Ice Age (LIA) in the Saint Lawrence valley. There were warmer springs and better growing conditions not only after the middle of the century but before as well, except for the cold 1810s.

A Complicated Climate, Past and Present

On the whole, cold winters don't seem to have garnered much attention, apart from the shock they tended to give newcomers. Gaultier noted that 1746 was a "horrible" winter. The cold snap of January 1859 generated the coldest recorded temperatures for southern Quebec and the Saint Lawrence valley region and prompted some attention in newspaper articles. Even so, the country was generally well adapted to the cold, with efficient stoves and plentiful supplies of firewood, at least outside the cities (Lambert 1810, p. 123). Residents adapted to the cold, taking advantage of the below-freezing temperature to preserve food. The frozen water surfaces of lakes and rivers and snow-covered fields provided a convenient means of transportation, superior to the muddy, rutted summertime roads. Cold winters were normal, though sometimes disagreeable, such as when strong winds whipped up snow crystals into a vicious, stinging cyclone of white particles known as a *poudrerie*, reducing visibility and confusing direction, keeping prudent travellers at home. Storms aside, winter was the season for paying calls, going to concerts, organizing various societies, and general socializing.

It is rather the springs, when supplies started to run low and the isolation from outside contact started to wear, and even more the summers, the all-important growing season that determined whether the harvests would be plentiful enough to avoid hunger and perhaps even have a surplus to sell or export, that caught the attention of the diarists and that were described in more detail. Detailed attention was given to the ice breakup, when the ships from abroad could finally navigate the Saint Lawrence River and break the winter isolation. The pleasure shown at an early snowmelt that would allow early access to the fields and early sowing, giving the grain time to mature, can be seen most clearly in the earlier records of Gaultier, McCord, and Sutherland.

The evidence of scorching summers, particularly in the Quebec City region, is particularly at odds with the notion of primarily cold and difficult conditions during the height of the Little Ice Age. The Little Ice Age was not, however, a time of uniformly cold conditions; the middle of the eighteenth century was a generally warmer period in between the cold seventeenth and nineteenth centuries. An increased level of seasonal changeability, with hot or dry seasons following immediately after cold or wet ones, may be as much a marker of the Little Ice Age as actual persistently cold conditions.

With the climate, as with many other aspects of life, we have a somewhat romanticised view of the past: crisp snowy winters, with sleighs and their jingling bells dashing across the snow, and mild, peaceful summers, spent tranquilly out in the fields. The reality is, as always, more complicated. While there were certainly periods of cold winters, especially in the later part of the nineteenth century, there were few, if any, months in the historical record where the temperature did not go above freezing at least once. As there was a general adaptation to using sleighs and ice roads in winter, seasons with many freeze–thaw cycles throughout the winter or when the winter river freeze-up or spring thaw was slow or vacillating produced frustration, travel restrictions, and in some years hardship as supplies ran low at the end of winter and transport of food and firewood was made difficult or impossible by mired roads.

Our perceptions of the weather and climate are shaped by popular culture, very recent memory, and our local environments. The winters of 2014 and 2015 were particularly cold in the Saint Lawrence valley region, with February 2015 being the coldest on record. As I write this in December of 2015, the current El Niño is being held responsible for an extremely warm start to the winter. We thus have had a record-breaking cold winter month, February, and a record-breaking warm winter month, December, in the same calendar year of 2015. Record snowfall amounts fell during the winter of 2007–2008, leading to several roof collapses from the accumulated weight of the snow. In 2011, a number of snowstorms led to extremely high snow accumulation in the Adirondacks to the south. When the waters from the spring snowmelt discharged via the Richelieu River to the Saint Lawrence were combined with an extremely wet spring, the result was sixty-nine days of flooding in the Richelieu valley. These recent events shape not only our memories but also our expectation of what the "normal" climate should be.

The increasingly urbanised Canadian population also tends to live in cities, which are not only warmer than the rural areas (where up to 40% of the Canadian population once lived as recently as the 1950s), but also where snow is removed from the urban environment to make movement by car, public transport, and foot easier. These are all a partial response to the nostalgic question of "where have the snows of yesteryear gone?"[12]

References

Anderson, K., 2005: *Predicting the Weather: Victorians and the Science of Meteorology*. University of Chicago Press, 331 pp.

Bethune, J., 1869: John Bethune's meteorological reports. McCord Family Fonds, Papers P001-835. McCord Museum Archives.

Cleghorn, R., 1833: Copies of weather diaries. McCord Family Fonds, Papers P001-839. McCord Museum Archives.

Coates, C. M., 2000: *The Metamorphoses of Landscape and Community in Early Quebec*. McGill-Queen's University Press, 231 pp.

Duhamel Dumonceau, H. L., 1746: Observations botanico-météorologiques faites à Québec par M. Gaultier pendant l'année 1745. *Mém. Acad. Roy. Sci.*, **1746**, 88–97.

Environment Canada, 1961: Hourly data report for May 31, 1961. Accessed 12/14/2015, http://climate.weather.gc.ca/climate_data/hourly_data_e.html?timeframe=1&Prov=&StationID=5934&type=line&MeasTypeID=dptemp&StartYear=1840&EndYear=2017&Year=1961&Month=5&Day=31#.

Financière agricole du Quebec, 2004a: L'état des cultures au Quebec pour le 1er juin 2004. Financière agricole du Quebec, Gouvernement du Quebec Rep. 2, 4 pp.

——, 2004b: L'état des cultures au Quebec pour le 30 novembre 2004. Financière agricole du Quebec, Gouvernement du Quebec Rep. 15 , 2 pp.

——, 2006a: L'état des cultures au Quebec pour le 10 octobre 2006. Financière agricole du Quebec, Gouvernement du Quebec Rep. 10, 3pp.

——, 2006b: L'état des cultures au Quebec pour le 6 juin 2006. Financière agricole du Quebec, Gouvernement du Quebec Rep. 2, 4 pp.

——, 2007: L'état des cultures au Quebec pour le 24 juillet 2007. Financière agricole du Quebec, Gouvernement du Quebec Rep. 5, 4 pp.

——, 2008: L'état des cultures au Quebec pour le 3 juin 2008. Financière agricole du Quebec, Gouvernement du Quebec Rep. 2, 5 pp.

---

12. A question that itself dates to 1533.

——, 2009a: L'état des cultures au Quebec ; Bilan 2008. Financière agricole du Quebec, Gouvernement du Quebec Rep. 15, 4 pp.

——, 2009b: L'état des cultures au Quebec pour le 2 juin 2009. Financière agricole du Quebec, Gouvernement du Quebec Rep. 3, 4 pp.

——, 2010: L'état des cultures au Quebec: Bilan 2009. Financière agricole du Quebec, Gouvernement du Quebec Rep. 15, 4 pp.

——, 2011a: L'état des cultures au Quebec: Bilan 2010. Financière agricole du Quebec, Gouvernement du Quebec Rep. 15, 4 pp.

——, 2011b: L'état des cultures au Quebec pour le 18 octobre 2011. Financière agricole du Quebec, Gouvernement du Quebec Rep. 12, 6 pp.

——, 2012: L'état des cultures du Quebec: Bilan 2011. Financière agricole du Quebec, Gouvernement du Quebec Rep. 15, 5 pp.

Gaultier, J.-F., 1745 unpublished manuscript: Journal des observations météorologiques de M. Gaultier à Kebec. Observatoire de Paris Manuscript, Doc. 64-5-B, 103 pp. [Available from Fonds Joseph-Nicolas Delisle, Observatoire de Paris, 61 Avenue de l'observatoire, Paris 75014, France.]

——, 1748 unpublished manuscript: Journal des observations météorologiques de M. Gaultier à Kebec depuis le 1 octobre 1747 jusqu'au 1 octobre 1748. Observatoire de Paris Manuscript, Doc. 64-6-A, 71 pp. [Available from Fonds Joseph-Nicolas Delisle, Observatoire de Paris, 61 Avenue de l'Observatoire, Paris 75014, France.]

——, 1754 unpublished manuscript: Journal des observations météorologiques de M. Gaultier à Kebec 1754. Meteorological Collection, Houghton Library, MS Can 42(1), 17 pp. [Available from Houghton Library, Harvard University, Cambridge, MA 02138.]

——, 1755 unpublished manuscript: Observations on Quebec, 1744–1754. Meteorological Collection, Houghton Library, MS Can 42(2). [Available from Houghton Library, Harvard University, Cambridge, MA 02138.]

Glackmeyer, L.-E., 1859 unpublished manuscript: Meteorological observations taken at Beauport near Quebec. Meteorology Account R6707-0-5-E. Library and Archives Canada.

Greer, A., 1985: *Peasant, Lord, and Merchant: Rural Society in Three Quebec Parishes, 1740–1840.* University of Toronto Press, 304 pp.

——, 1993: *The Patriots and the People: The Rebellion of 1837 in Rural Lower Canada.* University of Toronto Press, 385 pp.

Higgins, J., 2012: The 1846 Great Fire. Accessed 12/14/2015, http://www.heritage.nf.ca/articles/politics/st-johns-fire-1846.php.

Hough, F. B., 1872: *Results of a Series of Meteorological Observations Made under Instructions from the Regents of the University at Sundry Stations in the State of New York, Second Series, from 1850 to 1863, Inclusive.* Weed, Parsons and Company, 406 pp.

Lambert, J., 1810: *Travels through Lower Canada and the United States in the Years 1806, 1807, and 1808.* Richard Philips, 526 pp.

LeRoy Ladurie, E., 2004: Histoire humaine et comparée du climat : Canicules et glaciers XIIIe–XVIIIe siècles. Fayard, 740 pp.

McCord, J. S., 1835 unpublished manuscript: Meteorological tables. McCord Family Fonds, Papers P001-829. McCord Museum Archives.

——, 1836 unpublished manuscript: Scientific notebook. McCord Family Fonds, Papers P001-825. McCord Museum Archives.

——, 1838 unpublished manuscript: Letter to Dr. Skey. McCord Family Fonds, Papers P001-827. McCord Museum Archives.

——, 1840: Meteorological summary of the weather at Montreal, province of Canada (from 1836 to 1840). *Amer. J. Sci. Arts*, **41**, 330–331.

——, 1842a unpublished manuscript: Meteorological tables. McCord Family Fonds, Papers P001-830. McCord Museum Archives.

——, 1842b: Report of the meteorological observations made on the island of St. Helen's. Natural History Society of Montreal, Ed., Lovell, 18 pp.

——, 1843 unpublished manuscript: Notebook on climate and meteorology of North America. McCord Family Fonds, Papers P001-828. McCord Museum Archives.

——, 1847: Observations on meteorology. *Br. Amer. J. Med. Phys. Sci.*, **3**, 228–229.

McCord, T., and J. S. McCord, 1826 unpublished manuscript: McCord's meteorological register. Meteorological Records, 1798–1972, Container 1039, Envelope 320. Faculty of Arts and Sciences, Department of Meteorology, McGill University Archives.

Meyer, W. B., 2014: *Americans and their Weather*. Oxford University Press, 296 pp.

Moore, P., 2015: *The Weather Experiment: The Pioneers Who Sought to See the Future*. Farrar, Straus and Giroux, 395 pp.

Skakel, W., 1852 unpublished manuscript: Meteorological register 1842–1852. Meteorological Records 1798–1972, RG 2 Container 939 item 1. Faculty of Arts and Sciences, Department of Meteorology, McGill University Archives.

——, 1869 unpublished manuscript: Meteorological register 1862–1869. Meteorological Records 1798–1972, RG 2 Container 939 item 2. Faculty of Arts and Sciences, Department of Meteorology, McGill University Archives.

Slonosky, V. C., 2014: Daily minimum and maximum temperature in the St-Lawrence valley, Quebec: Two centuries of climatic observations from Canada. *Int. J. Climatol.*, **35**, 1662–1681, https://doi.org/10.1002/joc.4085.

——, 2015: Historical climate observations in Canada: 18th and 19th century daily temperature from the St. Lawrence valley, Quebec. *Geosci. Data J.*, **1**, 103–120, https://doi.org/10.1002/gdj3.11..

Smallwood, C., 1855 unpublished manuscript: The observations of Dr. Smallwood 1849–1855. Meteorological Records, 1798–1972, RG 32, Container 956, Envelope 322. Faculty of Arts and Sciences, Department of Meteorology, McGill University Archives.

——, 1862 unpublished manuscript: The observations of Dr. Smallwood 1856–1862. Meteorological Records, 1798–1972, RG 32, Container 955, Envelope 323. Faculty of Arts and Sciences, Department of Meteorology, McGill University Archives.

Spark, A., 1819 unpublished manuscript: Diary 1798–1819. Meteorological Records,

1798–1972, RG 2, Container 1039, Envelope 318–319. Faculty of Arts and Sciences, Department of Meteorology, McGill University Archives.

Sutherland, W., 1848 unpublished manuscript: The observations of Dr. Sutherland. Meteorological Records, 1798–1972, RG 32, Container 1039, Envelope 321. Faculty of Arts and Sciences, Department of Meteorology, McGill University Archives.

Wetter, O., and C. Pfister, 2013: An underestimated record breaking event—Why summer 1540 was likely warmer than 2003. *Climate Past*, **9**, 41–56, https://doi.org/10.5194/cp-9-41-2013.

Wien, Thomas. 1990. "Les Travaux pressants." Calendrier agricole, assolement et productivité au Canada au XVIIIe siècle. *Revue d'histoire de l'Amérique française* **43**, 535–58.

CHAPTER THIRTEEN

Extraordinary Seasons

Almost every season has its peculiarities and interesting properties; to describe the three centuries' worth of unusual seasons would strain the patience of even the most avid weather enthusiast. In the selection of unusual seasons in this chapter, I was mostly guided by weather episodes that the historical observers themselves considered extraordinary and thus described in detail. The information we have depends very much on the nature of the observer and the manuscripts and records that have been conserved. Thus, we have a thirty-eight-page letter from Jean-François Gaultier describing the hot, dry, and long summer of 1754, but nothing from Alexander Spark on what may possibly be the two warmest summers of the past two centuries and longer: 1807 and 1808. The cold snaps of the summers of 1844 and 1859 caught my attention in the graphs presented in the last chapter and are discussed in slightly more detail here. January of 1859, as described in the introduction, had the coldest temperatures recorded in the Saint Lawrence valley. In general, winters, even very cold ones, tended to be less remarked upon than summers, as summer was the most important season for crop growth and transport. There is a gap here between the time of Gaultier in the 1750s and the beginning of Spark's record in 1798, though further research holds promise of uncovering more records from ecclesiastical and military sources.

Climate of the Mid-Eighteenth Century: Gaultier's Seasons

The weather of the 1740s in Quebec City as recorded by Jean-François Gaultier was mostly warm and dry, particularly the summers, with the exception of the warm but rainy summer of 1744. The years 1743, 1746, and 1748 all had summer droughts. Gaultier recorded the apple harvest failing because the apples dried up and fell off the trees in July of 1743, 1745, and again in 1748. The drought of 1746 caused many of the smaller water sources such as streams and wells to dry up, but in 1747 the pastures were still green with cattle grazing outdoors until November 28. It is unfortunate that the detailed daily records of 1747 have not been recovered since 1747 was described as very warm, with an early spring and late fall. December of 1742 and the winters of 1747–1748 and 1749–1750 were described as long and cold, with an ice bridge forming on the Saint Lawrence River from Quebec City as far south as Montreal in 1749 (Duhamel Dumonceau 1750, p. 309). The winters towards the middle of 1740s, on the other hand, were extremely mild, with 1746 described as the mildest winter in living memory, with no ice road whatsoever established on the rivers and the first spring thaw occurring in early February. According to Gaultier's records, the mild winter and early spring led to an extremely early wheat harvest, with the grains ripe as early as August 4, 1746.

We have Gaultier's notes, if not his complete records, for ten summers between 1742 and 1754. Only two of these, 1744 and 1753, had weather poor enough to have an impact on the harvest, with 1753 having had a cold summer and continual cold rains. Gaultier reported widespread distress and food shortages following the disastrous harvest of 1753. There must also have been a bad harvest in 1742, just before Gaultier arrived in the colony, as he described 1742–1743 as a year of famine ("disette affreuse"; Duhamel Dumonceau 1743). In 1745, Gaultier noted that the harvest yields were 50% higher than in 1744, so presumably 1744 was also a bad year. The summer of 1745 was hot and dry, so much so that some of the nascent apples dried up and fell off the trees in July for lack of moisture. In August, Gaultier wrote that "the countryside is so beautiful and smiling that we hasten to praise its magnificence" (Gaultier 1745).[1] In summing up, he noted that "we had a very mild winter and excessive heat in summer . . . The years in which there are great heats are those which are ordinarily the most abundant, when the harvest is very good"[2] (Gaultier 1745).

---

1. "On ne tarissoit point en Eloge sur la magnificence de la Campagne . . . si belle et si riante" (Gaultier 1745).

2. "Nous avons eû un hyver très doux et des chaleurs pendant l'été qui ont été

The warm summer led Gaultier to speculate on a warming climate and human-induced climate change, in words which are by now familiar:

> The oldest inhabitants say that in earlier times, we couldn't start the wheat harvest until the 15th or 16th of September, and occasionally as early as the 12th, and that the wheat only started to ripen at this moment, and they never came to such perfect maturity as it does today, but since then much land has been cleared of forests, there is a great uncovering of the land and the wheat and other crops being more exposed to the Sun's rays come to maturity more quickly, so we can start the harvest earlier and there are grounds to hope that the more we clear the land in Canada, the more this land will become fertile and abundant, and the climate much milder.[3] (Gaultier 1745)

Here, again, we see a pattern that seems to recur in human thinking about climate change; warmer conditions tend to be thought of as being produced by human agency, whether for good or ill. Climate improvement was thought of as a positive and optimistic development, whereas current global warming is considered to be problematic. Interestingly, cold seasons don't seem to invoke a similar response; they tend to be seen primarily as natural events.[4]

The winter of 1745–1746 was again very mild, even more so than 1744–1745, and 1746 was another hot and dry summer, with not only grasses in the pasturage drying up and withering, but also many of the wells, water sources, and creeks (*beaucoup de sources* and *ruisseaux taris*; Duhamel Dumonceau 1747, p. 475). In 1747, a cold winter, an early spring with sowing earlier than normal, and a very hot summer led to another year of "good agricultural

---

excessives ... les années où il y a de grandes chaleurs sont celles qui sont ordinairement les plus abondantes, es où la récolte est très bonne" (Gaultier 1745).

3. "Les anciens du Pays dissent qu'autre fois on ne pouvait commencer La récolte des bleds que le 15 ou 16 Septembre, et que quelques fois dès le 12, et que les bleds ne commençoient a être muris que dans ce temps, es qu'ils ne venoient Jamais à une aussi parfait maturité qu'aujoudhuy, mais depuis qu'on a défriché beaucoup de Terre, et abbatu beaucoup de bois, il y a eû beaucoup de découvert et les bleds et autres productions de la Terre ayant été plus exposés aux rayons de Soleil sont venue plustost à maturité, aussi commence ton plustôt la récolte es il y a Lieu d'esperer que plus on défrichera de terre en Canada, plus ce pays deviendra fertile es abondant es que son Climat s'adoucira beaucoup" (Gaultier 1745).

4. An exception to this could be the 1960s, when increasingly cold conditions were linked to industrial pollution, although the cold was also speculated to be a harbinger of a coming (natural) ice age.

productivity" (Gaultier, 1748). The first frost was very late, with the fields still green and cattle grazing in the pastures until the first serious snow on November 28. April of 1748 was sunny, with the sowing again started early. In June and early July, the heat was so extreme it was thought that such high temperatures had never before been felt in Canada, a sentiment that would recur in the near future.[5]

Water sources, already low from last year's drought, started to dry up. Plants wilted and some trees had their leaves dry up, brown, and fall off (*brulées et grilles*; Gaultier 1748). There was some rain in mid-July, but drought set in by the end of July: "the wells, the sources, the smaller rivers were all completely dried up, even the tides were lower than usual, and smaller than had ever been seen before" (Gaultier 1748).[6] As in 1743 and 1745, grass withered and apples dried up and fell off the trees in midsummer, before they could ripen. The pattern continued into 1749 with a cold winter and hot summer. Despite these years of drought in the 1740s, 1754 would leave even more of an impression on Gaultier. He devoted an entire memoir to the summer of 1754, which we will examine in detail below.

The Summer of 1754

In his letter for 1754, Gaultier indicated that the season was so extraordinary that it merited a detailed monograph and, thanks to this, the summer of 1754 is perhaps the best-described season in the historical record. Gaultier devoted over sixteen pages of closely written manuscript and marginalia to the subject. At the beginning of his letter, Gaultier describes 1753 as having had a cold and wet summer, with a poor harvest in Quebec City, although the harvest in Montreal was said to have been exceptionally good. In 1753, the harvest had initially promised well, but a cold spring had led to a late sowing. This led to disaster when unexpectedly strong frosts, combined with continuous rains, occurred at the end of August and beginning of September:

> But as they sowed late they were surprised towards the end of August and in the beginning of September by heavy white frost, by cold and continual rains

---

5. "Chaleurs excessives; il a fait excessivement chaud jusqu'au 16 (Juillet) es qu'on n'avoit peut être jamais senti des chaleurs si grandes au Canada" (Gaultier 1748).

6. "Les puits, les sources, les petites rivieres etoient entirèment tarris, les mareés etoient même plus basses et beaucoup moins considérable qu'on ne les avoit jamais vues" (Gaultier 1748).

[so] that the wheat rusted, such that they had much difficulty in ripening and the harvest couldn't be started until the end of September, and even then with much difficulty.[7] (Gaultier 1755)

The poor harvest of 1753 in the region of Quebec City led to food shortages and near-famine conditions in the spring of 1754. Gaultier noted, however, that in Montreal, some 200 kilometers to the southwest, 1753 produced an exceptionally good harvest, with an early spring that enabled an early start to the planting and good weather throughout the summer:

> We can say that the air temperature of these seasons of spring and summer were so advantageous in this district of Montreal for all fruits of the Earth that there were abundant harvests of all sorts of goods, and that the grain harvest was so abundant that the granges couldn't hold all the harvest, and we were obliged to make very considerable piles outside. In a word, we don't recall, in living memory, of having ever seen such an abundant harvest in the district of Montreal.[8] (Gaultier 1755)

There was little widespread illness in Quebec City in 1753, Gaultier continued, except during November and December, "when nearly all the town and garrison became ill with a bilious complaint, which responded well to treatment. This illness was attributed to the rotten flour [from the rusted and diseased harvest]" (Gaultier 1755).

January 1754 started off very warm, with temperatures around the freezing mark and rain on January 4. By January 6, nearly all the snow which had previously fallen had melted, as had the river ice. This thaw posed a

---

7. "Mais comme ils avoient été semé tard ils furent surpris vers la fin d'aoust [et] dans tout le mois de septembre par les gelées blanches fortes par de pluyes froides et continuelles que les [bleds ont] rouilliés . . . de sorte qu'ils eûrent beaucoup de peine a murir et que la recolte ne peut être faite qu'à la fin de septembre es encore ce fût avec beaucoup de peine" (Gaultier 1755).

8. "On peut dire que la température de l'air de ces saisons du printemps et de l'été se comportait si avantageusement dans ce gouvernement de Montreal pour tous les biens de la Terre qu'on y eus des fruits de toutes espèces en abondance que la récolte des bleds y fut si abondante que les granges ne pouvaient pas contenir toute la récolte, on fût obligé des tats de gerbes dehors très considérables . . . on ne souvient pas de mémoire d'homme d'avoir jamais vû une récolte si abondante dans le gouvernement de Montreal en un mot" (Gaultier 1755).

major inconvenience in Canada in winter. Without a smooth layer of snow or river ice, the roads turned into an impassable mire, and it was impossible to transport logs from the forests as heating fuel. To compound the misery, all the meat that had been stored the previous autumn in natural freezers started to thaw, compounding the food shortages because of the poor harvest. "We have perhaps never seen such a mild January in Canada," commented Gaultier (1755). The warmth and lack of ice extended to the Labrador coast and up "towards Hudson's Bay" (though the geography at that time was uncertain and Hudson Bay was generally thought to be both much closer and farther east relative to Quebec City than it actually is) (Gaultier 1755).

February saw a return to cold and heavy snows, with "*des poudreries horribles*," episodes of blowing snow so fierce and blinding it was impossible to go outdoors (Gaultier (1755). A thaw set in again in March, with maple sap running and being collected as early as March 12, and continued on and off for the rest of the month as temperature swings saw pronounced freezing followed by thaws. Illness recurred, with pneumonias and putrid fevers, attributed once again to the spoiled wheat used to make bread, especially for the poorer citizens. Gaultier added a bitter complaint to the fact that the abundant grain in Montreal was confined by the authorities to that district to feed their residents, leaving those in the Quebec district to fall ill with their poor and rotten grains. The disconnect between the two neighbouring regions shows the extent to which they were considered two separate and independent administrative units, though the large-scale transport of grain over several hundred kilometers during a winter with poor roads and no firm river ice might have been difficult in any event. The snow was largely melted by April, which meant an early sowing season and hopes of a better harvest in 1754. Strawberry plants were in flower by May 2, and the first fruit was ripe by June 12. The lack of rain started to be worrying, however, and by June 17, the heat had already started to build:[9]

> The botanico-meteorological history of this month is so interesting that it must be described in as much detail as possible. We can see that the warmth was strong, and nearly continual, such that they gave rise to a singular phenomena in the vegetation. The cherry trees flowered a second time, even while the first fruits were still ripe on the tree, something which has perhaps

---

9. "Il fit le 17 [juin] une chaleur très considérable" (Gaultier 1755).

never been seen, especially in Canada, in such a way that the trees had ripe fruit and large bunches of flowers at the same time.[10] (Gaultier 1755)

The strong and continual heat continued to nourish the grain, though the heat and drought ruined the hay harvest in Montreal: "Never have the prairies in the district of Montreal been so dry, so arid, or have furnished so little"[11] (Gaultier 1755). The wheat harvest started on August 10, 1754 in Montreal and was again abundant, as much so as that of 1753; the harvest in Quebec started on August 17. The heat caused the grains of wheat to loosen and fall off as the stalks were harvested. To try to get around this problem, harvesting was done in the cooler temperatures of the night and early morning.

The drought continued on through August 1754 in the Quebec City district:

> The excessive heat dried up most of the water sources, wells, ponds and fountains. It completely dried up the smaller rivers such that the beds were dry by the 10th. Never had such small quantities of water been seen, nor such low levels everywhere in Canada, even in the St-Lawrence River.[12] (Gaultier 1755)

There was so little water that the cattle suffered greatly from thirst, diminishing their milk, an important source of protein in the habitant diet (Greer 1985) during the summer months, with some cattle even reportedly dying of thirst. Another devastating consequence of the heat and drought was forest fire:

> In the woods, through carelessness a voyager or hunter had not put out the fire he made in the night to cook his food. This fire made considerable ravages

---

10. "L'histoire botanico-météorologique de ce mois est trop intéressante pour qu'on ne la détaille pas le plus qu'on pourra. On y verra que les chaleurs y ont été trés vives, et presque continuelles, qu'elles ont occasionné un phenomene singulier dans la vegetation. Elles sont sans d'autre cause que les cerisiers à grappes ont fleuri une seconde fois dans le temps même que leur premiers fruits étoit a maturité, ce qu'un n'a peut-être jamais vu surtout en Canada de sorte que les arbres avoient des fruits murs, et de très belles grappes de fleurs en même temps" (Gaultier 1755).

11. "Jamais les prairies du gouvernement de Montreal n'avions été si seches, si arides, n'avoient jamais si peu fourni" (Gaultier 1755).

12. "Cette excessive chaleur a fait tarent la plus grande partie des sources, des puits, des mares et des fontaines. Elle a aussi desséché entièrement les petites rivières de sorte que leur lit étoit à sec le 10. Jamais in n'avoit vu les eaux en si petite quantité ni si basse partout en Canada même dans le fleuve St-Laurent" (Gaultier 1755).

and burned large extents of woods, entire fields of wheat and prairies on the edge of the forest. These arid forests inflamed such a large quantity of wood at once, and woods such as pine and fir heated the air to such a degree that an excessive heat was felt everywhere in the cities as in the country. The air was so hot and heavy at Quebec on the 11th of the month that it was hard to breathe, everywhere was the odour of smoke, of fire, of ash which formed a dust in the air and smelled of burnt turf . . . These fires obscured the Sun so much with their thick smoke that it was possible to look at the Sun as if through coloured glass. The air was so hot on the 20th at Quebec as much from the heat of the weather as from the fires in the woods that asthmatics found it difficult for several days.[13] (Gaultier 1755)

Finally, on August 23, 1754, it rained heavily enough and long enough to put out the fires as well as to replenish the water sources. A cooler spell in September didn't last long, though, as the heat and sunny weather returned in the middle of the month, with temperatures going up to 23°R (~26°–29°C). The drought left the mills without enough water power to grind the grain into flour. Although good for wheat, the prolonged drought damaged the fruit and vegetable harvest. The unusually prolonged heat produced a second flowering of fruits and berries and an autumn crop of strawberries and raspberries in October. Gaultier had never seen this happen before; a similar phenomenon happened in Europe during the exceptionally warm and dry summer of 1540 (Wetter and Pfister 2013).

None of the usual rain, snow, or frost occurred in October 1754, which instead saw summerlike conditions. Indeed, with the help of a few autumn

---

13. "Dans les bois par l'imprudence que quelque voyageur ou chasseur qui n'avoit pas éteint le feu qu'il avait allumé la nuit . . . pour faire sa cuisson. Ce feu fut des ravages considérables et a brulé de grandes étendues de bois, des près de bleds toutes entières et des prairies qui étoient sur le bord des forets. Ces forets qui avoyent aride couvre avoyent enflammé une si grande quantité de bois à la fois, et munie de bois épineux comme pins, sapins et puit avoyent échauffé l'air ou la température qu'on sentoit partout dans les villes comme dans les campagnes une chaleur excessive. L'air étoit si chaud si pesant si lourd a quebec Le 11 du mois qu'on avoit de la peine à respirer, on ne sentoit partout qu'une odeur de fumée, de feu, de cendre qui étoient une poudrerie dans l'air et qui avoyent une odeur de tourbe brulée . . . Ces feux obscurcissoient tellement l'ait es le soleil par les fumées épaisses qu'on pouvoit regarder aussi finement le soleil que si on avoit eû un verre coloré . . . L'air fut si échauffé le 20 à quebec tant par la chaleur du temps que par les feux qui étoient dans les bois que les asthmatiques s'en trouvèrent fort incommodés pendant quelques jours" (Gaultier 1755).

rains, the grass grew green and abundant enough to leave the cattle out in the fields to graze until November:

> This warmth, this mild weather, so agreeable and so charming, continued for the rest of the month to the point where the warmth makes the liquid in the thermometer rise to 20 or 21 degrees above the freezing point [23°–26°C] and the temperature so mild, so nice, and so agreeable that we would believe we were in late spring or summer.[14] (Gaultier 1755)

Gaultier didn't record many afternoon temperatures in 1754, making it difficult to place the warmth of 1754 in the context of his earlier observations. The harvest was two weeks later than in 1746 and a week later than in 1748, while the first snow didn't occur until November 25, nearly as late as the first snowfall of November 28, 1747.

There are a few other instances in the extant weather diaries describing hot, dry summers, though both are confined to short, one-line descriptions. In 1800, Alexander Spark noted on July 6 that the "therm[ometer] stood at the highest I have ever seen it observed in this country" (Spark 1819) when it recorded an afternoon temperature of 96°F (35°C); his own record would be surpassed, however, on July 9, 1803, with a temperature of 97°F (36°C), while his reading of 96°F (35°C) would be matched on July 12, 1807, and July 17, 1808, which are very likely to be to the two hottest summers on record to date in the Saint Lawrence valley. Spark further commented on July 17, 1800, that, as in 1754, it was "very dry," with the "woods on fire in many places" (Spark 1819).

The Summers of 1807 and 1808: Alexander Spark and John Lambert

Spark made very few comments on the weather, not even during the astonishingly warm summers of 1807 and 1808 nor during the famously cold summer of 1816 (the year without a summer), so inferences about the climate from his records have to be taken directly from his daily observations, with no other commentary such as that given by Gaultier on the state of agriculture or seasonal markers in the natural world. Spark's temperature

---

14. "Cette chaleur, ce temps si douce, si agréable et si charmante eut continué le reste du mois au point que la chaleur a fait monter la liqueur du thermomètre souvent à 20 et 21 degrés au-dessus de la congélation et la température a été si douce, si belle, et si agréable qu'on croyoit être à la fin du printemps ou dans l'été" (Gaultier 1755).

observations can be broken into two periods, before and after May 1807, when he moved to the manse next to the newly built Saint Andrew's Presbyterian Church on Rue Sainte-Anne. It is very unfortunate, from a climatological perspective, that this move coincided with an apparently very warm summer, as it makes it difficult to disentangle a truly warm summer from what could be the effects of a different thermometer location. Statistical analysis of Spark's measurements suggests they're consistent with modern observations with no particular discontinuities aside from the relocation in May 1807. Fortunately, there is some independent evidence from John Lambert's *Travels through Lower Canada and the United States in the Years 1806, 1807, and 1808*. In a section entitled "scorching summers," Lambert collated the maximum and minimum temperatures for Quebec City, along with the general weather conditions, for the summer months of 1807, reproduced in Table 13.1 below (Lambert 1810). Spark's records are missing for May 1807 and incomplete for June 1807, when the highest temperature he recorded was 83°F (28°C). Spark's highest temperature for July was 96°F (35.5°C) on July 12, with 95°F (35°C) recorded on July 27 and 88°F (31°C) recorded on August 31.

The summer of 1808 was once again very warm, with Spark's records showing it to be the warmest in the entire observational record for the Saint Lawrence valley region. Lambert wrote that

> the summer of 1808 was the hottest that has been known for several years in Canada. In the months of July and August, the thermometer was several times at 90 [°F; 32°C], and 95 [°F; 35°C] and one or two days it rose to 103°F [39°C], in the shade at Montreal and Three Rivers. At Quebec it was 101° or 102° [38.3°–38.9°C] . . . It appears that it was unusually hot about that time in England, and I suppose it was the same upon the Continent. (Lambert 1810, p. 132)

In fact, the summer temperatures for 1807 in central England (Manley 1953; Parker et al. 1992), Uppsala in Sweden (Bergström and Moberg 2002; H. Bergström 2013, personal communication), and central Europe (Auer et al. 2007; http://www.zamg.ac.at/histalp/) are among the warmest in these centuries-long records, with the summer of 1808 also very warm in many central European locations. In the Saint Lawrence valley, 1808 is the warmest summer (defined here as the summer half of the year, April 15 to October 14) on record with an estimated average temperature of 66.4°F (19.1°C). The next warmest summers, 1955 and 1825, had average maximum temperatures

Table 13.1. Lambert's synopsis of the summer of 1807 in Quebec City (Lambert 1810, p. 129).

	Lowest temperature (°F)	Highest temperature (°F)	Weather
May	20° (−7°C)	75° (24°C)	Continual rain
June	50° (10°C)	90° (32°C)	Rain in the first week, afterwards dry and warm
July	55° (13°C)	96° (36°C)	Dry and sultry
August	68° (20°C)	90° (32°C)	Fine warm weather with little rain
September	46° (8°C)	78° (25°C)	Fine mild weather

of 64.9 and 64.8°F (18.3 and 18.2°C), respectively, while the warmest summer of the twenty-first century so far, 2001, had an average summer temperature of 64.4°F (18.0°C). For a recent comparison, 2015 and 2016 had mean summer temperatures of 64.2 and 64.0°F (17.9 and 17.8°C). The ordering changes slightly when different measures are used. For example, Gaultier recorded mainly morning (minimum) temperatures in 1754, while many of the afternoon temperatures went unrecorded, so that year is excluded from mean temperature calculations. If we look at only minimum (that is, morning) temperatures, the value most often recorded in the historical diaries, 1820 is the warmest summer half-year with an average minimum temperature of 57.4°F (14.1°C) between April 15 and Oct 14, followed by 1825 then 1955. Because so many observations are missing, it is only when we look at the temperature values in much finer detail, such as August maximum temperatures, that the summers of 1807 and 1754 stand out as the two warmest on record: the average August maximum temperature is 85.1°F (29.5°C) for 1807 and 82.9°F (28.3°C) for 1754. The warmest comparable August mean maximum temperature in recent times was 82.0°F (27.8°C) in 2001. August 2016, widely reported as a hot summer, had a mean maximum temperature of 81.3°F (27.4°C).

Given these limits, both in terms of accuracy and completeness, of the temperature records, it's hard to classify with absolute confidence which were the warmest summers on record, but 1754, 1807, and 1808 are all outstanding as hot and dry summers, as seen from both the temperature readings and the contemporary remarks listed above. Again in 1846, Sutherland noted at the end of August that the "water [was] very low, small streams dried up in the country" (Sutherland 1848). The years 1870 and 1873 are also remarkable; Smallwood recorded thirteen months of above-average temperatures from December of 1869 to December 1870 (Smallwood 1873). As in 1808, and possibly in 1754, the unusually warm temperatures were not so much from excessively hot days and record-breaking temperatures but rather were caused by a sustained warmth held over many months. It is tempting to try to link these extraordinary seasons, as well as the mild winters, to global-scale, disruptive phenomena such as the El Niño–Southern Oscillation. While the recent warm summers of 1998, 2005, and 2010 all do follow after winters with pronounced negative values of the Southern Oscillation index (SOI), 1870 and 1873 follow months of mild to pronounced *positive* SOI values (La Niña conditions) in the previous winter months and, in fact, come *before* negative SOI values, indicating an El Niño event (Allan and Ansell 2006). Four of the eight years described by Gaultier and Louis-Edouard Glackmeyer as having mild autumns and late arriving snows in Quebec City might be linked to El Niño in 1744, 1747, 1846, and 1854, but the others, including 1754, are not. Of the warmest Decembers on record, 1891 is linked to a very strong El Niño, 2001 is linked to a neutral state, and December 1998 is linked to a moderate La Niña. The temptation to link mild, green Decembers to El Niño will have to be resisted.

Cold Nineteenth-Century Summers

The summer of 1816 was the coldest on record and is notorious in the Northern Hemisphere as the year without a summer (Harrington 1992; Brugnara et al. 2015; Baron 1992; Chenoweth 2009; Hamilton 1986; Klingaman and Klingaman 2013). Works both scholarly and popular have been devoted to analysing the volcanic explosion of Tambora in April of 1815 and the effect it had on climate around the world the following year. The miserable summer of 1816 inspired, among other literary works, Mary Shelley's *Frankenstein* and George Byron's poem "Darkness."

Spark recorded snow on June 6 and 7, 1816, describing the weather as "bleak." Although there were some warm days reaching as high as 30°C during July and August, the temperature didn't go above 20°C after August 18, and "bleak" and "haze" appear frequently in Spark's weather descriptions (Spark 1819). The summer after the eruption of Mount Pinatubo in 1992 was also cold and wet in Montreal and was the only summer on record without a single day breaking past 86°F (30°C).[15]

---

15. Technically speaking, the temperature as measured at the Montreal International (Dorval) airport did not reach 30°C between June 21 and September 21 2017,

While the cold summer of 1816 and its link to the Tambora explosion are well known, Spark's record shows that the climate started to cool even before 1816. The summer of 1809 was also cold, and the year 1809 is one of the coldest on record. An elusive volcanic explosion inferred from ice-core and tree-ring records has now been tentatively dated to December 1808, most likely in Indonesia (Guevara-Murua et al. 2014). This appears to have had a considerable impact on the climate in the Saint Lawrence valley, as the warmest summer on record in 1808 cooled rapidly into one of the coldest years in 1809. And although 1816 is considered the benchmark year for cold summers in North America, partly based on Spark's records (Brugnara et al 2015; Chenoweth 2009; Hamilton 1986), we can see from Glackmeyer's diary that 1844 had the latest recorded frost date in the Saint Lawrence valley (July 7; Fig. 12.7). The evolution of that summer cold snap as recorded by the various diarists in Quebec City and Montreal is shown in Table 13.2.

On July 4, 1844, Glackmeyer recorded a morning temperature of 45°F (7°C), describing the day as "Cold a.m. But beautiful after" (Glackmeyer 1859). July 5 and 6 were rainy but warmer, and on July 7 the morning temperature was 48°F (9°C), with "hoar frost this morning, fine & clear all day" as the day's description (Glackmeyer 1859; Table 13.2). Meanwhile in Montreal, John Bethune also recorded cold mornings, with 44°F (7°C) on July 4 and 49°F (9°C) on the morning of July 8, though with no additional weather descriptions (Bethune 1869). William Sutherland noted "cool—cold during the night" on July 4, and "cool & dusty" July 7 (Sutherland 1848).[16] Alexander Skakel, living in the heart of the city of Montreal, recorded 7:00 a.m. temperatures of 51°F (10°C) and 53°F (12°C), both days being listed as "fair" (Skakel 1852).

The summers of the late 1840s and 1850s continued to be on the cool side. "Fine April weather," John Samuel McCord grumbled sarcastically on May 26, 1847. June continued "cold, raw, and miserable." By the end of June 1847, McCord worried that "this continual rain must entirely destroy the crops." (McCord 1852). The next month, the 1847 typhus epidemic hit Montreal along with a heat wave.

---

but temperatures above 30°C were recorded earlier in June 2017 and after September 21 2017.

16. Sutherland's recorded temperature values are 64° and 66°F for 7:00 a.m. and 3:00 p.m., respectively, but these values would seem to be suspect given the other values.

Table 13.2. Frost in July 1844.

| | Quebec City | | | | Montreal | | | |
| | Glackmeyer (1859) | | Bethune (1869) | | Sutherland (1848) | | Skakel (1852) | |
	Temperature (°F/°C)	Weather	Temperature (morning) (°F/°C)	Weather	Temperature (morning) (°F/°C)	Weather	Temperature (°F/°C)	Weather
Jul 4	45/7	Cold a.m. but beautiful after	44/7	—	62/17	Cool—cold during the night	51/11	Fair
Jul 5	68/20	Most agreeable, showers at night	50/10	—	64/18	North wind barometer falling	59/15	Showers
Jul 6	60/16	Rained all night, cleared up & fine	62/17	—	68/20	Rain during the night wind strong during the day	57/14	Fair
Jul 7	48/9	Hoar frost this morning, fine & clear all day	67/19	—	66/19	Cool & dusty	53/12	Fair
Jul 8	66/19	Fine & warm all day	49/9	—	65/18	—	57/14	Fair

June and July 1859

The nadir of the cold mid-nineteenth-century summers hit in 1859, with two cold snaps: the first in the beginning of June and the second in early July. The frost hit first in Montreal on June 4 and again in Quebec City on June 10. This was the same year Glackmeyer recorded −40°F (−40°C) and Charles Smallwood recorded −43.6°F (−42°C) on January 10, the coldest temperature on record for the Saint Lawrence valley (see the introduction). The cold snap and "great frost" in Montreal on the night of June 4, 1859, caused considerable worry for the gardens and crops. A newspaper clipping in McCord's diary recounts the damage caused: "Gardens suffered more than fields. Melons, cucumbers and tomatoes went by the board." In an effort to prove wrong "the oldest inhabitant . . . that venerable authority [who] would have it that Saturday night, the 4th June 1859, was the coldest that had taken place in that month on over a quarter of a century," the newspaper printed the coldest June days on record since 1840: temperatures of 28.1° and 28.2°F (−2.2 and −2.1°C) had been recorded on June 11, 1842, and June 1, 1843, respectively; June 4, 1859, was 30.2°F (−1.0°C). Nevertheless, the article conceded that "still, it was low enough and frost sufficient even on the 4th June to make that date a remarkable one in the annals of climatology" (McCord 1852).

On June 11, 1859, the morning temperature in Quebec City was only 38°F (3°C) and described as "very cold" in the weather remarks (Table 13.3). The morning temperature of June 5 was 41°F (5°C), and the temperature was 40°F (4°C) on June 9, with a "strong gale" on June 6 (Glackmeyer 1859). Bethune recorded 40°F (4°C) in Montreal on June 5, a "strong gale" on June 6, 41°F (5°C) the morning of June 9, and 39°F (4°C) the morning of June 11. Temperatures rose as high as 82°F (28°C) by the afternoon of June 15 (Bethune 1869). Smallwood's readings confirm this sequence, with a reading of 40.2°F (4.5°C) June 11 (Smallwood 1862).

The rest of the month continued to be variable and unsettled. By June 14, it was "sultry and muggy," with thunderstorms in Quebec City on June 15, but June 18 was again "cold, raw and damp" in Montreal (Glackmeyer 1859; McCord 1859). McCord noted that June 23 was "the first day we could spend out of doors," and that by June 28 it was "fearfully hot;" a thermometer in the deeply shaded woods read 88°F (31°C). June 29 was "superlatively hot," but a fall of 41°F during the night led to a renewed cold spell (Table 13.4) (McCord 1859).

Although the recorded temperature in July reached as low as it had in June and was even colder than in 1844, there doesn't appear to have been any frost recorded in this second cold spell in July 1859. Tables 13.2 and 13.3

Table 13.3. Temperature and weather on June 5–11, 1859.

	Quebec City		Montreal			
	Glackmeyer (1859)		Bethune (1869)		McCord (1859)	
	Morning temperature (°F/°C)	Weather	Morning temperature (°F/°C)	Weather	Minimum temperature (°F/°C)	Weather
Jun 2	59/15.1	—	60/15.5	Showers	—	—
Jun 3	60/15.5	—	39/3.9	Showers p.m.	—	—
Jun 4	40/4.4	—	43/6.1		—	—
Jun 5	40/4.4	Cloudy	40/4.4		(unavailable as McCord was travelling as circuit court judge)	Cold
Jun 6	40/4.4	Fine	45/7.2	Strong gale	—	Cold
Jun 7	49/9.4	Fine	50/10.0		—	Fine but cold
Jun 8	56/13.3	Rain	60/15.5	rain	56/1.3	Cold rain and high wind
Jun 9	44/6.6	Fine	41/5.0	—	—	A fine day but chilly
Jun 10	44/6.6	Rain all day Froze at night ice formed on the pools of water	47/8.3	rain	—	Alas! Never had a worse day of rain and cold wind. Very cold, blowing hard
Jun 11	38/3.0	Very cold & cloudy all day	39/3.9	—	—	
Jun 12	41/5.0	Fine	40/4.4	—	41/5.0	A fine bright sunshine

show that the thermometers alone aren't necessarily a good indicator of potentially damaging cold and frosts; frosts were recorded in the weather observations of July 7, 1844, in Quebec City and June 4, 1859, in Montreal, but the temperatures recorded by our observers stayed above 40°F (4°C). Other meteorological factors, such as humidity and wind, also play a role in determining ground frosts. Nevertheless, 4°C in July remains an outstandingly chilly night.

Winters

As mentioned in chapter 12, winters, even very cold ones, seldom received as much attention as summers. Gaultier noted on a number of occasions

Table 13.4. A second cold snap in July 1859.

	Quebec City		Montreal			
	Glackmeyer (1859)		Bethune (1859)		McCord (McCord 1859)	
	Morning temperature (°F/°C)	Weather	Morning temperature (°F/°C)	Weather	Minimum Temperature (°F/°C)	Weather
Jun 29	68/20	Showers	73/22.8	Showers PM with thunder	(Travelling as circuit court judge)	A superlatively hot day! Thunderstorms evening, night. Fall of therm[ometer] 41° as observed by Mr. Hackett.
Jun 30	53/11.7	Fine	51/10.5	—	—	Cold rainy morning
Jul 1	50/10.0	Fine	53/11.7	—	52/11.1	Cool morning, fine day
Jul 2	56/13.3	Showers all day	60/15.5	Thundershowers	52/11.1	Cloudy, threatening rain
Jul 3	54/12.2	Fine	55/12.8	Rain AM	52/11.1	Raining, still cold and comfortless
Jul 4	44/6.7	Fine	45/7.2	—	41!!/5 [exclamations in original]	Bright cold morning. I scarcely suppose there can be found a precedent of so cold a night between 3 & 4 July
Jul 5	45/7.2	Fine	50/10.0	—	51/10.5	A fine, pleasant, hazy day

in January and February 1743 that all the mercury had contracted into the bulb of his thermometer; it was so cold that he literally couldn't measure the temperature with his instruments designed in France. Gaultier's recording of, or more precisely his difficulty in recording, cold temperatures in Quebec City spurred Joseph-Nicholas Delisle's research into expanded thermometer scales in France. Gaultier's original manuscript letters are conserved among the Joseph-Nicholas Delisle papers at the Paris Observatory, and it is thanks to this connection that so many of Gaultier's records survive today.

The year 1743 continued with violent and cold winds, with up to 30 (Parisian) feet (approximately 9.75 meters) of snow in the woods. All in all, Gaultier found it difficult to imagine more ghastly weather than the conditions he encountered during his first Canadian winter (Fig. 12.2). The following winter, 1743–1744, still had a handful of occasions when the temperature was so cold the mercury contracted into the thermometer bulb but fewer than the previous winter. The thermometers constructed in Paris were slowly expanding to meet the requirements of recording the Canadian climate. The winters of 1744–1745 and 1745–1746 were both very mild, with little ice on the Saint Lawrence river. After the warm and dry summer of 1746 (see above), the winter of 1746–1747 was rendered "terrible by the length and violence of the cold" (Gaultier 1748).[17] An ice bridge formed across the Saint Lawrence during the winters of 1749 and 1753 (e.g., Houle and Moore 2008).

From Fig. 12.2, we can see that there were a series of cold winters in the early 1800s (1809, 1815, 1817, 1818, and 1821 all stand out as exceptionally cold), but winters were milder for much of the 1830s to the 1850s; McCord described the alternating freeze-up and thaws of the early 1840s in considerable detail. Rather, as seen in the previous chapter, warm winters and winters that alternated between above- and below-freezing temperatures were more inconvenient than cold snowy ones, as transport became more difficult both over land roads, which became mired in mud, and by water, when the ice was not solid. Rivers that were neither entirely frozen over, and so could not serve as roads, nor entirely ice free, and thus safe for boats, were so dangerous as to be unusable, and to those living on the island of Montreal, a dangerously isolating barrier.

McCord and the Unusual Winters of 1841–1842 and 1842–1843

McCord (1843) noted that the winter of 1841–1842 was one of "constant alternation of frost and rain throughout." In December 1841, he reported "rain & mild weather for last fortnight—the snow which fell about that period is all washed off and the season as open as in November. No ice." On December 10, it was "raining as hard as in midsummer, the thermometer ranges today from 35 to 40°[F] (2–4°C)." In January 1842, "the first half of this month (January) was tolerably cold but from the 18th to the end, thaws and rain prevailed." New Year's Day had seen the temperature plunge to −2°F (−19°C) in town [−4°F (−20°C) in the countryside] and then rise to 38°F (3°C) by 9:00 a.m. January 2, 1842. "The consequence of this unsettled and unusual weather was soon visible . . . in the unhealthy state of the public health," wrote

---

17. "L'hyver de 1746 avoit été terrible par la longueur et par la violence du froid" (Gaultier 1748).

McCord in his notebook. "At this date, 9 Feb, the city is full of sickness [and] influenza" (McCord 1843).

The following winter "proved a most extraordinary one," according to McCord, who regretted that his appointment as district court judge had obliged him to suspend his personal observations:

> The winter set in suddenly on the 30 Novr and from then to the 6 Jan'y the falling of snow and continual cold was uninterrupted. At that period there could not have been less than between 4 and 5 feet (1.2–1.5 metres) of snow through the District of Montreal. The 6 Jan'y was a lovely day, but in the next the rain set in and in the course of a week the fields were bare everywhere. (McCord 1843)

On January 31, "we had a heavy gale of wind rain and mild weather, which washed all the District . . . of every vestige of snow" (McCord 1843). This alternating between snow-covered and muddy roads made travel difficult. McCord was making his rounds of the countryside as district judge during this period, switching between wagons and sleighs as snow fell and then thawed. Cold weather returned on February 5, 1842:

> The weather from the 5 February continued cold and in fact a new winter set in and continued cold throughout the month, and into the present month of March . . . This day, St Patrick's, we are again enclosed by a severe snow storm from the NE which is raging with great fury. The roads, almost blocked up by the storm of the 14, will now be almost impassible. (McCord 1843)

By April there was "an average of 4 feet [of snow] everywhere, and in some drifts 10 to 11 feet . . . The country from St Charles to the St Lawrence so covered with snow that the tops of the pickets alone are visible above the snow"(McCord 1843). McCord's account of that winter ends with snow squalls on April 9, 1843. The late 1850s saw a return of colder temperatures, culminating in 1859. The late 1860s to the 1890s then saw a number of colder winters. Smallwood spent considerable effort in writing up the effects of an incredible number of snowstorms during the winter of 1868–1869.

Smallwood and the Snows of 1869

A decade after the extremely cold temperatures of January 1859, the snowy season of 1868–1869 caught Smallwood's attention. He published several articles "on the heavy snows of the winter of 1868–69," both in Canada (Smallwood 1869a) and in the United Kingdom (Smallwood 1869b). "The unprecedented large amount of snowfall" was "by far the greatest amount from available records" (Smallwood 1869b, p. 387). Winter set in on December 7, 1868, with the ice road from Montreal across the Saint Lawrence good for crossing by December 25. October and November 1868 had received 23.20 inches (58.9 centimeters) of snow between them, while fourteen days of snow in December brought another 27.96 inches (71 centimeters). Fifteen days of snow in January 1869 added 28.09 inches (71.39 centimeters), while seventeen days of snow in February brought an incredible 73.76 inches (187.3 centimeters) for a total snowfall of 13.6 feet (4.1 meters) by the end of February. On February 3, it snowed from 7:00 a.m. to 9:00 p.m., bringing a total of 13.9 inches (35.3 centimeters) to the city. Even in snow-hardened Montreal, the "high wind carried loose snow with great violence into enormous drifts of 15–20 feet, rendering travel by rail entirely suspended and sleigh very difficult . . . snow shoes were generally resorted to" (p. 389).

A second storm in February lasted from 3:15 p.m. on February 14 until 2:15 p.m. on February 16, giving another 14.9 inches (37.8 centimeters) of snow to the area. A third storm just over a week later, on February 23, added yet another 11.15 inches (28.3 centimeters) of snow. At 4:00 a.m. on March 18, another storm hit, this time leaving 8.82 inches of snow. By this point the accumulated snowfall over the winter led to a thick blanket of snow and enormous drifts, which blocked the railways.

Railroad trains were snowed up for several days; passengers suffered from "hunger as well as delay," as the temperature dropped to 8°F (−13°C) (Smallwood 1869b, p. 390). Drifts formed "immense tunnels, or snow-banks on either side some 20 feet (6 m) high" which "no amount of power used by three or four heavy steam locomotion could force through" (p. 390).

The rescue engines themselves were stranded, and in the end "swarms of men" armed with shovels had to come to the rescue of stranded passengers and engineers alike (Smallwood 1869b, p. 390). Until that winter, the mean snowfall of the past twenty years, according to Smallwood's records, had been 79.50 inches (201.9 centimeters), while the highest monthly snowfall depth until that winter had been 45.74 inches (116.2 centimeters) in January 1866. By way of comparison, Smallwood mentioned that the "winter of 1745 [was] said to be very mild with only 24 inches (61 cm) of snow" throughout the entire winter (p. 391). The winter of 1826–1827 had no snow until January 17, 1827; the snow that fell from January 17 to 18, 1827, was "the heaviest fall of snow on record when 60 to 70 inches (about 150 to 175 cm) of snow fell" (Smallwood 1869a, p. 63).[18]

In the *Canadian Naturalist* for 1869 Smallwood (1869c, p. 117) noted that the "meteorological year" of 1868 (December 1867 to November 1868) had been unusually dry, despite the unusually high snowfall amounts in October and November, with only thirty-one days of rain throughout the year (compared to an average of 73 days; Smallwood 1858 p. 200). He noted that

> very few observations of a reliable kind on the rain and snowfall have been recorded for Montreal, but the few to which we have had access would give the mean annual amount of rain somewhat above 36 inches, or about double the quantity which fell during the past year (1868). This unusual dryness was also felt in Great Britain and on the continent of Europe. (Smallwood 1869c, p. 117)

Even as late as the 1860s, scientists were still struggling to get reliable records and calculate the averages for precipitation.

July of 1868 is singled out for its "extreme heat" and, in fact, July 1868 is among the warmest Julys of the two-and-a-half-century-long record (Smallwood 1869c p. 116). Once again, Smallwood compared his results to those obtained in previous years by McCord and Archibald Hall; he found that his annual-mean temperature for the year of 42.45°F (5.8°C) agreed "exactly with that observed by the late Mr. Justice McCord" but was 2°F lower than that obtained by "the late Dr. Hall" (Smalloood 1869c, p. 116). There is no explanation given for this discrepancy. Smallwood didn't consider it important to give the years used in either McCord's or Hall's calculations of the mean annual temperature. We see the idea of year-to-year or decade-to-decade changes in the climate that could affect the mean are no longer considered to be important by Smallwood. Climate was now considered constant, with short-term fluctuations but no longer-term changes, at least in historical time periods of centuries to millennia.

Although 1872 has one of the highest recorded temperatures for the summer half of the year and 1870 is one of the warmest years on record, in 1871 it was the cold snaps that Smallwood found remarkable. From January 22 to 27, 1871, the thermometer stayed continually below 0°F (−18°C), while

---

18. This is not quite accurate on Smallwood's part. According to newspaper clippings, snow had fallen on December 31, 1826, and light snow fell in early January (F. Ponari 2016, personal communication).

a second cold snap in February saw the thermometer dip down to −28°F (−33°C). December 1871 also saw an usual cold snap, with winter having set in early on November 29 "with unusual severity, and somewhat earlier than usual, causing severe losses to shipping" (Smallwood 1871, p. 338).

The seasons discussed here are, for the most part, those that struck the individual observers as unusual. The repeated cold spells during the 1810s, 1840s, and late 1860s and 1870s made it difficult for the nineteenth-century climatologists to have a general sense of whether the climate was warming, cooling, or staying about the same. This variability probably helped to dismiss the earlier, optimistic ideas of climatic improvement through anthropogenic land changes. After years of debate, the ups and downs of the nineteenth century saw the idea of a stable, unchanging climate, at least on the scale of historical time periods, become more established. Although there continued to be interest and discussion on short term, decadal scale climatic cycles by Eduard Bruckner (Stehr and von Storch 2000) and Jean Mascart (Mascart 1925), climatology in the twentieth century became more about gathering long enough time series to fully describe the statistics of this stable climate. The idea of short-term changes over decades to centuries, rather than gradual changes over millenia, was one that would not be fully articulated again until the 1960s.

References

Allan, R., and T. Anell, 2006: A new globally complete monthly historical gridded mean sea level pressure dataset (HadSLP2): 1850–2004. *J. Climate*, **19**, 5816–5842, https://doi.org/10.1175/JCLI3937.1.

Auer, I., and Coauthors, 2007: HISTALP—historical instrumental climatological surface time series of the Greater Alpine Region. *Int. J. Climatol.*, **27**, 17–46, https://doi.org/10.1002/joc.1377. [http://www.zamg.ac.at/histalp].

Baron, W., 1992: 1816 in perspective: The view from the northeastern United States. *Climate since A.D. 1500*, R. S. Bradley and P. D. Jones, Eds., Routledge, 124–144.

Bergström, H., and A. Moberg, 2002: Daily air temperature and pressure series for Uppsala (1722–1998). *Climatic Change*, **53**, 213–252, https://doi.org/10.1023/A:1014983229213.

Bethune, J., 1869: John Bethune's meteorological reports. McCord Family Fonds, Papers P001-835. McCord Museum Archives.

Brugnara, Y., and Coauthors, 2015: A collection of sub-daily pressure and temperature observations for the early instrumental period with a focus on the "year without a summer" 1816. *Climate Past*, **11**, 1027–1047, https://doi.org/10.5194/cp-11-1027-2015.

Chenoweth, M., 2009: Daily synoptic weather map analysis of the New England cold wave and snowstorms of 5 to 11 June 1816. *Historical Climate Variability and Impacts in North America*, L.-A. Dupigny-Giroux and C. J. Mock, Eds., Springer, 107–121.

Duhamel du Monceau, H. L., 1744: Observations Botanico-Météorologiques faites à Québec par M. Gautier, pendant l'année 1743. *Mém. Acad. Roy. Sci.*, **1744**, 135–155.

——, 1747: Observations botanico-météorologiques faites en Canada par M. Gaultier. *Mém. Acad. Roy. Sci.*, **1747**, 466–488.

——, 1750. Extraits des observations botanico-météorologiques faites à Québec pendant l'année 1749, par M Gautier, Médécin du Roi en Canada." *Mém. Acad. Roy. Sci.*, **1750**, 309–10.

Gaultier, J.-F., 1745 unpublished manuscript: Journal des observations météorologiques de M. Gaultier à Kebec. Observatoire de Paris Manuscript, Doc. 64-5-B, 103 pp. [Available from Fonds Joseph-Nicolas Delisle, Observatoire de Paris, 61 Avenue de l'observatoire, Paris 75014, France.]

——, 1748 unpublished manuscript: Journal des observations météorologiques de M. Gaultier à Kebec depuis le 1 octobre 1747 jusqu'au 1 octobre 1748. Observatoire de Paris Manuscript, Doc. 64-6-A, 71 pp. [Available from Fonds Joseph-Nicolas Delisle, Observatoire de Paris, 61 Avenue de l'Observatoire, Paris 75014, France.]

——, 1755 unpublished manuscript: Observations on Quebec, 1744–1754. Meteorological Collection, Houghton Library, MS Can 42(2). [Available from Houghton Library, Harvard University, Cambridge, MA 02138.]

Glackmeyer, L.-E., 1859: Meteorological observations taken at Beauport near Quebec. Meteorology Account R6707-0-5-E. Library and Archives Canada.

Greer, A., 1985: *Peasant, Lord, and Merchant: Rural Society in Three Quebec Parishes, 1740–1840*. University of Toronto Press, 304 pp.

Guevara-Murua, A., C. A. Williams, E. J. Hendy, A. C. Rust, and K. V. Cashman, 2014: Observations of a stratospheric aerosol veil from a tropical volcanic eruption in December 1808: Is this the Unknown ~1809 eruption? *Climate Past*, **10**, 1707–1722, https://doi.org/10.5194/cp-10-1707-2014.

Hamilton, K., 1986: Early Canadian weather observers and the "year without a summer." *Bull. Amer. Meteor. Soc.*, **67**, 524–532, https://doi.org/10.1175/1520-0477 (1986)067<0524:ECWOAT>2.0.CO;2.

Harrington, C. R., Ed., 1992: *The Year Without a Summer? World Climate in 1816*. Canadian Museum of Nature, 576 pp.

Houle, D., and J.-D. Moore, 2008: Les ponts de glace sur le fleuve Saint-Laurent: Un indicateur de la sévérité des hivers entre 1620 et 1910. *Nat. Can.*, **132**, 75–80.

Klingaman, W. K., and N. P. Klingaman, 2013: *The Year Without a Summer: 1816 and the Volcano that Darkened the World and Changed History*. St. Martin's Press, 338 pp.

Lambert, J., 1810: *Travels through Lower Canada and the United States in the Years 1806, 1807, and 1808*. Richard Philips, 526 pp.

Manley, G., 1953: The mean temperature of central England, 1698–1952. *Quart. J. Roy. Meteor. Soc.*, **79**, 242–261, https://doi.org/10.1002/qj.49707934006.

Mascart, J., 1925: *Notes sur la variabilité des climats*. Audet, 382 pp.

McCord, J. S., 1843 unpublished manuscript: Notebook on climate and meteorology of North America. McCord Family Fonds, Papers P001-828. McCord Museum Archives.

——, 1852 unpublished manuscript: Diary – March 28 1847–April 28 1852. McCord Family Fonds, Papers P001.B1, Item 0416. McCord Museum Archives.

——, 1859 unpublished manuscript: Diary – January 1 1859–December 31 1859. McCord Family Fonds, Papers P001.B1, Item 0410. McCord Museum Archives.

Parker, D. E., T. P. Legg, and C. K. Folland, 1992: A new daily central England temperature series, 1772–1991. *Int. J. Climatol.*, **12**, 317–342, https://doi.org/10.1002/joc.3370120402.

Skakel, W., 1852 unpublished manuscript: Meteorological register 1842–1852. Meteorological Records 1798–1972, RG 2, Container 939, item 1. Faculty of Arts and Sciences, Department of Meteorology, McGill University Archives.

Smallwood, C., 1855 unpublished manuscript: The observations of Dr. Smallwood 1849–1855. Meteorological Records, 1798–1972, RG 32, Container 956, Envelope 322. Faculty of Arts and Sciences, Department of Meteorology, McGill University Archives.

——, 1858: The meteorology of the vicinity of Montreal. *Proc. 11th Meeting of the American Association for the Advancement of Science*, Montreal, Canada, American Association for the Advancement of Science, 197–204.

——, 1862 unpublished manuscript: The observations of Dr. Smallwood 1856–1862. Meteorological Records, 1798–1972, RG 32, Container 955, Envelope 323. Faculty of Arts and Sciences, Department of Meteorology, McGill University Archives.

——, 1869a: On the great snow falls of 1869. *Can. Nat.*, **4**, 62–64.

——, 1869b: On the heavy snowstorms of the winter 1868–69 in the province of Quebec, Dominion of Canada. *Proceedings of the Meteorological Society*, Vol. 4, J. Glaisher, Ed., 387–391.

——, 1869c: Meteorological report of the year 1868. *Can. Nat.*, **4**, 115–118.

——, 1871: Meteorological results for Montreal for the year 1871. *Can. Nat.*, **6**, 334–339.

——, 1873 unpublished manuscript: Meteorological register 1868–1873. Meteorological Records, 1798–1972, RG 32, Container 956, Envelope 326. Faculty of Arts and Sciences, Department of Meteorology, McGill University Archives.

Spark, A., 1819 unpublished manuscript: Diary 1798–1819. Meteorological Records, 1798–1972, RG 2, Container 1039, Envelope 318–319. Faculty of Arts and Sciences, Department of Meteorology, McGill University Archives.

Stehr, N., and H. von Storch, Eds., 2000: Eduard Bruckner—The sources and consequences of climate change and climate variablity in historical times. Springer, 338 pp. doi.org/10.1007/978-94-015-9612-1.

Sutherland, W., 1848 unpublished manuscript: The observations of Dr. Sutherland. Meteorological Records, 1798–1972, RG 32, Container 1039, Envelope 321. Faculty of Arts and Sciences, Department of Meteorology, McGill University Archives.

Wetter, O., and C. Pfister, 2013: An underestimated record breaking event—Why summer 1540 was likely warmer than 2003. *Climate Past*, **9**, 41–56, https://doi.org/10.5194/cp-9-41-2013.

Biographical Sketches

The following are brief biographical sketches of many of the people mentioned in this book. It is meant to situate the people mentioned in the text. It is not a comprehensive overview of all meteorologists or climatologists of the period.

QUEBEC CITY WEATHER OBSERVERS

Jean-François Gaultier (1706–1756). Gaultier was born in Rouen, Normandy, in northern France and educated in medicine in Paris, where he came into contact with some of the most renowned botanists and natural philosophers of his day at the Jardin du Roi, where medical students were instructed in the therapeutic uses of plants. These connections, notably to the Jussieu brothers and Duhamel Dumonceau, helped him obtain the prestigious appointment of royal physician in Quebec City from 1742 to 1756. He was the first person to keep systematic instrumental meteorological observations in Canada. Detailed records were sent every year to his correspondents at the Académie Royale des Sciences. Gaultier's temperature observations, especially of the cold winter temperatures in Canada, were of crucial importance to the development of thermometers in the eighteenth century (Boivin 1974).

James Thompson (1733–1830). Thompson studied civil engineering at some point in his youth. He participated in the siege of Louisburg and the capture of Quebec in 1759. He supervised repairs and dealt with military and government construction for all of the British colony of Canada until 1825. During the 1770s and 1780s, he kept a diary in which he noted down the weather and how it affected his civil engineering works (Rioux 2003, Chapman and McCulloch 2010).

Alexander Spark (1762–1819). Spark was a Scottish clergyman who founded the first Presbyterian Church in Canada: Saint Andrew's in Quebec City. A product of the Scottish Enlightenment, Spark did much to promote education in Canada. He kept meteorological records from 1798 to 1819 (Campbell 1887; Lambert 1983, 1984).

Louis-Edouard Glackmeyer (1793–1881). Glackmeyer was a notary in the Quebec City region whose records cover the 1844 to 1859 period. His father had been a bandmaster from Germany who settled in Quebec in 1776. Glackmeyer was politically active and did much to reform the profession of notaries. He was interested in music and botany as well as in meteorology (Vachon 1982).

William Ward, Royal Engineers (unknown–1867). William Cuthbert Ward served in the Napoleonic Wars. He was promoted to lieutenant colonel in 1837, to colonel in 1851, and to major general in 1858. Ward was the senior engineering officer in Fort York (Toronto) during the construction of the Toronto Observatory and later was the senior officer in charge of meteorological observations in Quebec City (Royal Engineers 2016).

MONTREAL WEATHER OBSERVERS

Thomas McCord (1750–1824). McCord was a Protestant Irish business man and public office holder. Thomas McCord's father, John McCord, settled in Quebec in 1764 and engaged in trade. Thomas McCord became a businessman in 1770 and was a citizen of some importance by the 1790s. He left for Ireland in 1796 for business purposes on what was intended to be a brief visit, but political unrest and rebellion prolonged his stay until 1805 when he returned to Montreal as a general merchant. He was a member of the elected assembly, justice of the peace, and police magistrate and was a political force behind the establishment of a regular paid police force in 1818.

Starting in 1813, McCord worked with his young son John Samuel to record temperatures in Montreal (Senior 1987).

Alexander Skakel (1776–1846). A Scottish educator who taught at the Classical and Mathematical School and later at the Royal Grammar School, Alexander Skakel was one of Montreal's principal educators in the early nineteenth century. Skakel taught John Samuel McCord and Archibald Hall as well as many other citizens of Montreal in the early nineteenth century. He was a founding member of the Montreal Natural History Society. Skakel is likely to have started keeping meteorological observations with his brother William in the early 1820s. Fragments (probably copies) of the observations survive for 1842–1852 and 1862–1868 (Frost 1988).

William Skakel (unknown–1863). William was Alexander Skakel's brother, who is presumed to have continued Alexander's meteorological record after Alexander's death in 1846.

Robert Cleghorn (1778–1841). Cleghorn was a Scottish botanist who operated Montreal's first commercial nursery, Blink Bonny Gardens, north of the city. Fragments of his records, including daily temperatures, survive for the period 1829 to 1833 (McGuire 2010).

John Bethune (1791–1872). Bethune was an Anglican clergyman, the son of a Presbyterian minister and a refugee from the American Revolution. He was educated by Bishop Strachan in Upper Canada and was influenced by Strachan to become an Anglican, rather than a Presbyterian, clergyman: the first to be educated and ordained in Canada. Bethune was assigned to be a minister of Christ Church and rector of the Anglican parish in Montreal in 1818. He was the principal of McGill College from 1835 to 1846, a contentious period during which he came into conflict with other educational authorities in Montreal. Bethune succeeded nonetheless in starting the construction of the McGill University campus. He remained at Christ Church Anglican parish for the rest of his life, becoming dean in 1854. He kept a meteorological diary, noting the minimum and maximum temperature, barometric pressure, wind, and weather from 1838 to 1860 (Cooper 1972).

John Samuel McCord (1801–1865). One of the most prominent meteorologists of the first half of the nineteenth century, John Samuel McCord was considered by Smallwood to be the "Pioneer of Canadian Meteorology"

(Smallwood 1860, p. 309). As a child and young man, John Samuel contributed to and may even have been the instigator, under the influence of his teacher Alexander Skakel, of the meteorological journal kept with his father from 1813 to 1826. He studied law and was a member of the Montreal Militia. He was deeply involved in the Natural History Society of Montreal (NHS), acting as secretary in the 1830s and 1840s. McCord instigated the first bihourly weather observations kept by the military on Saint Helen's Island from 1839 to 1841. His involvement in the government's military action to quell the Rebellions of 1837–1838 may have led to his semi-exile from Montreal by means of his appointment as a district court judge, which involved considerable travel and significantly curtailed his scientific activities (Young 2014). In his quest to determine whether the climate had changed, and whether humans were responsible for any changes in climate, McCord collected and analysed numerous historical records of weather and climate. These were conserved among his scientific papers by his son, David Ross McCord, and are housed in the archives of the McCord Museum of Canadian History. McCord's own observations, along with those he collected from other observers, now form the core of the documents used in modern studies of the climatic history of the Saint Lawrence valley region.

Archibald Hall (1812–1868). A native Montrealer, Hall studied under Skakel as a child and as a young man became one of Skakel's natural philosophy demonstrators during Skakel's evening classes and public lectures for adults. Hall studied medicine at McGill before completing his medical degree in Edinburgh. He was appointed to the McGill Medical Faculty in 1835, where he remained for the rest of his life. He was also associated with the Montreal General Hospital. He had an interest in zoology and was awarded the Natural History Society's silver medal for his memoir on the mammals and birds of Montreal. He edited two medical journals, the *British American Journal of Medical and Physical Science* and later the *British American Journal*. Hall kept detailed meteorological records that he published in these journals, which he also sent to Joseph Henry as part of the Smithsonian volunteer weather observing network (Canada Medical Journal 1868; Bensley 1976).

Charles Smallwood (1812–1873). Born and educated in Birmingham, Charles Smallwood studied medicine in England at the University College of London, graduated with an MD and emigrated to Canada in 1833. In 1834, he settled in Saint Martin on Ile Jésus in the Saint Lawrence River, just to the north of Montreal (present-day Laval), where he "acquired one of the largest medical practices in the Country" (Smallwood 1861). He built his own observatory where he kept not only detailed meteorological records but also observations on terrestrial magnetism, atmospheric electricity, and ozone. Smallwood developed various self-recording instruments. He received some government support for his observatory, and in 1862 moved the observatory to McGill University, where he was appointed honorary professor of meteorology. Astronomical observations for timekeeping also made the McGill Observatory the principal centre in Canada for the determination of time and longitude. Smallwood was also a member of the Natural History Society of Montreal and the American Association for the Advancement of Science (Marshall 1972; Marshall and Bignell 1969).

William Sutherland (Sunderland) (1815–1875). Sutherland graduated from McGill Medical School in 1836 and spent several years on the Niagara frontier before returning to Montreal in the early 1840s. Sutherland participated in a bilingual medical school, giving lectures once in French and once in English as well as operating a free dispensary for poorer patients. Sutherland also edited a medical journal with Francis Badgely. Sutherland was appointed to the McGill Medical Faculty in 1849. William Sutherland's meteorological and personal diary for 1844–1848 is conserved in the McGill University Archives (Canada Medical Record 1875; Canada Medical and Surgical Journal 1875; Roland and Bernier 2000).

Clement Henry McLeod (ca. 1850–1917). One of the first engineering graduates of McGill in 1874, as an undergraduate McLeod had been allowed to live in the main college building so as to be able to take readings at the McGill Observatory. After Smallwood's death in November of 1873, McLeod was given charge of the observatory. He trained briefly under Kingston in Toronto. McLeod was eventually appointed supervisor of the observatory as well as professor of the Department of Civil Engineering and vice dean of the Faculty of Science at McGill. He and his students undertook important observations in the telegraphic determination of longitude in 1883, which resulted in more accurate measurements for the entire continent (Bignell 1962).

HUDSON BAY COMPANY OBSERVERS
There were many others not mentioned in this book.

William Wales (1734–1798). Wales was a mathematician, astronomer, and scientific explorer who was sent with Joseph Dymond to Churchill, on the western shore of Hudson Bay, to observe the 1769 transit of Venus (Carlyle and Howse 2004; Wulf 2012). He spent the winter of 1768–1769 at Fort Churchill preparing for the transit and kept a meteorological journal during his stay.

Thomas Hutchins (unknown–1790). Hutchins was a surgeon first employed by the HBC at York Factory in 1766. He worked together with Andrew Graham, master of the post of Fort Severn, Ontario, inspired by William Wales' year in York Factory in 1768–69. Hutchins was interested in meteorology, performing careful and detailed experiments on the congelation of mercury, which won him the Royal Society's prestigious Copley medal in 1783, and keeping meteorological and magnetical observations (Houston et al. 2003; Williams 2003; Binnema 2014).

Samuel Hearne (1745–1792). Born in London, Hearne joined the Royal Navy at the age of eleven as servant to a captain during the Seven Years' War. During the winter of 1768–1769 he learned astronomical and navigational skills from William Wales, posted to the Prince of Wales Fort by the Royal Society to observe the Transit of Venus in June 1769. Hearne retired to London in 1787, and encouraged by Wales and Lapérouse, published his travel journals and notes of his explorations in the northwestern continental interior, including scientific and ethnographic details of interest to scientists and explorers (Houston et al. 2003; Mackinnon 2003; Binnema 2014).

John Siveright (1779–1856).[1] Siveright was born in Scotland and, at the age of 19, entered the fur-trade as an apprentice clerk with the Montreal-based new North-West Company. After an incidence of violence between the North-West Company, the Hudson's Bay Company, and the Red River settlement in 1816, Siveright was posted to Sault Sainte Marie (which he labelled in his journals as Saint Mary's Falls) from 1821 to 1823 and then to Fort Coulonge on the Ottawa River from 1823 to 1832. After the coalition between the North-West Company and the Hudson's Bay Company, Siveright was retained as a clerk by the Hudson's Bay Company. Following the directives that had been put in place in the 1810s, Siveright kept meticulous weather records, copies of which were retained in McCord's scientific papers. He was put in charge of the Timiskaming district in 1843, was briefly a chief factor in Montreal, and retired to Scotland in 1849 (Galbraith 1985; Arthur 1985).

---

1. Thanks to James Gordon at the Hudson's Bay Company Archives, Archives of Manitoba.

KINGSTON WEATHER OBSERVER

James Williamson (1806–1895). Born and educated in Edinburgh, where he studied mathematics and astronomy, Williamson was recruited to be professor of mathematics and natural philosophy at Queen's College, Kingston, in 1841. He lectured in a wide range of subjects, including natural history and chemistry. With the help of local fundraising, an Observatory was built at Kingston, which Williamson directed for the following three decades. As well as astronomical observations, he determined the latitude and longitude of Kingston, and collected meteorological observations. The Kingston observatory became a part of the Meteorological Service of Canada in 1876 (Jarrell 2003).

TORONTO WEATHER OBSERVERS (ROYAL ARTILLERY)

John Henry Lefroy (1817–1890). Lefroy's military education was at the Royal Military Academy in Woolwich, where he met Charles Younghusband. Although initially hoping to be accepted into the Corps of Engineers, he was posted instead to the Royal Artillery, with further study in astronomy at the Royal Engineer's school. In 1839, Lefroy was selected by Sabine to go to Saint Helena in the South Atlantic to take magnetic observations. He was trained in magnetic observations by Humphrey Lloyd in Dublin. He remained at Saint Helena until 1842, when he was transferred to the Toronto Observatory after Charles Riddell was granted medical leave and resigned the post of director. Before taking up his duties in Toronto, Lefroy spent nearly two years on a voyage of scientific discovery in the Canadian Northwest, which established his reputation as a geoscientist. At Fort Chipewyan in Alberta he made hourly magnetic observations, with his companion Bombardier William Henry, for over four months, with readings taken every two minutes during episodes of high magnetic activity. These were later described by Humphrey as the most important contributions to the study of the geomagnetism ever made. Upon his return to Toronto late in 1844, he took up the directorship of the observatory. He was elected to the Royal Society in 1848 and, as a founding member, acted as both vice president and later president of the Canadian Institute. He returned to England in 1853, where he continued his scientific work as a part of his military career. He was appointed governor of Bermuda and was also closely involved in the Crimean War. He returned to Canada for the 1884 meeting of the British Association for the Advancement of Science (BAAS), where he served as president of the geographical section (Vetch and Stearn 2004; Whitfield and Jarrell 1982; Lefroy 1895; Toronto Public Library 1967, p. 95).

Charles J. Buchanan Riddell (1817–1903). An officer and geologist, Riddell entered the Royal Military Academy at Woolwich and graduated into the Royal Artillery in 1834. His first tour of duty was in Quebec from 1835 to 1837. He was chosen to serve as superintendent of the Canadian Magnetic and Meteorological Observatory under Sabine and was the acting officer in charge of establishing the observatory. It was Riddell, acting under Bayfield's geological advice, who made the decision to relocate the observatory to Toronto. He was invalided back to England in 1841, where he was given the position of assistant superintendent of the military magnetic observatories. In this post, he continued to help Sabine in the collation and analysis of magnetic observations. He was elected to the Royal Society in 1842 (James and Baigent 2004; Toronto Public Library 1967, p. 95).

Charles Younghusband (1821–1899). Described by Lefroy as "the youngest and smallest officer in the service" (Lefroy 1895, p. 19), Charles Younghusband followed his father into the Royal Artillery. Younghusband unexpectedly became acting director of the Toronto Observatory at the age of twenty when Riddell was sent home on medical leave and Lefroy, the appointed director, was on a two-year scientific expedition of the Canadian Northwest. He was later attached to Sabine's office and became a member of the Royal Society in 1852.

TORONTO OBSERVATORY (CIVILIAN)

George Kingston (1816–1886). Kingston was born in Portugal, the son of an English wine merchant, and was sent to school in England. He entered the Royal Navy and was awarded the gold medal for mathematics at the Naval College. He left the navy at the age of twenty-six and enrolled at Cambridge, where he earned an MA in mathematics. He emigrated to Canada where he became the head of the Naval College in Quebec City and in 1855 took the positon of professor of the Department of Meteorology at King's College, York (later the University of Toronto), and director of the Toronto Observatory. The latter had passed into civil control of the provincial government of Upper Canada (Ontario) when the Royal Artillery withdrew from the observatory in 1854. With the cooperation of Egerton Ryerson, Kingston instituted a network of observers at twelve grammar schools across Ontario. Kingston and other meteorologists campaigned for the formation of a Canadian national meteorological service. They succeeded in having the Meteorological Service of Canada (MSC) inaugurated in 1871 under the Department of the Marine and Fisheries for the purposes of issuing storm warnings. Thereafter, Kingston took an active role in developing Canada's official meteorological service, corresponding with volunteer observers, distributing instruments, and collecting observations nationwide (Thomas 1982; Toronto Public Library 1967, p. 95).

OTHERS (CANADA)

Jacques Cartier (1491–1557). Cartier made three voyages to North America in 1534, 1535, and 1541. In 1534, he sailed around the Gulf of Saint Lawrence, which he mapped. In 1535, he discovered the Saint Lawrence River and sailed up it; the river was so long no one among the Native Americans Cartier encountered in the east had seen its source. Cartier concluded that it represented the fabled passage to the Orient most European explorers of the time were seeking. Cartier and his expedition arrived at the island of Montreal in October 1535, which was at the time the site of the village of Hochelaga. He returned to Quebec where the expedition experienced a severe winter; twenty-five men died of scurvy before a cure of white spruce bark was communicated to them by the Iroquois. Cartier documented the physical and social geography of the Saint Lawrence valley. He returned in 1541 when he and Jean-François Roberval were to set up a colony near the site of present-day Quebec City; this attempt failed and Cartier returned to France in 1542 (Trudel 2014a).

Samuel de Champlain (ca. 1570–1635). Champlain was the French explorer and geographer who, in 1608, founded Quebec City, the first colony in New France. He was an excellent draughtsman who produced early maps of New France. His journals contain descriptions of the landscape and climate and were referred to in nineteenth-century documents about climate change. Considered the founding father of New France, there is extensive literature about Champlain and his founding of Canada (Trudel 2014b).

Michel Sarrazin (1659–1734). Sarrazin first arrived in New France as a naval surgeon in 1685. In 1686, he was appointed surgeon major of the colonial troops. After an illness, he returned to France in 1694 and spent three years studying to become a physician in Paris, where he studied at the Jardin du

Roi under Tournefort. Sarrazin returned to New France in 1697 and from this time forward became a natural philosopher as well as a doctor, dissecting animals and collecting botanical plants and seeds that he forwarded to the Jardin du Roi; his work was incorporated in Tournefort's textbook on botany. Sarrazin's medical competence was held in high esteem, and some of his works on illnesses were known to La Galissonière. Sarrazin was Tournefort's academic correspondent and, after Tournefort's death, that of Réaumur. Sarrazin was also in contact with the botanists at the Jardin des Plantes in Paris, including Vaillant and the Jussieus as well as the president of the Académie Royale des Sciences, the Abbé Bignon. Vaillant prepared a "catalogue of the plants of Canada" from Sarrazin's work, which, however, was never published (Rousseau 2014).

Roland-Michel de La Galissonière (1693–1756). La Galissonière was educated in Paris and joined the navy as a midshipman in 1710. He often sailed in ships supplying France's American colonies. In 1747, he was entrusted with the role of commandant general in New France while the appointed governor general, La Jonquière, was being held prisoner by the British. While La Galissonière would have preferred to remain in the navy, it was deemed necessary for him to take command of New France because of the war with England; he was later commended for his devotion to duty. La Galissonière threw himself into the defence and development of the colony but was at the same time a reluctant governor and in 1749 was granted permission to leave. While in New France he had set up a system of collecting specimens at the posts and forts throughout the continent, which were forwarded to Gaultier in Quebec City and from there to the Jardin du Roi in Paris, a system described in detail by Kalm. After 1750, La Galissonière was in a position to further develop his interest in science, and he corresponded with Duhamel Dumonceau, Gaultier's contact with the Académie Royale des Sciences, and with Bernard de Jussieu at the Jardin du Roi (Taillemite 1974).

James McGill (1744–1813). Born in Glasgow, McGill attended the University of Glasgow and emigrated to Canada sometime before 1766. He came from a family of traders and set up as a trader in the North American interior, eventually becoming a partner in the North-West Company. He settled in Montreal in 1775. He was a loyalist during the American Revolution and was appointed Justice of the Peace in 1776. He built a trading empire based on fur and supplying military posts. His summer home in Montreal was a property on the southern slope of Mount Royal named "Burnside"; this later became the land on which the eponymous university McGill endowed in his will was built. The current departments of Geography, Atmospheric and Oceanic Sciences and Mathematics and Statistics of McGill University are located in the 12-storey Burnside Hall on the site of McGill's original summer house (Cooper 2003).

Alexander MacKenzie (1764–1820). A legendary figure of Canadian exploration, MacKenzie was born in Scotland, and emigrated to New York with his father as a child of ten. The American Revolution broke out soon after, and MacKenzie's father, commissioned in the British Army, died in 1780. Alexander was sent to school and safety in Montreal in 1778. A year later, he joined the fur trade and became a partner in the North-West Company. In 1789, MacKenzie was sent as head of an exploring party to investigate the river now named after him, with its source in Great Slave lake; it was thought the river would go through Alaska to the Pacific, but instead the MacKenzie river flows for over one thousand kilometers north to the Arctic Ocean. In 1793, MacKenzie and his party went west, descended the Peace river, crossed the Rockies following Amerindian paths, and reached Bella Coola. Two days later they reached the Pacific, and completed the first crossing of the North American continent north of Mexico. His journeys to the Arctic and Pacific oceans added enormous, previously unknown, landmasses to European maps. MacKenzie was knighted in 1802 (Lamb 2003).

James Somerville (1775–1837). Born in Scotland, Somerville studied at the University of Glasgow, receiving his degree in arts in 1792 and in divinity in 1799. He was offered a position teaching the children of Scottish merchants in Quebec City, where he was befriended by Alexander Spark. He organised a school which was later overseen by Daniel Wilkie, after Somerville was invited to become the minister of the St. Gabriel Street Church in Montreal in 1802. He was ordained by John Bethune (the father of the John Bethune discussed in chapter 5) in 1803. With a life-long interest in science and learning, especially geology and meteorology, Somerville was one of the leading forces behind the establishment of the Montreal Natural History Society in 1827 (Campbell 1887; McDougall 2003).

Daniel Wilkie (1777–1851). Wilkie was a Presbyterian clergyman who studied at the University of Glasgow. He relished theological controversy but supported the moderate faction of the Church of Scotland that joined Christian belief to an Enlightenment education. He emigrated to Quebec as a teacher in 1803 and soon opened his own school. He offered courses in geography as well in grammar and mathematics; he also provided evening classes for adults. Wilkie supported general education. In 1810, he opened the Classical and Mathematical School, which attracted the sons of the French Canadian and British elites as well as those of military officers. His school suffered from financial instability and lack of discipline, though his reputation remained solid. Although offered a position as a clergyman or teacher both in the Presbyterian and Anglican denominations, Wilkie refused them and stayed in Quebec City, where he assisted Alexander Spark at Saint Andrew's Presbyterian Church. Wilkie was the owner and editor of the bilingual Quebec newspaper *The Star/L'étoile*, in which he aimed to provide nonpartisan and full accounts of political debates and court decisions (Lambert 1985).

William Logan (1798–1875). One of the giants of Canadian science, Logan was born in Montreal and was a pupil of Alexander Skakel until the age of sixteen, when he was sent to Edinburgh to complete his education. He attended Edinburgh University for a year as a medical student. He was elected to the Geological Society of London in 1837. On a visit to North America in 1840, he met the British geologist Charles Lyell, who introduced him in turn to John William Dawson. In 1841, the Natural History Society of Montreal and the Literary and Historical Society of Quebec petitioned for a geological survey of Canada; Logan applied and received the position of provincial geologist. Logan, with exceptional energy and skill, devoted decades to surveying Canada, founding the Geological Survey of Canada. His extraordinary contributions include the identification of evidence of glaciation, the finding of animal fossils in Cambrian rocks, and the organization of the Canadian contribution to the Great Exhibit in London of 1851, when he was also elected as a member of the Royal Society and to the Paris Exposition of 1855, where he was presented the cross of the Légion d'honneur. His *Geology of Canada* remains a basic reference (Campbell 1887; Winder 2003, Zeller 2009).

John William Dawson (1820–1899). Dawson was born in Pictou, Nova Scotia. His parents were Scottish immigrants. Dawson's religious convictions played an important role in his life as a scientist and educator. While educated in the classics and science at the Pictou Academy, he studied geology and natural history on his own. He attended the University of Edinburgh in the 1840s, and in 1842 he guided Charles Lyell on a tour of the geology of the Albion coal mines near Pictou. He trained in the United Kingdom as an exploration geologist, the first in British North America. He was elected a fellow of the Geological Society of London in 1854. He was unexpectedly offered the positon of principal of McGill and did much to build up the university from the dilapidated state it had fallen into and turn it into a thriving institution. He was appointed a member of the Royal Society in 1862 (Sheets-Pyenson 1996; Eakins and Eakins 1990).

BRITISH ARMY OFFICERS IN CANADA

John Colborne, Commander-in-Chief (1778–1863). Colborne was an army officer who fought in the Napoleonic Wars in Spain and Portugal and was commander of the Coldstream Guards during the Battle of Waterloo. He was governor of Upper Canada (Ontario) from 1828 to 1836, at a time when it was still very much a frontier society. He was appointed commander in chief of Canada in 1836 and was the military leader of Canada during the 1837 and 1838 Rebellions (Wilson 1976; Seymour 2004).

George Landmann, Royal Engineers (1780–1854). As Landmann's father was a professor of artillery at the Royal Military Academy at Woolwich in the United Kingdom, Landmann was raised and educated in a military milieu. He was commissioned as a second lieutenant in the Royal Engineers at the age of 15 and promoted to first lieutenant at seventeen, when he was posted to the Canadas. While in Quebec, Landmann received a package containing cowpox vaccine, which he used to vaccinate the children of a fellow officer in the first smallpox vaccination in Canada. His later career encompassed the Peninsular War in Spain (Tunis 2003).

Richard Bonnycastle, Royal Engineers (1791–1847). Bonnycastle was the son of the mathematics professor at the Royal Academy of Woolwich; his family was described as being "exceptional in that his middle-class status came from his father's intellectual achievements in a military setting. The Royal Military Academy was the first tertiary school in the English-speaking world to furnish advanced engineering and scientific courses to prospective officers of the artillery and the engineers. Academy graduates dealt not only with heavy weapons and fortifications but also with roads, harbours, and canals, with cartographical, meteorological, and geological observations, and with drawing and painting because of the need to sketch for military purposes" (Raudzens 1988). Bonnycastle and Colonel John By, who was in charge of building the Rideau Canal and the town at its northern terminus, which later became Canada's capital Ottawa, "are the outstanding examples in Upper Canada of constructive imperial military officers" (Raudzens 1988).

Bonnycastle contributed much to both the civil society and the military defence of the Canadian colony. His building of Fort Henry and his defence of Kingston prevented an invasion from the United States during the Rebellions of 1837–1838. He was among those who petitioned Colborne for the establishment of an observatory in Canada to help develop the sciences in the colony. Bonnycastle wrote extensively on the state of the Canadas and Newfoundland, especially their physical geography, during his tour of duty as Commanding Engineer in these colonies. He and his family settled in Kingston (Raudzens 1988; Chichester and Lunt 2004).

William Kelly, Royal Navy (dates unknown). A naval surgeon who joined Bayfield in 1829 and remained with the surveying expedition for the next twenty years, Kelly was also a keen naturalist who took meteorological and astronomical observations with Bayfield. He corresponded with McCord on the climate of Lower Canada and published his findings, along with his analysis of the observations kept at the Quebec Citadel by Mr. Watt, in the *Transactions of the Literary and Historical Society* (McKenzie 1976).

Henry Bayfield, Royal Navy (1795–1885). Born in northern England, Bayfield joined the navy when he was eleven years old. He saw action in the Napoleonic Wars and on Lake Champlain during the War of 1812. From 1816 onward, he served primarily as a naval surveyor and, during his lifetime, accomplished the charting and coastal survey not only of the Great Lakes and western Saint Lawrence River, but also the vast area of the Gulf of Saint Lawrence and the Strait of Belle Isle. The thousands of islands and inlets along the Great Lakes made the hydrographical survey a time-consuming task, often carried out under difficult circumstances.

His work as a naval surveyor gave Bayfield great insight into the geological formations and physical geography of the Saint Lawrence valley as well as experience of the weather in these coastal regions, as he often camped out in rough conditions. The surveying team consisted of Bayfield, Bayfield's assistant Midshipman Collins, and later William Kelly. They carried an array of instruments with them, including barometers and chronometers for estimating altitude and longitude. It was upon Bayfield's recommendation that the magnetic observatory was moved from Saint Helen's Island near Montreal to Toronto to avoid the natural magnetism of the underlying rocks of the Saint Lawrence valley. Bayfield spent his winters in Quebec City where he prepared his maps and participated in the meetings of the Literary and

Historical Society of Quebec (LHSQ), publishing his geological findings in the LHSQ's *Transactions*. As astronomical observations formed a large part of his surveying work, Bayfield was also a fellow of the Royal Astronomical Society. Bayfield spent the rest of his life in the Canadian colonies (British North America), moving to Charlottetown when his surveying work brought him to the Gulf of Saint Lawrence (McKenzie 1976, 1982).

FRENCH SCIENTISTS

Blaise Pascal (1623–1662). The son of a local administrator from Clermont, Pascal was a mathematical prodigy, teaching himself from the works of Euclid and developing theories of conic sections as a teenager. He invented a calculating machine able to perform arithmetic mechanically. With his brother-in-law Florin Périer, Pascal tested Torricelli's theories on the weight of the atmosphere by measuring the height of the mercury column in a barometer with increasing altitude as Périer ascended the Puy de Dôme in 1646. This was described as the first verification of a theory in a controlled scientific experiment by the collection of instrumental measurements. He proved the existence of a vacuum above the mercury contained in a barometer; until then it was thought to be impossible for a vacuum to exist. This paved the way for the development of the air pump. Pascal, in letters and discussions with contemporary mathematician Pierre Fermat, cofounded the field of probability. He is described by Napier Shaw as "one of the best writers and profoundest thinkers France has ever produced" (Napier Shaw and Austin 1932, p. 119; Larousse 1994, p. 442; Oxford University Press 1999; Attali 2000).

Jean-Dominique (Giovanni Domenico) Cassini (1625–1712). Cassini was appointed professor of astronomy in Bologna at the age of twenty-five, where he calculated the rotational periods of Jupiter and Mars. He also calculated the movements of the moons of Jupiter, which allowed another scientist, Ole Romer, to estimate the speed of light. Cassini was recruited by Louis XIV's minister Colbert in 1669 to establish an astronomical observatory in Paris and became a French citizen in 1673. Among his discoveries were the moons of Saturn and the gap in the rings of Saturn. His work on the parallax of Mars allowed him to estimate the size of the solar system, and he also undertook fundamental observations on the size and shape of Earth. His calculations disagreed with the theoretical estimates of Newton, and the quest to measure the size and determine the shape of Earth would become a recurring theme in French science over the next century. The 1997 spacecraft sent to observe Saturn and Jupiter was named the *Cassini* space probe in recognition of Jean-Dominique Cassini's pioneering observations of these planets and of the solar system in general (Larousse 1994, p. 133; Oxford University Press 1999; Hall 1983; Hahn 1971).

Philippe de La Hire (1640–1718). La Hire was the son of Laurent de La Hire, a professor of the Royal Academy of Painting and Sculpture and a portrait painter of some renown. Philippe de La Hire's early training was as an artist, and he travelled to Italy in his youth. There he became interested in geometry and perspective, and soon became more interested in mathematics than painting. He returned to Paris in 1664 and worked on conic sections. He was elected to the Académie Royale des Sciences in 1678 and was assigned to work with Jean Picard on surveying. Their aim was to produce more accurate maps, a task that would occupy the scientists at the Paris Observatory for the next century. His work in cartography linked La Hire to the observatory and to Cassini and Maraldi. La Hire also worked on the project to supply water to the gardens of Versailles, then under construction, and so began the first systematic recording of precipitation in Paris. La Hire worked in many areas of the sciences, including magnetism, falling bodies, astronomy, and optics. He developed instruments, including meteorological instruments such as thermometers and barometers at the observatory. Every year he published a summary of the monthly rainfall totals and other measurements such as the extremes of pressure and temperature. His son, Gabriel Philippe de La Hire (1677–1719), worked with his father at the observatory and also published meteorological reports (Bobis and Lequeux 2012; O'Connor and Roberts 2008).

Joseph Pitton de Tournefort (1656–1708). Professor of botany at the Jardin du Roi and member of the academy, Tournefort did much to establish the science of botany, developing the concept of the genus and investigating the uses of plants in medicine. He wrote several treatises on the plants of the Parisian region and travelled to the Levant (Middle East) on a voyage of scientific exploration from which he brought back a number of botanical specimens. His textbook, *Elements of Botany*, described his system of plant classification which later inspired Linnaeus. He classified thousands of botanical specimens. He was killed in a carriage accident while transporting plants in Paris; the street where this accident took place is now named after him (Larousse 1994, p. 547).

Giacomo Filippo Maraldi (1665–1729). The nephew of Jean-Dominique Cassini, Maraldi was an important astronomer, mathematician, and geoscientist working at the Paris Observatory. He worked on a catalogue of fixed stars, studied Mars extensively, and thus discovered that the Martian ice caps were not aligned with the rotational poles. He also realised that the corona visible during an eclipse belonged to the sun and not the moon. He was one of the geoscientists who measured the arc of the meridian between Dunkirk and Perpignan, which helped determine the size and shape of Earth. He kept meteorological observations and wrote extensively on the distribution of atmospheric pressure around the world (Larousse 1994, p. 389; Oxford University Press 1999).

Sébastien Vaillant (1669–1722). A botanist and student of Tournefort, Vaillant became the director of the Jardin du Roi and established the first hothouse in France. A member of the Académie Royale des Sciences, he worked on a botanical catalogue of plants that was published after his death (Larousse 1994, p. 555).

Jacques Cassini (1677–1756). The son of Jean-Dominique, Jacques Cassini carried on his father's work. He is most remembered for his work in cartography and surveying, including the determination of the length of the arc of a meridian between Dunkirk and Perpignan with his cousin Giacomo Filippo Maraldi. Appointed as a student astronomer to the Académie Royale des Sciences in 1692, he was also elected to the Royal Society in 1696, corresponding with Isaac Newton and Edmond Halley (Larousse 1994, p. 133; Editors of Encyclopædia Britannica 2017; Hahn 1971).

Antoine Jussieu (1686–1758). Jussieu studied at Montpellier, renowned for its reputation in medicine, and became a physician. He travelled through Spain, Portugal, and southern France collecting plants, which he brought with him to Paris. He practised medicine in Paris and became a member of the Académie Royale des Science in 1715. He took over Tournefort's position as professor in the Jardin des Plantes in Paris upon the latter's unexpected death. His most notable work was on the uses of plants, particularly in medicine (Larousse 1994, p. 322).

René Antoine Ferchault de Réaumur (1683–1757). A scientific polymath, Réaumur made significant contributions in a number of fields, including metallurgy and meteorology, although he is best remembered today for founding the field of entomology. His work on thermometers developed not only a thermometer with consistent units and a calibration method that made measurements from different instruments intercomparable but was also known for specifying the physical conditions for the placement of the thermometer to measure air temperature and limit the influence of direct and reflected solar radiation (Larousse 1994, p. 478).

Joseph-Nicholas Delisle (1688–1768). Delisle was a professor of mathematics at the Collège de France in Paris and a naval geographer. He was elected to the Académie Royale des Sciences in 1714 as a student of Maraldi's. In 1725, he travelled to Saint Petersburg to establish an astronomical observatory and stayed for 25 years. He developed the thermometer and temperature scale used by Gaultier, and many of Gaultier's original letters can be found with his scientific papers. He organised the international efforts to observe the transit of Venus to help determine the distance of the sun from Earth, constructing a map that showed where the best locations for observation would be. He used his numerous contacts and his membership in the Royal Society to ensure international scientific cooperation despite the international tensions of the Seven Years' War (Larousse 1994, p. 182; Wulf 2012).

Bernard Jussieu (1699–1777). Brother to Antoine, Bernard also studied medicine at Montpellier and became a member of the academy. He founded a method of classification based on embryonic characteristics. He became a second demonstrator at the Jardin du Roi upon Sébastien Vaillant's promotion to first demonstrator. He designed a botanical garden at Versailles (Larousse 1994, p. 322).

Henri-Louis Duhamel Dumonceau (1700–1782). Duhamel Dumonceau initially studied law in Paris, but while in Paris became interested in botany and the work of the Jussieu brothers at the Jardin du Roi. He wrote a treatise on saffron, which brought him to the notice of the Académie Royale des Sciences and Maurepas, minister of the navy. Duhamel Dumonceau worked on naval problems for Maurepas and with Buffon prepared a report on the strength of different types of wood. He remained interested in practical botany and forestry for the rest of his life. He was the first to be appointed inspector of the navy. He established a school for naval architecture and engineering, the École de la Marine. Duhamel Dumonceau was Gaultier's principal correspondent at the academy and submitted Gaultier's letters and weather diaries for publication, along with similar weather observations he himself kept at his estate in France (Payne 2002).

Joseph Jussieu (1704–1779). The youngest Jussieu brother, Joseph formed part of the 1739 French expedition to Peru to measure the length of the meridian at the equator and determine the shape of Earth. He remained in South America for 35 years, sending specimens to his brothers in Paris, including quinine for the treatment of malaria (Larousse 1994, p. 322; Ferreiro 2011).

Georges Louis Leclerc, comte de Buffon (1707–1788). A native of Burgundy, Buffon studied with Jesuits in Dijon, with a particular interest in mathematics, before obtaining his degree in law and travelling in France and Europe. He was nominated to the Académie des Sciences in 1733 as a physicist. The pressing need for wood as a raw material led Buffon to be appointed to work on arboriculture, where he met and worked with Duhamel Dumonceau; the experience was not happy. Buffon turned towards botany, and was appointed intendant of the Jardin du Roi in 1739, a post that had been expected to go to the more experienced Dumonceau. Buffon, however, made good use of his position and spent the rest of his life writing a 36-volume opus on Natural History. He was an opponent of Linnaeus' classification theory, believing instead in a slow transformation of species, and thus an early evolutionist. Buffon was passionately concerned with the communication of science, and his works had an immense influence on the development of natural science in the eighteenth century (Larousse 1999; Laissus 2007; Hoquat 2007).

César François Cassini (1714–1785). Son of Jacques Cassini, César François Cassini also became, in turn, director of the Paris Observatory. His scientific contributions were in the area of geodesy and cartography; under his direction, the complete survey of France was undertaken, and his map of France was the most detailed produced at that time. He verified the length of the meridian in 1744 (Larousse 1994, p. 133).

Louis de Bougainville (1729–1811). Bougainville showed an early aptitude for mathematics and in his twenties published a treatise on integral calculus. Bougainville was elected to the Royal Society of London in 1756 and to membership in the Académie Royale des Sciences in 1789. He served as secretary to Gaston-Pierre-Charles de Lévis-Mirepoix, ambassador extraordinaire, who travelled to London in 1754 in an attempt to find a diplomatic solution to the skirmishes between the English and French in the Ohio valley that eventually precipitated the Seven Years' War. Bougainville entered the military at the relatively late (for the time) age of 21; in 1756, he was named aide-de-camp to Louis-Joseph de Montcalm when the latter was given command of the French regulars in Canada. After fighting in numerous battles in North America, Bougainville was dispatched to France in 1758 to report on the situation in the colony. He returned to Canada in 1759 and commanded troops in the defence of Quebec City during the summer of 1759. After the fall of Quebec, Bougainville spent the following winter and summer of 1760 fighting in defence of the colony and was taken prisoner and sent to France after the surrender of Montreal. In 1766, he left on an expedition of exploration and discovery around the world, exploring South America, sailing through the Strait of Magellan, and stopping in Tahiti. His descriptions of the South Pacific, and especially Tahiti, created a sensation and developed the idea of the "noble savage" and state of innocence versus civilisation, which would be later expounded upon by Rousseau. He continued his naval career until 1792. He was imprisoned during the Reign of Terror but was eventually released and in 1795 became a member of the Institut de France, which in some measure replaced the Académie Royale des Sciences. He was appointed to the Bureau des Longitudes in 1799 (Taillemite 1983).

Antoine Laurent de Lavoisier (1743–1794). Lavoisier was a keen student of Bernard Jussieu. At 23 he won a prize from the Académie Royale des sciences on a treatise on urban lighting. He was appointed a member of the Académie at 25. He revolutionised chemistry by recognizing that mass is conserved in chemical reactions, and by disproving the phlogiston theory. His pioneering work on combustion and the nature of heat and, in conjunction with British scientist Joseph Priestly, on the discovery of oxygen, set the foundation for modern chemistry. He was executed during the Terror after the French revolution (Larousse 1994; Oxford University Press 2016).

Jean-Dominique Cassini (1748–1845). The last of the Cassinis to be director of the Paris Observatory, Jean-Dominque Cassini completed the work of his father on the map of France. He was also involved in the ongoing attempt to accurately measure the longitudinal difference between the Paris and London Observatories. He was imprisoned during the French Revolution and much of his work in cartography was taken over by the new republic in the Bureau des Longitudes (Larousse 1994, p. 133).

Aimé Bonpland (1773–1858). Born in La Rochelle, Bonpland originally trained as a naval surgeon, before embarking with Alexander von Humboldt on a years-long voyage to the Americas, chiefly South America. On his return to France he was the intendant of Malmaison, the imperial residence of Empress Josephine during the Napoleonic era (Larousse 1999).

Louis Pasteur (1822–1895). A biologist and chemist, Pasteur studied at the École Normale Supérieure and was appointed to the Faculty of Science in 1854. "Undoubtedly the most important medical scientist working in the 19th century" (Oxford University Press 2016), his most important contribution was the development of germ theory. In his studies of microorganisms, he definitively disproved the theory of spontaneous generation. He discovered the source of anthrax and developed a vaccination against it. He fought for the recognition of the role of microorganisms in causing disease and the importance of sterilization in the prevention of bacterial infections (Larousse 1994, p. 443).

Jean Mascart (1872–1935). Mascart was a French astronomer and mathematician with an interest in the history of science and in the mathematical analysis of climatic data. He was the director of the Lyon observatory.

BRITISH SCIENTISTS AND PHILOSOPHERS

Isaac Newton (1642–1727). Newton is widely regarded as one of the, if not the, most important figures of the scientific revolution. "It is hardly necessary to say that Newton's work supplied all the sciences with mathematical principles and calculus" (Napier Shaw and Austin 1932, p. 121). His work on motion and forces, and, in particular, his development of mathematical notations to describe these motions and forces, laid the foundation for all subsequent work in physics. In the atmospheric sciences, he contributed work on optics, work on temperature scales, and laws of cooling, fluids, and sound transmission (Westfall 2004; Napier Shaw and Austin 1932, p. 121).

William Derham (1657–1735). Derham was a clergyman as well as a natural historian and member of the Royal Society. As rector of Upminster in Essex, he kept detailed meteorological records, which he published in the *Philosophical Transactions of the Royal Society* starting in 1697. He was in correspondence with others across Europe who also kept daily observations and used the records they sent or published, including those published by the scientists of the Royal Observatory in Paris, to compare weather conditions across Europe. He was especially interested in atmospheric pressure and its variations. The most notable of his weather analyses was his work describing and identifying the successive waves of cold during the "Great Frost" winter of 1708, which devastated much of western Europe. He also used measurements of air pressure with a barometer to infer elevation. He delivered the Boyle lectures of 1711 and 1712, which were later published as his popular work on natural theology *Physico-Theology*. He was a close friend of John Ray and knew many of the other scientists of this time, including Isaac Newton and Edmond Halley. He was appointed chaplain to the Prince of Wales in 1716, after which he spent part of his time in Windsor, moving his instruments there. He also became interested in astronomy during his later years (Smolenaars 2004).

David Hume (1711–1776). A Scottish philosopher, Hume (Robertson 2004) is one of the best remembered figures of the Scottish Enlightenment. He wished to develop a new basis for all knowledge and human understanding based on the "science of man" (Robertson 2004), and proposed a philosophy of radical scepticism. He wrote extensively on moral and political philosophy. He upheld the idea of climatic determinism, which linked national personality characteristics to climate.

Adam Smith (1723–1790). A philosopher associated with the mid-eighteenth century Scottish Enlightenment. His most influential works have been in the fields of morality and economy.

James Hutton (1726–1797). Hutton, described as the "first British geologist" (Napier Shaw and Austin 1932, p. 125) devoted his life to natural history and was the Scottish scientist who first propounded the idea of "uniformitarianism" in his influential work *Theory of the Earth*. He held that the surface of Earth is being continually eroded, the sediments compacted and uplifted through continual action, so that Earth's surface was continually being recycled. This could only take place over long periods of time; Hutton was among the first to realise the immense periods of time that geological processes implied: "we find no vestige of a beginning—no prospect of an end" (Repcheck 2003, p. ix) Hutton also kept meteorological observations; one of his most well-known early scientific publications was the *Theory of Rain*. According to Napier Shaw, he is "said to have given us the wet-and-dry-bulb thermometers" (Napier Shaw and Austin 1932, p. 125).

Daines Barrington (ca. 1728–1800). Daines Barrington was the fourth son of Viscount Barrington and trained as a lawyer and legal scholar. However, his main occupation was as an antiquarian and natural historian. He was elected a fellow of the Royal Society. One of his notable contributions to science was the publication of *The Naturalist's Journal*, a set of printed forms consisting of one week per page, with vertical columns for the methodical recording of weather, wind, plants in flower, and other natural phenomena. Barrington sent a copy of the *Journal* to Gilbert White and encouraged White's observations and the eventual publication of White's *The Natural History of Selborne*, which was proven to be one of Charles Darwin's early inspirations. Barrington was also an enthusiastic supporter of Arctic voyages and of the search for the Northwest Passage to the Orient by way of the Arctic Ocean north of Canada. He mobilised the Royal Society's support of voyages of discovery, including Captain James Cook's third voyage to the Pacific. His varied interests and enthusiasms did not always lead to meticulous research, and there were some doubts as to his accuracy and credibility in his antiquarian work. His combination of classical scholarship and natural history, however, led him to combine the two in his examination of ancient literature for evidence of climatic change (Miller 2004).

Edward Jenner (1749–1823). Edward Jenner was a surgeon who invented the vaccination process against smallpox. An earlier and riskier method of preventing smallpox by introducing matter from smallpox pustules into the skin of healthy people in the hopes of giving them a milder, protective form of the disease had been introduced into the United Kingdom by Lady Mary Wortley Montague. Jenner noticed that people infected with cowpox rarely caught smallpox and in the 1790s, after receiving his medical degree from Saint Andrew's in Scotland, started his vaccination trials (Baxby 2004).

John Dalton (1766–1844). Dalton's most significant contribution to science is his theory of the atom. A Quaker, Dalton's family was not well off and he entered into domestic service at the age of ten. Given his aptitude for scholarship, he was encouraged to become a schoolmaster; when this failed, he worked as an agricultural labourer. He eventually returned to teaching and was encouraged to keep a meteorological diary, which he did until the day of his death: a neighbour claimed that he could set his watch by the moment when Dalton opened his window to read his thermometer. Dalton's 1793 publication of *Meteorological Reflections and Essays* concerned not only his observations but also his theories on air, water, vapour, and rain, which led to his theory of the atom. His meteorological interests also led to a quantitative investigation into the hydrological cycle, and it is from this interest in rain and water vapour that he developed his theories on the dynamics of fluids and gases, today known as "Dalton's Law." His major breakthrough in the theory of the atom was to recognise that gases combined in fixed ratios, which led to the idea that each element has a characteristic atomic weight. It was through these quantitative methods of measuring the weights of gases that individual elements could be identified. Compounds, including compound gases such as carbon dioxide, were a combination of different elements in strict proportions. Dalton's ideas lead to a revolution in atomic chemistry. Although he completed an "astonishing amount of scientific work" (Napier Shaw and Austin 1932, p. 132) in his lifetime, "the subject he had always cared about more than any other" (Greenaway 2004) remained meteorology.

Luke Howard (1772–1864). A Quaker chemist, Howard is remembered today as one of the founders of the science of meteorology and climatology. His two great contributions were the naming and classification of clouds and his long-standing observations on urban climatology published in *The Climate of London*. His work on a simple classification schema for clouds led to a better understanding of their formation and role in precipitation. Howard was elected a member of the Royal Society in 1821 and was a founding member of the first Royal Meteorological Society. His personal meteorological observations and work on *The Climate of London* are the first known publications in English of the urban heat island effect, demonstrating that cities have a noticeably warmer climate than the surrounding countryside. He dedicated a later edition of *The Climate of London* to John Dalton (Burton 2004a; Hamblyn 2001; Napier Shaw and Austin 1932, p. 133).

John Lambert (1775–ca. 1811). Lambert travelled to Lower Canada in 1806 with his uncle James Campbell to investigate the potential for growing and producing hemp in the colony, a vital wartime resource for shipping. He produced a memoir with observations and statistics on the colony, accompanied by illustrations based on his watercolor sketches (Baigent 2004a).

John Frederic Daniell (1790–1845). Born in London and educated at King's College London, Daniell was perhaps one of the best-known meteorologists of his day. He was appointed a fellow of the Royal Society at twenty-three and developed meteorological instruments, including his eponymous hygrometer, which was in use around the world for much of the nineteenth century. His hygrometer enabled better understanding of the role of humidity and water vapour in atmospheric processes. He published an influential collection of meteorological essays in 1823 (Napier Shaw and Austin 1932 p. 138).

John Herschel (1792–1871). Herschel was widely acclaimed as one of the most eminent British scientists of the nineteenth century. Son of Court Astronomer William Herschel (the discoverer of Uranus), John Herschel was an accomplished mathematician who worked to bring continental mathematical ideas to Britain. He and his father, along with his aunt Caroline Herschel, made significant discoveries in stellar astronomy as well as planetary astronomy, cataloguing nebulae and developing theories of stellar evolution. John Herschel completed his father's work of cataloguing nebulae in the Northern Hemisphere and spent from 1833 to 1838 in South Africa cataloguing the stars as seen from the Southern Hemisphere. While he was in South Africa, he wrote a guide to making meteorological observations. He also issued a call for hourly observations to be recorded around the globe on four specific days a year: the equinoxes and solstices. Both his guide to observations and his call for worldwide hourly observations were widely circulated among meteorologists. His observations of double stars and the calculations of their orbits provided an important confirmation of the universality of Newton's laws of gravity. He emerged in the 1820s as "Britain's first modern physical scientist" (Crowe 2004), and his writings on the scientific method influenced a generation of Victorian scientists, including Charles Darwin, Michael Faraday, and John Samuel McCord. Herschel played a founding role in the establishment of the Royal Astronomical Society, serving three times as its president and was twice awarded the society's gold medal. He was also a member of the Royal Society and of the British Association for the Advancement of Sciences (Crowe 2004; Napier Shaw and Austin 1932, p. 139).

Charles Daubeny (1795–1867). An Oxford professor with an interest in chemistry, geology, and botany, Daubeny was a founding member of the British Association for the Advancement of Science. Although his main interest was in chemistry, he also published on climate and reported on the progress of meteorology in North America after his visit in 1838.

Charles Lyell (1797–1875). Lyell was an influential geologist whose main theory was that the processes in action today could explain all of the geological past: there was no need to invoke past catastrophes to explain the origin of features such as mountains. His work, along with that of Hutton, invoked vast stretches of time and changed thinking about the age of Earth. He explained climate change as a consequence of changes in physical geography (Rudwick 2004).

Charles Darwin (1809-1882). Possibly the most famous nineteenth-century scientist today, Darwin developed the theory of evolution by natural selection. Many of the foundational observations for his theory came from his five-year voyage to South America as the gentleman companion of the captain of the HMS *Beagle*, Robert FitzRoy (Desmond et al. 2004).

James Croll (1821-1890). A Scottish geologist who was largely self-educated, Croll's work on applying mathematical principles to geology brought quantitative methodology to the earth sciences. The work for which he is most remembered is his book *Climate and Time*, which laid the foundation for the astronomical theory of glaciation (Kushner 2004).

Robert Henry Scott (1833-1916). Scott studied experimental physics and engineering in Dublin, moving to Berlin after completing his studies. There he worked under Heinrich Wilhelm Dove, one of the most eminent meteorologists and climatologists of the nineteenth century. He translated Dove's *Laws of Storms*, dedicating it to Robert FitzRoy. After FitzRoy's death in 1865, Scott was appointed director of the Met Office by a committee appointed by the Royal Society. It is likely that his close friendship with Edward Sabine, then president of the Royal Society, helped secure Scott's position. Scott was also secretary of the International Meteorological Committee from 1873 to 1900 and worked tirelessly to establish the international cooperation so vital in meteorology. It was through Scott that George Kingston received many of the meteorological instruments that furnished the observing posts throughout Canada (Burton 2004b; Anderson 2005).

BRITISH ARMY OFFICERS

Edward Sabine, Royal Artillery (1788-1883). Edward Sabine came from a long line of military officers and was enrolled in the Royal Military Academy of Woolwich at the age of fourteen. He served in the Napoleonic Wars and in North America, at Quebec and on the Niagara Peninsula, during the War of 1812. After the wars, he became involved in scientific explorations, notably on Ross's 1818 expedition to find the Northwest Passage and later on Parry's 1819-1820 Arctic expedition. Sabine became especially interested in magnetic and meteorological phenomena while on these voyages. He dedicated most of his long life as a soldier scientist to establishing observatories and analysing data from these two fields. Sabine was also interested in geodesy, the measurement of the exact shape of Earth, and participated with John Herschel in one of many efforts to determine the difference in longitude between the Paris and Greenwich Observatories. Sabine was elected a member of the Royal Society in 1818 and served as its president from 1861 to 1871. He also participated in the BAAS, serving as general secretary for twenty years and president in 1852. During the 1830s, Sabine worked to build a coalition between the Royal Society, the BAAS, the military, and parliament to support his magnetic crusade, a project to which he would devote the next thirty years. Sabine worked with the Royal Engineers, and later with the Army Medical Department, to establish regular meteorological observations in all their foreign and colonial stations. His work in analysing the vast amounts of data generated by the observatories led to pioneering research in statistical methods to discover periodic phenomena in the earth sciences. For example, he discovered that the eleven-year solar sunspot cycle was also reflected in geomagnetic observations (Good 2004; Napier Shaw and Austin 1932, p. 137).

William Reid, Royal Engineers (1791–1858). Reid was the son of a Church of Scotland minister. He was educated at the Edinburgh Academy and entered into the Royal Military Academy in 1806, where he learned surveying. He served in the Peninsular War, where he survived being wounded three times. In 1814 and 1815, he was part of the British Army sent to North America during the War of 1812. After the peace of 1816, he returned to Woolwich as an adjutant of the Royal Sappers and Miners, later a part of the Royal Engineers. He was serving in Barbados in the early 1830s when he witnessed hurricanes and became interested in the origins of storms, then a matter of controversy in the United States. Reid was in contact with William Redfield, one of the protagonists in the controversy, with whom he collaborated and to whom he sent his observations of storms in the British West Indies. He was governor of Bermuda from 1839 to 1846. Reid presented Redfield's ideas to the BAAS meeting in 1838, where Herschel gave them his support. Reid was elected a fellow of the Royal Society in 1838 on the strength of his work on storms. While developing his theory of storms, he followed Herschel's philosophy of developing laws from gathering data rather than by hypothesizing without collating facts. He prepared a set of sailing rules for ships to avoid the worst effects of cyclones and hurricanes when caught on the open ocean. Reid was elected vice president of the Royal Society in

1849. He also served as a commanding royal engineer in Woolwich (Baigent 2004b; Napier Shaw and Austin 1932, p. 297).

Robert FitzRoy, Royal Navy (1805–1865). FitzRoy entered the Royal Naval College at Portsmouth at twelve years old and was the first to graduate with a perfect score in his exams. While on a surveying mission in South America in 1828, FitzRoy was appointed interim captain of the HMS *Beagle*, a position that was confirmed upon returning to the United Kingdom. FitzRoy was an able captain and surveyor and was appointed captain of a second survey of South America. On this second voyage, Charles Darwin accompanied FitzRoy as his guest and as a gentleman naturalist. After an unsuccessful stint as governor of New Zealand, FitzRoy was appointed chief of a new department to collect and collate meteorological and oceanographic information for the Board of Trade, with the aim of reducing shipping losses. As both a man of science and an experienced sailor, FitzRoy instituted regular observations of the weather, wind, and ocean temperature at sea. From this information, he produced detailed charts of wind direction distribution for different sectors of the oceans as well as publications intended for general use such as his *Barometer and Weather Guide* (FitzRoy 1858). He instituted storm warning signals at sea ports, which were deduced from synoptic analyses of weather observations, and published the first weather forecasts in newspapers. FitzRoy inaugurated several aspects of modern meteorology from marine instruments to wind and weather charts. He is considered the founder of the United Kingdom's Met Office (Anderson 2005; Moore 2015; Barometer World 2016; Napier Shaw and Austin 1932, p. 149).

SWEDISH SCIENTISTS

Anders Celsius (1701–1744). A native of Sweden, Celsius was appointed professor of astronomy to Uppsala University in 1730. In 1737, he participated in the voyage organised by the French Académie des Sciences to Lapland to measure the shape of the earth. Celsius investigated magnetic phenomena and aurora. In 1742, he devised a thermometer scale with "100" as the freezing point and "0" as the boiling point of water. This scale, reversed, has been adopted as the metric centigrade thermometer scale (Larousse 1994).

Carl Linnaeus (1707–1778). A pastor's son from Sweden, Linnaeus was a physician and botanist who discovered more than 100 new species in his travels, first around Lapland and later in Europe, including France where he met the Jussieu brothers. He pioneered the field of taxonomy and developed the system of cataloguing all living things with two names, a generic (genus) name and a specific (species) name still used in taxonomy today. Royal physician and botanist, he founded the Stockholm Academy and spent the rest of his life as a professor in Uppsala. His extraordinary work in collecting and cataloguing botanical specimens made him one of the most respected and influential natural philosophers of the eighteenth century (Oxford University Press 2016; Hoquat 2007).

Pehr (Petter) Kalm (1716–1779). A student of Linnaeus and of Anders Celsius, Pehr Kalm was the son of a Finnish clergyman and a Scottish mother, Catherine Ross. Although from a poor background, he attended the University of Åbo. He developed an interest in practical botany applied to agriculture, and attended the University of Uppsala in 1740. Linnaeus was interested in the potential of native North American plants to expand the pool of potential crops which could be grown in the northern climate of Sweden. In 1747, he chose Kalm for a voyage to North America. Kalm's travel journal contained detailed descriptions of his time in Canada, its physical geography, and social customs. He spent several weeks with Jean-François Gaultier collecting specimens (Jarrell 1979).

AMERICAN SCIENTISTS

Hugh Williamson (1735–1819). Williamson (Elliott 1979, p. 275–276) was an eighteenth-century polymath and one of the signatories of the Declaration of Independence. Born in Pennsylvania, he studied theology in Connecticut and medicine in Edinburgh and Utrecht. While in Europe he became friends with Benjamin Franklin, then living in London, and conducted experiments in electricity with Franklin. After some time spent as a preacher, he became a professor of mathematics at the College of Philadelphia. He started a medical practice after obtaining his medical degree and later became a member of legislature and then of Congress. His most influential scientific work was his *Observations on the Climate in Different Parts of America*, published in 1811.

James Espy (1785–1860). Espy started his career as a school teacher and then as a school principal, including a period of teaching mathematics and classics at the Franklin Institute in Philadelphia. In 1836, he gave up teaching to concentrate on meteorology. Espy was a meteorologist connected at various times with the U.S. War Department, the U.S. Navy, and the Smithsonian.

Espy investigated the role of heat in the development and maintenance of storms. Although his main thesis was incorrect, his work on the thermodynamics of convection was an important contribution to meteorological theory. He also worked to develop meteorological networks in the United States and promoted the use of the telegraph in weather forecasting (Elliott 1979, 88–89; Fleming 1990).

William Redfield (1789–1857). Redfield was never formally educated. After working as a saddlemaker, he became involved in railroad design and engineering. In 1821, he noticed a centripetal pattern in the direction of fallen trees after the passage of a storm (a hurricane) in western Massachusetts and developed a theory of the movement of storms as a progressive whirlwind. This insight won him considerable acclaim, including that of mariners. Redfield collected over ten years of observations on West Indies hurricanes and mariners' reports, which he used to construct synoptic maps showing the progression of storms (Elliott 1979, p. 215; Napier Shaw and Austin 1932, p. 296).

Joseph Henry (1797–1878). Born in Albany, Henry came from an impoverished background and worked his way through college. He attended the Albany Academy, studying medicine before switching to engineering in 1825. He was appointed professor of mathematics and physics at Albany and later professor of natural philosophy at Princeton. He discovered the principal of electromagnetic induction, independently of Michael Faraday, although Faraday was the first to publish and receive credit for the discovery. His invention of the electrical relay to overcome resistance was critical for the development of the telegraph. In 1846, he was appointed secretary of the Smithsonian Institution and as such developed a large network of weather observers. When the meteorological network was paired with the telegraph, Henry was able to map daily weather conditions (Oxford University Press 2016; Larousse 1994).

John Disturnell (1801–1877). Disturnell was a prolific travel writer whose works include statistical information about the geography of the places he visited (Williams-Mystic 2017).

Alexander Bache (1806–1867). A great-grandson of Benjamin Franklin, Bache was educated at the Military College of West Point and became a military engineer. During his career, he was a professor of natural history and chemistry at the University of Pennsylvania, the president of Girard

College, and superintendent of the U.S. Coastal Survey. He was interested in geophysics and terrestrial magnetism and established the first magnetic observatory in the United States in 1840. Bache made astronomical, marine, and magnetic observations an integral part of the Coastal Survey's work and applied the technology of the telegraph to the determination of longitude (Elliott 1979, p. 19–20; Fleming 1990).

Louis Agassiz (1807–1873). Agassiz was a Swiss natural philosopher with an early interest in zoology and fossils and was known as the foremost expert in fossil fish of his time. He is chiefly remembered today for his pioneering work in establishing the existence of the Great Ice Ages of the past, with continental ice sheets covering much of North American and parts of Europe. He showed the role ice and glaciers played in erosion and geology. Agassiz was appointed professor of zoology and geology at Harvard in 1847 and settled permanently in the United States. His insistence on first-hand investigation and field work changed the way much of natural history was studied. Agassiz was also notable for accepting evolution but rejecting Darwin's theory of natural selection (Oxford University Press 2016).

Arnold Guyot (1807–1884). A Swiss geoscientist, Guyot studied in Neuchâtel, Strasbourg, and Berlin. He taught in Paris in the late 1830s, was a professor of history and physical geography in Neuchâtel, and emigrated to the United States in 1848. He was appointed to the chair of geology and physical geography at Princeton in 1854. Working with the Smithsonian Institution, he organised meteorological stations in the eastern United States (Oxford University Press 2016).

Elias Loomis (1811–1889). Loomis studied at Yale and in Paris and became a professor of mathematics and physics in Ohio, later moving to New York and then returning to Yale as a professor in 1860. Many of his scientific works on a wide range of subjects, including meteorology and storms, came to be considered classics in their fields (Napier Shaw and Austin 1932, p. 143).

Lorin Blodget (1823–1901). Blodget became a volunteer contributor to Joseph Henry's Smithsonian network at the age of twenty and was hired in 1851 as a calculator to compile weather and climate statistics for the Smithsonian and for the Army Medical Department, which sent meteorological reports to the Smithsonian. The project did not go well, and, after he was dismissed, Blodget used his results to publish a highly influential work on the climatology of North America, without crediting the Smithsonian for the data collection or his employment. Blodget also worked for the Pacific Railroad Survey, determining altitudes by measuring atmospheric pressure. He was a member of the American Philosophical Society and the American Association for the Advancement of Science (Elliott 1979, p. 32–33; Fleming 1990, p. 110–111).

Edward Norton Lorenz (1917–2008). Born in Connecticut, Lorenz studied mathematics at Dartmouth College, Harvard, and the Massachusetts Institute of Technology (M.I.T). During the Second World War, he applied his knowledge of mathematics to weather forecasting in the United States Army Air Corps. After the War, he held the position of meteorologist at M.I.T, later becoming professor of meteorology there. He was among the pioneers in using computers to model the mathematical equations governing atmospheric dynamics. In 1961, he discovered the computer output changed dramatically when slightly different values were used as starting points (known as sensitivity to initial conditions). This discovery led in turn to the realization that the nonlinear equations governing the weather give rise to a system in which similar conditions recur repeatedly, but not in a predictable way. Through this Lorenz discovered the application of chaotic theory, earlier discovered by Henri Poincaré in 1880s, to weather; his discovery set off a scientific revolution (O'Connor and Robertson 2008; Oxford University Press, 2016).

OTHER SCIENTISTS MENTIONED IN THIS BOOK

Evangelista Torricelli (1608–1647). Torricelli was educated in Rome and, at Galileo's invitation, moved to Florence. After Galileo's death in 1642 he was appointed the grand ducal mathematician and professor of mathematics in Florence. He discovered that the reason why artesian wells could never be raised more than about 32 feet was that the atmosphere had weight and exerted pressure on the water. The water could only rise until it was counteracted by the weight of the atmosphere. Repeating the experiment with mercury instead of water, Torricelli noticed that the level of mercury varied over time and invented the barometer in 1644. He also worked on the development of a thermometer (Oxford University Press 2016; Larousse 1994, p. 546; Napier Shaw and Austin 1932, p. 119).

Gabriel Fahrenheit (1686–1736). Born in Danzig, Fahrenheit went to Amsterdam as a young man, where he quickly found himself immersed in making physical apparatus. He moved between Holland and England, becoming skilled in glass-blowing. In 1709, he began to make thermometers which could be standardised, and thus intercompared. Before this level of technical skill was accomplished, there was no way of ensuring that the amount of liquid contraction or expansion which represented a degree was the same from one instrument to another, so that the readings of different instruments could not be directly compared to one another. He set as his zero point the coldest possible freezing temperature obtained with a mixture of salt and ice. In 1715, he replaced the alcohol in his eponymous thermometers with mercury, giving the thermometer named after him its definitive form (Larousse 1994; Oxford University Press 2016).

Theodore Augustin Mann (1735–1809). A Catholic priest and monk from England, Mann was the son of a land surveyor from Yorkshire who was taught scientific methods and principles by his father. In his early life, he devoted himself to many fields of study, including physics and theology. He was appointed to the Royal Academy of Brussels and to the Royal Society, and was nearly appointed bishop of Quebec after Canada became part of the British Empire. His work included compiling early meteorological records for the region of the Netherlands. Mann was later appointed to organise the Palatine Society of Mannheim, the first organised attempt to take simultaneous international meteorological observations. His immense output of scientific, historical, educational, and popular works gave him a lasting reputation in the Netherlands. His treatises on climate change influenced McCord (Arblaster 2004).

Alexander von Humboldt (1769–1859). Born to a noble Prussian family, Alexander von Humboldt and his brother Willheim received a thorough education by tutors at home before completing their studies at the University of Göttingen. Alexander was interested in geology and soon after graduation was appointed director of mines in Franconia. He travelled to Paris where he made the acquaintance of Aimé Bonpland; they determined to go on a voyage of scientific discovery and were given extremely rare permission by the Spanish authorities to visit the Spanish colonies in America, which were closed to visitors at that time. During his voyage to South America he made extensive geophysical observations. He pioneered the use of graphical methods to summarise scientific observations, including the use of isothermal lines to reveal patterns in the geographic distribution of temperature. Upon his return, he spent many years in Paris, the scientific capital of the world. He was one of the last scientists to have comprehensive knowledge of many fields of science, from anatomy to magnetism, and was arguably the most famous and influential scientist of the nineteenth century. He inspired a generation of scientists, including Herschel, Sabine, and McCord with his integrated view of the world in his masterpiece, multivolume work *Cosmos* (Larousse 1994, p. 307; Wulf 2015; Napier Shaw and Austin 1932, p. 133).

Christophe Buys-Ballot (1817–1890). The son of a Dutch minster, Buys-Ballot was a professor of mathematics at the University of Utrecht. He organised the meteorological observation system of the Netherlands and founded the Netherlands Meteorological Institute in 1854, one of the earliest national meteorological services. He also, as director of the Netherlands Meteorological Institute, developed the first forecasting service and storm warning system in Europe in 1860. He developed the law relating wind direction to pressure distribution taught to every meteorologist, that the wind direction is always oriented with the low pressure to the left in the Northern Hemisphere (right in Southern Hemisphere), and that the closer together the isobars, the stronger the force of the wind (Oxford University Press 2016; Larousse 1994, p. 124; Napier Shaw and Austin 1932, p. 303).

Heinrich Wilhelm Dove (1803–1879). The German professor Dove was one of the most prominent meteorologists of the mid-nineteenth century. After an education in natural philosophy at Breslau and Berlin, he was appointed professor of physics in Königsberg and Berlin. In 1849, he was appointed the director of the Prussian Meteorological Institute, founded in 1847. Dove was considered by Humboldt to be the founder of contemporary meteorology. Dove's work emphasised the Humboldtian view of climate and its relationship to biogeography, and his outlook was statistical rather than the more thermodynamical atmospheric physics. His compilations of observations made him the "founder of the entire superstructure of accurate climatological knowledge" Napier Shaw and Austin 1932, p. 290). Dove also published a theory of storms based on the conflict between tropical and polar currents, and based on his collection of historical meteorological data published monthly maps of isotherms from 1729 onward (Bernhardt 2004; Napier Shaw and Austin 1932).

Eduard Bruckner (1863–1927). Born in Jena, Bruckner studied in Dorpat (now Estonia), Dresden, and Munich, was appointed professor of geography at Bern, and taught in Halle before moving to Vienna. He promoted the view of climate as dynamic rather than static on historical time scales, developing the idea of a 35-year periodicity in climate, and investigated the impact of climatic fluctuations on human societies (Stehr and von Storch, 2000).

INSTRUMENT-MAKERS

John Newman (ca. 1783–1860). Newman was a highly reputed instrument-maker, whose clients included John Ross (for his polar expedition), Joseph Henry of the Smithsonian Institution, Humphry Davy, and Isembard Kingdom Brunel. In his earlier career, he supplied equipment and acted as laboratory assistant to the Royal Institution, where chemist Humphry Davy often gave demonstrations. He made standard barometers for many observatories, including one for the Royal Society in 1822 that became a well-known reference instrument for calibrating other barometers. His surviving instruments testify to the high quality of his work and support his well-deserved reputation as the finest instrument-maker of his day. Following his retirement, his premises were taken over by Negretti and Zambra (Dawes 2004).

Alexander Adie (1775–1858) and son **John Adie (1805–1857).** Alexander Adie was a well-known instrument-maker in Edinburgh, with a special interest in meteorology. The Adies were well patronised by professors from the University of Edinburgh and by the Stevenson family of engineers and writers, including Robert Stevenson, engineer, and his son Thomas Stevenson, inventor of the Stevenson screen, and father of the writer Robert Louis Stevenson. Alexander Adie kept a meteorological register with instruments of his own construction, and he designed and patented a marine barometer that was both sensitive to changes in air pressure and of sturdy construction (Morrison-Low 2004a, b; Locher 2007).

Enrico Angelo Ludovico Negretti (1818–1879) and **Joseph Warren Zambra (1822–1887).** "The world's most famous instrument makers," Negretti and Zambra of London were the foremost meteorological instrument-makers of the second half of the nineteenth century. Their maximum thermometers were particularly prized as having solved the difficulty of getting the marker of the maximum temperature to stay at the maximum point reached by the mercury column by placing the marker inside the thermometer tube. They also pioneered the use of enamel on the back of the thermometer, which enabled a thinner thread of mercury to be used, making the thermometer more sensitive (Collins, 2016).

References

Anderson, K., 2005: *Predicting the Weather: Victorians and the Science of Meteorology*. University of Chicago Press, 331 pp.

Arblaster, P., 2004: Mann, Theodore Augustine (1735–1809). *Oxford Dictionary of National Biography*. Oxford University Press, accessed 01/29/2016, http://dx.doi.org/10.1093/ref:odnb/17948.

Arthur, E., 1985: Sieveright, John. *Dictionary of Canadian Biography*. Vol. 8. University of Toronto/Université Laval, accessed 01/29/2016, http://www.biographi.ca/en/bio/sieveright_john_8E.html.

Attali, J., 2000: *Blaise Pascal, ou, Le génie français*. Fayard, 538 pp.

Baigent, E., 2004a: Lambert, John (b. c.1775, d. in or after 1811). *Oxford Dictionary of National Biography*. Oxford University Press, accessed 01/27/2016, http://dx.doi.org/10.1093/ref:odnb/15940.

——, 2004b: Reid, Sir William (1791–1858). *Oxford Dictionary of National Biography*. Oxford University Press, accessed 01/27/2016, http://dx.doi.org/10.1093/ref:odnb/23345.

Barometer World, 2016: Admiral FitzRoy. Accessed 02/29/2016, http://www.barometerworld.co.uk/admiral-fitzroy/index.htm.

Baxby, D., 2004: Jenner, Edward (1749–1823). *Oxford Dictionary of National Biography*. Oxford University Press, accessed 01/27/2016, http://dx.doi.org/10.1093/ref:odnb/14749.

Bensley, E. H., 1976: Hall, Archibald. *Dictionary of Canadian Biography*. Vol. 9. University of Toronto/Université Laval, accessed 02/21/2018, http://www.biographi.ca/en/bio/hall_archibald_9E.html.

Bernhardt, K.-H., 2004: Heinrich Wilhelm Dove's position in the history of meteorology of the 19th century. Accessed 01/28/2016, http://www.meteohistory.org/2004polling_preprints/docs/abstracts/bernhardt_abstract.pdf.

Bignell, N., 1962: Official time signal: 100 years. *McGill News* (summer), 16–22.

Binnema, T., 2014: *Enlightened Zeal: The Hudson's Bay Company and Scientific Networks, 1670–1870*. University of Toronto Press, 459 pp.

Bobis, L., and J. Lequeux, 2012: *L'Observatoire de Paris 250 Ans de Science*. Gallimard Observatoire de Paris.

Boivin, B., 1974: Jean-Francois Gaultier. *Dictionary of Canadian Biography*. Vol. 3. University of Toronto /Université Laval, accessed 02/21/2018, http://www.biographi.ca/en/bio/gaultier_jean_francois_3E.html.

Burton, J., 2004a: Howard, Luke (1772–1864). *Oxford Dictionary of National Biography*. Oxford University Press, accessed 01/27/2016, http://dx.doi.org/10.1093/ref:odnb/13928.

——, 2004b: Scott, Robert Henry (1833–1916). *Oxford Dictionary of National Biography*. Oxford University Press, accessed 01/27/2016, http://dx.doi.org/10.1093/ref:odnb/37944.

Campbell, R., 1887: *A History of the Scotch Presbyterian Church, St. Gabriel Street, Montreal*. W. Drysdale, 807 pp.

Canada Medical and Surgical Journal, 1875: The Late William Sutherland M.D. *Can. Med. Surg. J.*, **3**, 377–380.

Canada Medical Journal, 1868: The Late Archibald Hall, M.D., L.R.C.S.E. *Can. Med. J.*, **4**, 429–432.

Canada Medical Record, 1875: The Late William Sutherland M.D. *Can. Med. Rec.*, **3**, 473–474.

Carlyle, E. I., and D. Howse, 2004: Wales, William (bap. 1734, d. 1798). *Oxford Dictionary of National Biography*. Oxford University Press, accessed 01/27/2016, http://dx.doi.org/10.1093/ref:odnb/28457.

Chapman, E., and McCulloch, I., 2010: *A Bard of Wolfe's Army: James Thompson, Gentleman Volunteer, 1733–1830*. Robin Brass Studio, 361 pp.

Chichester, H. M., and J. Lunt, 2004: Bonnycastle, Sir Richard Henry (1791–1847). *Oxford Dictionary of National Biography*. Oxford University Press, accessed 01/29/2016, http://dx.doi.org/10.1093/ref:odnb/2856.

Colburn, 1858: *Colburn's United Service Magazine*. H. Colburn, 311.

Collins, P., 2016: Negretti and Zambra. Accessed 02/29/2016, http://www.barometerworld.co.uk/negrettiandzambra/index.htm

Cooper, J. I., 1972: Bethune, John (1791–1872). *Dictionary of Canadian Biography*. Vol. 10. University of Toronto/Université Laval, accessed 10/08/2014, http://www.biographi.ca/en/bio/bethune_john_1791_1872_10E.html.

Cooper, J. I., 2003: McGill, James. *Dictionary of Canadian Biography*. Vol. 5. University of Toronto/Université Laval, accessed 10/8/2017, http://www.biographi.ca/en/bio/mcgill_james_5E.html.

Crowe, M. J., 2004: Herschel, Sir John Frederick William, first baronet (1792–1871). *Oxford Dictionary of National Biography*. Oxford University Press, accessed 01/27/2016, http://dx.doi.org/10.1093/ref:odnb/13101.

Dawes, H., 2004: Newman, John (bap. 1783, d. 1860). *Oxford Dictionary of National Biography*. Oxford University Press, accessed 01/27/2016, http://dx.doi.org/10.1093/ref:odnb/39365.

Desmond, A., J. Moore, and J. Browne, 2004: Darwin, Charles Robert (1809–1882). *Oxford Dictionary of National Biography*. Oxford University Press, accessed 01/27/2016, http://dx.doi.org/10.1093/ref:odnb/7176.

Eakins, P. R., and J. S. Eakins, 1990: Dawson, Sir John William. *Dictionary of Canadian Biography*. Vol. 12. University of Toronto/Université Laval, accessed 01/29/2016, http://www.biographi.ca/en/bio/dawson_john_william_12E.html.

Editors of Encyclopædia Britannica, 2017: Jacques Cassini. *Encyclopædia Britannica*, accessed 02-09-2017, https://www.britannica.com/biography/Jacques-Cassini.

Elliott, C. A., 1979: *Biographical Dictionary of American Science*. Greenwood Press, 360 pp.

FitzRoy, R., 1858: *Barometer and Weather Guide*. Board of Trade, London, 25 pp.

Fleming, J. R., 1990: *Meteorology in America, 1800–1870*. Johns Hopkins University Press, 264 pp.

Frost, S. B., 1988: Skakel, Alexander. *Dictionary of Canadian Biography*. Vol. 7. University of Toronto/Université Laval, accessed 10/08/2014, http://www.biographi.ca/en/bio/skakel_alexander_7E.html.

Galbraith, J. S., 1985: Simpson, Sir George. *Dictionary of Canadian Biography*. Vol. 8. University of Toronto/Université Laval, accessed 01/29/2016, http://www.biographi.ca/en/bio/simpson_george_8E.html.

Good, G. A., 2004: Sabine, Sir Edward (1788–1883). *Oxford Dictionary of National Biography*. Oxford University Press, accessed 01/27/2016, http://dx.doi.org/10.1093/ref:odnb/24436.

Greenaway, F., 2004: Dalton, John (1766–1844). *Oxford Dictionary of National Biography*. Oxford University Press, accessed 01/27/2016, http://dx.doi.org/10.1093/ref:odnb/7063.

Hahn, R., 1971: *The Anatomy of the Scientific Institution: The Paris Academy of Sciences, 1666–1803*. University of California Press, 433 pp.

Hall, A. R., 1983: *The Revolution in Science 1500–1750*. Longman, 373 pp.

Hamblyn, R., 2001: *The Invention of Clouds: How an Amateur Meteorologist Forged the Language of the Skies*. Picador, 403 pp.

Helferich, G., 2011: *Humboldt's Cosmos*. Tantor, 384 pp.

Holmes, R., 2008: *The Age of Wonder: How the Romantic Generation Discovered the Beauty and Terror of Science*. Harper, 552 pp.

Hoquat, T., 2007: *Buffon/Linné: Éternels rivaux de la biologie?* Dunod, 221 pp.

Houston, S., Ball, T., and Houston, M., 2003: *Eighteenth-Century Naturalists of Hudson Bay*. McGill-Queen's University Press, 333 pp.

James, T. E., and E. Baigent, 2004: Riddell, Charles James Buchanan (1817–1903). *Oxford Dictionary of National Biography*. Oxford University Press, accessed 01/27/2016, http://dx.doi.org/10.1093/ref:odnb/35747.

Jarrell, R. A., 1979: Kalm, Pehr. *Dictionary of Canadian Biography*. Vol. 4. University of Toronto/Université Laval, accessed 02/05/2016, http://www.biographi.ca/en/bio/kalm_pehr_4E.html.

Jarrell, R. A., 2003: Williamson, James. *Dictionary of Canadian Biography*. Vol. 12, University of Toronto/Université Laval, accessed 10/07/2017, http://www.biographi.ca/en/bio/williamson_james_12E.html.

Jones, J., 2004: Hutton, James (1726–1797). *Oxford Dictionary of National Biography*. Oxford University Press, accessed 01/27/2016, http://dx.doi.org/10.1093/ref:odnb/14304.

Kushner, D., 2004: Croll, James (1821–1890). *Oxford Dictionary of National Biography*. Oxford University Press, accessed 01/27/2016, http://dx.doi.org/10.1093/ref:odnb/6744.

Laissus, Y., 2007: *Buffon: La nature en majesté*. Gaillimard, 127 pp.

Lamb, W. K., 2003: MacKenzie, Sir Alexander. *Dictionary of Canadian Biography*. Vol. 5. University of Toronto/Université Laval, accessed 10/7 2017, http://www.biographi.ca/en/bio/mackenzie_alexander_5E.html.

Lambert, J. H., 1983: Alexander Spark. *Dictionary of Canadian Biography*. Vol. 5. University of Toronto/Université Laval, accessed 04/18/2012, http://www.biographi.ca/en/bio/spark_alexander_5E.html.

——, 1984: *One Man's Contribution: Alexander Spark and The Establishment of Presbyterianism in Quebec 1784–1819*. St. Andrew's Presbyterian Church, 26 pp.

——, 1985: Wilkie, Daniel. *Dictionary of Canadian Biography*. Vol. 8. University of Toronto/Université Laval, accessed 04/18/2012, http://www.biographi.ca/en/bio/wilkie_daniel_8E.html.

Larousse, 1994: *Inventeurs et Scientifiques: Dictionnaire de Biographies*. Larousse, 692 pp.

Lefroy, J. H., 1895: *Autobiography of General Sir John Henry Lefroy*. C. Lefroy, Ed., Pardon and Sons, 342 pp.

Linsley, S. M., 2004: Noble, Sir Andrew, first baronet (1831–1915). *Oxford Dictionary of National Biography*. Oxford University Press, accessed 01/27/2016, http://dx.doi.org/10.1093/ref:odnb/35243.

Locher, F., 2007: The observatory, the land-based ship and the crusades: earth sciences in European context, 1830–50. *Br. J. Hist. Sci.*, **40**, 491–504.

Marshall, J. S., 1972: Charles Smallwood. *Dictionary of Canadian Biography*. Vol. 10. University of Toronto/Université Laval, accessed 04/18/2012, http://www.biographi.ca/en/bio/smallwood_charles_10E.html.

——, and N. Bignell, 1969: Dr. Smallwood's weather observatory at St. Martin's. *Nat. Can.*, **96**, 483–490.

McDougall, E. A. K., 1988: Somerville, James (1775–1837). *Dictionary of Canadian Biography*. Vol. 7. University of Toronto/Université Laval, accessed 01/29/2016, http://www.biographi.ca/en/bio/somerville_james_1775_1837_7E.html.

McGuire, S., 2010: Robert Cleghorn: Nurseryman & man of culture. Accessed 01/29/2016, http://montrealhistory.org/tag/cleghorn/.

McKenzie, R., 1976: Admiral Bayfield, Pioneer Nautical Surveyor. *Fish. Mar. Serv. Spec. Publ.*, **32**, 13 pp.

——, 1982: Bayfield, Henry Wolsey. *Dictionary of Canadian Biography*. Vol. 11. University of Toronto/Université Laval, accessed 01/29/2016, http://www.biographi.ca/en/bio/bayfield_henry_wolsey_11E.html.

Miller, D. P., 2004: Barrington, Daines (1727/8–1800). *Oxford Dictionary of National Biography*. Oxford University Press, accessed 01/27/2016, http://dx.doi.org/10.1093/ref:odnb/1529.

Moore, P., 2015: *The Weather Experiment: The Pioneers who Sought to See the future*. Farrar, Straus and Giroux, 395 pp.

Morgan, H. R., Ed., 1946: Parish register of Brockville and vicinity, 1814–1830. *Papers and Records*. Vol. 38. Ontario Historical Society, 77–108.

Morrison-Low, A. D., 2004a: Adie, Alexander James (1775–1858). *Oxford Dictionary of National Biography*. Oxford University Press, accessed 01/27/2016, http://dx.doi.org/10.1093/ref:odnb/48378.

——, 2004b: Adie, John (1805–1857). *Oxford Dictionary of National Biography*. Oxford University Press, accessed 01/27/2016, http://dx.doi.org/10.1093/ref:odnb/48379.

Napier Shaw, W., and E. Austin, 1932: *Manual of Meteorology: Meteorology in History*. Vol. 1. Cambridge University Press, 343 pp.

O'Connor, J. J., and E. F. Roberts, 2008: Philippe de la Hire. School of Mathematics and Statistics, University of St. Andrews, accessed 02/06/2016, http://www-history.mcs.st-andrews.ac.uk/Biographies/La_Hire.html.

O'Connor, J. J., and Robertson, E. F., 2008: Edward Norton Lorenz. School of Mathematics and Statistics, University of St. Andrews, accessed 11/05/2017, http://www-history.mcs.st-andrews.ac.uk/Biographies/Lorenz_Edward.html.

Oxford University Press, 2016: *A Dictionary of Scientists*. Oxford University Press, accessed 02/04/2016 http://dx.doi.org/10.1093/acref/9780192800862.001.0001.

Payne, B., 2002: Henri-Louis Duhamel Dumonceau. Accessed 01/22/2016, http://www.mobot.org/mobot/osgl/author.asp?creator=Henri-Louis+Duhamel+du+Monceau.

Raudzens, G. K., 1988: Bonnycastle, Sir Richard Henry. *Dictionary of Canadian Biography*. Vol. 7. University of Toronto/Université Laval, accessed 01/29/2016, http://www.biographi.ca/en/bio/bonnycastle_richard_henry_7E.html.

Repcheck, J., 2003: *The Man Who Found Time: James Hutton and the Discovery of the Earth's Antiquity*. Perseus, 247 pp.

Rioux, C., 2003: Thompson, James. *Dictionary of Canadian Biography*. Vol. 6. University of Toronto/Université Laval, accessed 10/7/2017, http://www.biographi.ca/en/bio/thompson_james_6E.html.

Robertson, J., 2004: Hume, David (1711–1776). *Oxford Dictionary of National Biography*. Oxford University Press, accessed 01/27/2016, http://dx.doi.org/10.1093/ref:odnb/14141.

Roland, C. G., and J. Bernier, 2000: *Secondary Sources in the History of Canadian Medicine*. Vol. 2. Wilfrid Laurier University Press, 278 pp.

Rousseau, J., 2014: Sarrazin, Michel. *Dictionary of Canadian Biography*. Vol. 2. University of Toronto/Université Laval, accessed 10/08/2014, http://www.biographi.ca/en/bio/sarrazin_michel_2E.html.

Royal Engineers, 2016: History: A brief history of the Corps of Royal Engineers, accessed 1/3/16, http://www.army.mod.uk/royalengineers/26315.aspx.

Rudwick, M., 2004: Lyell, Sir Charles, first baronet (1797–1875). *Oxford Dictionary of National Biography*. Oxford University Press, accessed 01/27/2016, http://dx.doi.org/10.1093/ref:odnb/17243.

Senior, E. K., 1987: McCord, Thomas. *Dictionary of Canadian Biography*. Vol. 6. University of Toronto/Université Laval, accessed 01/29/2016, http://www.biographi.ca/en/bio/mccord_thomas_6E.html.

Seymour, A. A. D., 2004: Colborne, John, first Baron Seaton (1778–1863). *Oxford Dictionary of National Biography*. Oxford University Press, accessed 01/27/2016, http://dx.doi.org/10.1093/ref:odnb/5835.

Sheets-Pyenson, S., 1996: *John William Dawson: Faith, Hope, and Science*. McGill-Queen's University Press, 304 pp.

Smallwood, C., 1860: Contributions to meteorology reduced from observations taken at St. Martin, Isle Jesus. *C.E. Can. J.*, **27**, 308–312.

——, 1861: Letter to Henry Morgan. Container 10. McGill University Archives.

Smolenaars, M., 2004: Derham, William (1657–1735). *Oxford Dictionary of National Biography*. Oxford University Press, accessed 01/27/2016, http://dx.doi.org/10.1093/ref:odnb/7528.

Stehr, N., and H. von Storch. Eds., 2000: Eduard Bruckner—The sources and consequences of climate change and climate variability in historical times. Springer, 338 pp. http://dx.doi.org/10.1007/978-94-015-9612-1.

Taillemite, É., 1974: Barrin de La Galissonière, Roland-Michel, Marquis de La Galissonière. *Dictionary of Canadian Biography*. Vol. 3. University of Toronto/Université Laval, accessed 02/06/2016, http://www.biographi.ca/en/bio/barrin_de_la_galissoniere_roland_michel_3E.html.

——, 1983: Bougainville, Louis-Antoine de, Comte de Bougainville. *Dictionary of Canadian Biography*. Vol. 5. University of Toronto/Université Laval, accessed 02/06/2016, http://www.biographi.ca/en/bio/bougainville_louis_antoine_de_5E.html.

Thomas, M. K., 1982: Kingston, George Templeman. *Dictionary of Canadian Biography*. Vol. 11. University of Toronto/Université Laval, accessed 01/29/2016, ttp://www.biographi.ca/en/bio/kingston_george_templeman_11E.html.

Toronto Public Library, 1967: *Landmarks of Canada: A Guide to the J. Ross Robertson Canadian Historical Collection in the Toronto Public Library*. Vol. 1. Toronto Public Library, 383 pp.

Trudel, M., 2014a: Cartier, Jacques (1491–1557). *Dictionary of Canadian Biography*. Vol. 1. University of Toronto/Université Laval, accessed 01/29/2016, http://www.biographi.ca/en/bio/cartier_jacques_1491_1557_1E.html.

——, 2014b: Champlain, Samuel de. *Dictionary of Canadian Biography*. Vol. 1. University of Toronto/Université Laval, accessed 01/29/2016, http://www.biographi.ca/en/bio/champlain_samuel_de_1E.html.

Tunis, B. R., 2003: Landmann, George Thomas. *Dictionary of Canadian Biography*. Vol. 8. University of Toronto/Université Laval, accessed 10/07/2017, http://www.biographi.ca/en/bio/landmann_george_thomas_8E.html.

Vachon, C., 1982: Glackmeyer, Louis-Edouard. *Dictionary of Canadian Biography*. Vol. 11. University of Toronto/Université Laval, accessed 01/29/2016, http://www.biographi.ca/en/bio/glackmeyer_louis_edouard_11E.html.

Vetch, R. H., and R. T. Stearn, 2004: Lefroy, Sir John Henry (1817–1890). *Oxford Dictionary of National Biography*. Oxford University Press, accessed 01/27/2016, http://dx.doi.org/10.1093/ref:odnb/16343.

Westfall, R. S., 2004: Newton, Sir Isaac (1642–1727). *Oxford Dictionary of National Biography*. Oxford University Press, accessed 01/27/2016, http://dx.doi.org/10.1093/ref:odnb/20059.

Whitfield, C. M., and R. A. Jarrell, 1982: Lefroy, Sir John Henry. *Dictionary of Canadian Biography*. Vol. 11. University of Toronto/Université Laval, accessed 01/27/2016, http://www.biographi.ca/en/bio/lefroy_john_henry_11E.html.

Williams, G., 2003: Hutchins, Thomas. *Dictionary of Canadian Biography*. Vol. 4. University of Toronto/Université Laval, accessed 10/7/2017, http://www.biographi.ca/en/bio/hutchins_thomas_4E.html.

Williams-Mystic, 2017: Disturnell, John. Accessed 2/29/2016,, http://sites.williams.edu/searchablesealit/d/disturnell-john/.

Williamson, H., 1811: *Observations on the Climate in Different Parts of America*. New York, T. and J. Swords, 199 pp.

Wilson, A., 1976: Colborne, John, Baron Seaton. *Dictionary of Canadian Biography*. Vol. 9. University of Toronto/Université Laval, accessed 02/04/2017, http://www.biographi.ca/en/bio/colborne_john_9E.html.

Winder, C. Gordon, 1972: Logan, Sir William Edmond. *Dictionary of Canadian Biography*. Vol. 10. University of Toronto/Université Laval, accessed 10/7/2017, http://www.biographi.ca/en/bio/logan_william_edmond_10E.html.

Womersley, D., 2004: Gibbon, Edward (1737–1794). *Oxford Dictionary of National Biography*. Oxford University Press, accessed 01/27/2016, http://dx.doi.org/10.1093/ref:odnb/10589.

Wulf, A., 2012: *Chasing Venus: The Race to Measure the Heavens*. Alfred A. Knopf, 304 pp.

——, 2015: *The Invention of Nature: Alexander von Humboldt's New World*. Knopf, 473 pp.

Young, B., 2014: *Patrician Families and the Making of Quebec*. McGill-Queen's University Press, 452 pp.

Zeller, S., 2009: *Inventing Canada: Early Victorian Science and the Idea of a Transcontinental Nation*. McGill-Queen's University Press, 372 pp.

Index

Académie Royale des Sciences, 5, 7, 9–12, 17, 23, 51, 127, 142, 169, 212, 278, 284–288, 296
Academy of Natural Sciences, 200
Acadia College, 185
Adie, Alexander, 74, 303
Adie, John, 74, 303
Adie, Patrick, 190–191
Agassiz, Louis, 8, 154, 161, 299
Age of Discovery, 126
Age of Empire, 141
Albany Academy, 298
Albany Institute of New York, 70, 183
albedo, 33
Albert the Great, 28
Albion Mines, 185, 280
Allan, R., 254
American Academy, 162
American Association for the Advancement of Science, 5, 68, 162, 166, 177, 184, 199–200, 213, 273, 300
  Smallwood at 1857 meeting, 173–176
American Civil War, 149, 174, 184–185

American Journal of Science, 170
American Philosophical Society, 162, 166, 300
American Revolution, 8, 32, 49, 52, 54, 65, 80, 87, 148–149, 278–279
Amherst, Jeffery, 7
Ampère, André-Marie, 144
Anderson, Katherine, 12, 126, 167–168, 179, 187, 294, 296
Anglican Church, 80–81, 158–159, 271
Anglican Diocese of Christ Church, 80–81, 177
Ansell, T., 254
Anthrax, 289
Arblaster, P., 301
Aristotle, 28
Army Medical Corps, 136
Army Medical Department, 165, 295, 299
Arthur, E., 274
Ashe, Edward, 186
astronomy, 11–13, 52, 283–285, 289, 293
atmospheric ozone research, 165, 169, 172
atomic theory, 9

311

Attali, J., 283
Auer, I., 252
Austin, E., 169, 283, 289–290, 292–293, 295–296, 298–300, 302

Bache, Alexander, 71, 154, 161, 298–299
Badgely, Francis, 273
Baigent, E., 276, 292, 296
Banks, Joseph, 50
barometer, 8–9, 127–130, 172, 189–191, 290
Barometer World, 296
Baron, W., 254
Barr, 50
Barrie, D., 11, 142
Barrington, Daines, 30–36, 40, 42, 44–46, 290–291
Barrington, Viscount, 290
Battle of Waterloo, 281
Baxby, D., 291
Bayfield, Henry, 66, 70, 129, 131, 136, 147–148, 276, 282–283
Beaufort, Francis, 70
Beaujeu, Daniel-Hyacinthe Liénard de, 23
Belin, William, 76
Benn, C., 158
Bensley, E. H., 272
Bergström, H., 252
Bernárdez, P., 41
Bernhardt, K.-H., 302
Bernier, J., 273
Bethune, John, 6, 8, 57, 66–67, 79, 177, 210, 212, 214–215, 218, 220, 222f, 228, 228f, 271
  weather journal, 80–82, 255, 256t, 257, 258t, 259t, 279
Bianchi, G. G., 42
Bignell, Nancy, 170, 197, 273
Binnema, T., 52, 274
Bishop's Medical Facility, 199
Bishop's University, 170
"blackbulb" thermometer, 137
Blacken, Clarence, 28–29
Bleho, M., 228f
Blink Bonny Gardens, 271

Blodget, Lorin, 78–79, 299–300
Board of Trade, 296
Bobis, L., 284
Boileau, René, 76
Boivin, Bernard, 24, 269
Bompas, William, 188
Bond, W. C., 162
Bonnycastle, Richard H., 66, 94, 136, 281–282
Bonpland, Aimé, 126, 288–289, 301
botany, 23–24, 68, 277–278, 284
Bougainville, Louis de, 50, 142, 287–288
Britain, 125, 289–294
  rivalry with France, 142–146
British American Journal, 75, 170, 178, 272
British American Journal of Medical and Physical Sciences, 93, 170, 178, 272
British Army, 184, 207, 281–282, 294–296
  Medical Department, 169
British Association for the Advancement of Science, 5, 66, 68, 71, 184, 275, 293, 295
British Board of Trade, 167
British Empire, 125–126, 137, 141–146, 169, 174, 178, 184, 212–213, 301
British North America, 6–8, 54, 65, 72, 183–185, 281, 283
British Royal Navy, 167, 274, 282
Bruckner, Eduard, 264, 302–303
Brugnara, Y., 254–255
Brunel, Isembard Kingdom, 303
Bunn, J., 188
Bureau des Longitudes, 143, 288
Burgess, Agness, 178
Burton, J., 292, 294
Buys-Ballot, Christophe, 187, 302
By, John, 137, 281
Byron, George, 254

Caesar, 35
Cambridge University, 276
Campbell, James, 292
Campbell, Robert, 54–55, 66–67, 80, 270, 279–280

Canada Medical and Surgical Journal, 273
Canada Medical Journal, 177–179, 272
Canada Medical Record, 273
Canadian Confederation, 8
Canadian Department of Marine and Fisheries, 168, 197, 277
Canadian Dominion Land Survey, 204
Canadian Historical Climate Date Rescue Project, 207–208
Canadian Institute, 162, 171, 275
Canadian Magnetic and Meteorological Society, 276
Canadian Naturalist, 263
Canadian Naturalist and Geology, 170–171
Canadian–U.S. tensions, 183–184
Carlyle, E. I., 274
Carmichael, Charles, 189
Carpmael, C., 160
Cartier, Jacques, 95, 277
cartography, 11, 285
Cassidy, D. C., 34
Cassini, César François, 287
Cassini, Giovanni Domenico. *See* Cassini, Jean Dominique
Cassini, Jacques, 12, 285, 287
Cassini, Jean Dominique, 12, 142–143, 283–284, 285, 288
Cassini space probe, 284
Caswell, Alexis, 174
Cawood, John, 142–143, 145–146
Celsius, Anders, 296–297
Champlain and St. Lawrence Railroad, 215
Champlain, Samuel de, 6, 29, 91, 95, 277
chaos theory, 9, 129–130, 300
Chapmand, E., 270
Charlevoix, 95
Chartrand, R., 7, 18, 24
Chenoweth, M., 254–255
Cherriman, John, 186
Chichester, H. M., 282
cholera, 165, 168–160, 173, 175–176
Christ Church, 271
Christie, William, 203
Church of Scotland, 279, 295
City University of New York, 162

Clarke, Lawrence, Jr., 185
Classical and Mathematical School of Montreal, 78, 271, 279
Cleghorn, Robert, 6, 67–68, 210, 212, 220, 222f, 271
climate amelioration, 87–97
　Lambert on, 88–90
　McCord and Kelly on, 90–96
climate change, 27–48, 87–97, 125, 291, 294
　agricultural indicators of autumn, 230–233
　agricultural indicators of spring, 227–228
　agriculture in colonial Canada, 218–220
　autumn weather, 229–230
　Celtic forests, 32–33
　cities and the river, 208–211
　cycles, 34
　evaluating eighteenth-century theory, 39–41
　evaluating improvement theory, 44–45
　Gaultier on, 245
　history of theories, 45–46
　human-induced, 29–31
　ice melt and road breaks, 214–215
　improvement theory, 27–48, 88, 90–96
　in classical literature, 34–38
　Kelly on, 90–96
　McCord on, 74–76, 90–96
　modern studies, 41–44
　other seasons, 236–238
　shipping season, 216–218
　since AD 588, 38–39
　spring in Quebec, 220–227
　St. Lawrence River, 213–218
　three centuries of observations, 207–241
　transport and communication changes, 211–213
　wheat and hay crops, 233–236
　winter roads, 215–216
climate improvement theory, 27–48, 88, 90–96, 245
climate networks, 11–12, 184–186
climatic determinism, 290

Index　313

Coates, Colin, 29, 219, 235
Colborne, John, 72, 131–133, 148, 281, 282
Colburn, H, 148
Cole, Gunman, 157
Collège de France, 286
Collège de Montréal, 60
Collège de Québec, 7
College of Philadelphia, 297
College of Physicians and Surgeons of Lower Canada, 170
Collins, P., 303
Companie Franche de la Marine, 7
Connor, E., 12
Cook, James, 50, 291
Cooper, Amy, 157–158
Cooper, Gunner, 157–158
Cooper, J. I., 271, 279
Corps of Royal Engineers, 125, 129, 133, 135–138, 148, 166, 169, 184–185, 270, 275, 281, 295–296
Cotte, Louis, 168
Coulomb, Charles-Augustin, 142
Craigie, William, 185
Crimean War, 275
Croll, James, 8
Cross, James, 294
Crowe, M. J., 293
Cuvier, Georges, 24

Dalhousie, Countess, 66
Dalton, John, 9, 166, 292
"Dalton's Law," 292
Daniell, John Frederic, 52, 166, 292–293
Dark Ages, 44
Dartmouth College, 300
Darwin, Charles, 8, 68, 126, 167, 291, 293–294, 296, 299
data homogenization, 134
Daubeny, Charles, 66, 71, 146, 293
Davy, Humphry, 303
Dawes, H., 303
Dawson, John William, 67, 200–202, 280–281
deforestation, 27–48, 74–75, 87–97, 245

DeGreef, D., 189
Degroot, Dagomar, 29
Delaney, John, 185, 193
Delisle, Joseph-Nicholas, 13–14, 18, 20–22, 259, 286
Denison, W., 135
Derham, William, 12, 127, 289–290
Desauniers, Francois, 76
Desmond, A., 294
Diodorus of Sicily, 35–36
Disturnell, John, 95–96, 298
Dove, Heinrich Willhelm, 61, 67, 78–79, 294, 302
Duhamel Dumonceau, Henri-Louis, 14, 17–20, 27–28, 219, 244, 245, 269, 278, 286–287
Duhamel, S., 18
Dymond, Joseph, 273
dysentery, 169

Eakins, J. S., 281
Eakins, P. R., 281
East India Company, 146
École de la Marine, 286
École Normal Supérieure, 289
Edinburgh Academy, 295
Edinburgh Philosophical Journal, 78–79, 178
Editors of Encyclopædia Britannica, 285
Einstein, Albert, 51
El Niño, 237, 254
electricity, 171–172
electromagnetism. *See* magnetism
Elliott, C. A., 297–300
energy balance, 44–45
Enlightenment science, 8–12, 88
  Jardin du Roi, 10–11
  Paris Observatory, 11
  weather and climate networks, 11–12
Environment and Climate Change Canada, 222f
Environment Canada, 222
Esclangon, E., 13
Espy, James, 73, 135, 166, 297–298

Euclid, 283
extraordinary seasons, 243–267
   cold summers, 254–258
      Gaultier, 244–251
      McCord, 260–261
      Smallwood, 261–264
      Spark and Lambert, 251–254
   winters, 258–264

Fahrenheit, Gabriel, 13, 300–301
Faraday, Michael, 142, 293, 298
Fastré thermometer, 187
Feldman, T. S., 34
Ferguson, Archibald, 80
Ferguson, Rebecca, 177
Fermat, Pierre, 283
Fidler, Peter, 52
Filippo, Giacomo. *See* Philippe, Jean
Financière agricole du Québec, 219, 222, 228
FitzRoy, Robert, 51, 126, 129–130, 167, 187, 197, 294, 296
Fleming, J. R., 29–30, 32, 56, 73, 75, 79, 136, 166, 169, 172, 185, 298–300
Florut, 36
fluid dynamics, 9
Fourier, Joseph, 166
four-pole magnetic theory, 144–145
France, 51, 125, 168, 283–289
   rivalry with Britain, 142–146
Francis I, 31
Franklin Institute, 71, 297
Franklin, Benjamin, 297–298
French Ancien Régime, 51
French Empire, 126, 169, 207, 212
French Revolution, 12, 142, 168, 288
Frost, S. B., 78, 271
Fyson, D., 58

Galbraith, J. S., 274
Galileo Galilei, 300
Galissoniére, Roland-Michel Barrin de La, 18, 23, 278

Gaultier, Jean-Francois, 4–5, 10–11, 13–14, 17–25, 27–30, 32, 40, 49, 60, 73–74, 78, 92, 168–169, 190–192, 207, 209–210, 212, 218–221, 222*f*, 224, 226*f*, 227, 228*f*–230*f*, 231, 231*f*–232*f*, 233–234, 235*f*, 237, 269, 278, 286, 297
   and John Lambert, 88–89
   death of, 24–25
   medicine, 19–20
   meteorology, 20–24
   weather journal, 243–251, 253–254, 258–260
Gauss, Carl Friedrich, 144
Gauvin, J.-F., 15
geodesy, 11, 204
Geography, 50–52, 126–128, 282, 286
Geological Society of London, 280
Geological Survey of Canada, 280
geological-scale climate change, 8, 128–131, 264
geology, 55, 66–68, 125, 127, 280–283, 285, 290, 293–294
geomagnetism. *See* magnetism
Gerbi, A., 10
Gibbon, E., 40
Girard College, 298–299
Glacken, C., 30, 32
Glackmeyer, Louis-Edouard, 3, 6, 80, 210, 212–213, 216, 218, 220–222, 222*f*, 226*f*, 227–230, 228*f*–232*f*, 235, 270
   weather journal, 254–255, 256*t*, 257, 258*t*–259*t*
Golinski, J., 30, 52
Good, G. A., 128, 161, 184, 295
Grace, William, 157
Graham, Andres, 274
Graham, Gunner, 154–155
Grand Trunk Railway, 197
Great Exhibit of London, 280
"Great Frost," 39, 290
Great Ice Age, 8, 40, 161
Greenaway, F., 292
Greenwich Observatory, 295
Greer, Allan, 218–220, 234–235
Gresham College, 39

Index   315

Grosjean, M., 42
Guelph Agricultural College, 193
Guevara-Murua, A., 255
Gulf Stream, 33
Gunn, Donald, 185
Guyot, Arnold, 154, 161, 299

Hadley circulation cell and radiation theory, 33
Hahn, R., 284–285
Hall, Archibald, 6, 55, 67–68, 78–80, 145, 183, 185, 199, 263, 271–272, 284
    medical meteorology, 169, 171–172, 177–178
Hall, Jacob, 177
Halley, Edmond, 127, 285, 290
Hallman, E. S., 133, 137
Hamblyn, R., 70, 76, 292
Hamilton, K., 254–255
Hannibal, 31
Hansteen, Christopher, 144
Harbour Commission of Montreal, 203
Harrington, C. R., 254
Harris, W. S., 132
Hartt, C. Fred, 185
Harvard Observatory, 162
Harvard University, 299, 300
Hearne, Samuel, 53, 274
Helferich, G., 127
Helmsley, Professor, 185
Henry, George, 55
Henry, Joseph, 70, 78, 142, 154, 161–162, 183–185, 198, 201, 272, 298–299, 303
    medical meteorology, 172, 177–178
Henry, William, 154, 275
Herodian, 35
Herodotus, 35–36, 39
Herschel, Caroline, 128, 293
Herschel, John F. W., 66, 68, 71–72, 133, 135, 137, 144–147, 160, 166, 293, 302
    instructions for observations, 128–131, 185, 295
Herschel, William, 128, 293
Hippocrates, 19

Hire, Philippe de La, 11
HMS *Beagle*, 126, 167, 294, 296
Hocquart, Indendant Gilles, 219
Hoe, Professor, 185
Holmes, R., 24, 50, 125, 128
Hoquat, T., 287, 297
Hough, F. B., 217*f*
Houle, D., 260
Houston, S., 53
Houston, T., 274
Howard, Luke, 52, 76, 292
Howse, D., 274
Hudson's Bay Company, 52–53, 72, 154, 185, 188
    observers, 273–274
Humboldt, Alexander von, 50–51, 76, 94–95, 125, 141, 144–145, 160–161, 288, 301–302
    contributions of, 126–128
Humboldt, Willheim, 301
Hume, David, 55, 290
hurricanes, 166, 295–296, 298
Hutchins, Thomas, 53, 274
Hutton, James, 8, 55, 290, 294
hydrological cycle, 291–292
hygrometers, 292–293

Imbrie, J., 8
Imbrie, K., 8
industrialization, 209–211
insolation, 44–45
Institut de France, 12
instrument makers, 303
International Meteorological Committee, 294
International Meteorological Society, 187
Iroquois people, 277
isolines, 127
isotherms, 51
Issar, A., 41, 43

Jackson, Andrew, 154
James, T. E., 276

Janković, V., 52, 128, 166
Jardin du Roi, 9–12, 17, 23–24, 51, 269, 277–278, 284–287
Jarrell, R. A., 18, 275-276, 297
Jenner, Edward, 8, 291
Jesuit Collège de Montréal, 92
Johnston, James, 154, 160
Johnston, John, 159
Jones, Charles, 157, 160
Jones, J., 290
Jones, T. P., 71
Jonquière, 278
Josephine, Empress, 289
Journal of the Canadian Institute, 171
Journal of the Franklin Institute, 71
Jussieu, Antoine, 10, 17, 269, 285–287, 297
Jussieu, Bernard, 10, 17, 269, 278, 286–288, 297
Jussieu, Joseph, 10, 269, 278, 286–287, 297
Justinian, 35, 39

Kalm, Pehr, 18–24, 90, 272, 278, 297
Kelly, William, 66, 70, 75–77, 129, 131, 136, 207, 209, 282
  climate improvement theory, 90–96
Kew barometers, 189–191
King's College, 151–152, 185, 276, 292
Kingston Observatory, 275
Kingston, George, 4, 6, 200–203, 273, 276–277, 294
  Meteorological Service in Canada, 185–194
Kington, J., 12, 19, 34, 51, 168
Kirby, William, 24
Klingaman, N.P., 254
Klingaman, W. K., 254
Kuntz, H., 68, 70, 72, 78
Kupperman, Karen O., 29, 33, 51
Kushner, D., 294

l'Académie Royale des Sciences, Lettres et Beaux Arts of Belgium, 200

l'Observatoire Physique Central of St. Petersburg, 199–200
La Hire, Gabriel Philippe, 284
La Hire, Laurent de, 284
La Hire, Philippe de, 143, 284
  thermometer, 22
La Niña, 254
Laclerc, George-Louis, 32
Laissus, Y., 287
Lamb, H. H., 40–44, 279
Lambert, John, 50, 55–57, 236, 270, 280, 292
  on climate amelioration, 88–89
  weather journals, 251–254, 253*t*
Lamontagne, R., 18, 23
land cultivation, 27–48
Landmann, George, 50, 136–137, 281
Lapérouse, 274
large-scale mapping, 3–6
Larousse, 283–289, 296, 298, 300–302
Lavoisier, Antoine Laurent de, 288
Le Petit Séminarie, 60
Leclerc, Georges Louis, 24, 275–276, 287
Lefroy, John Henry, 6, 50, 52, 71–72, 94, 133, 147, 184, 276
  Northwest Magnetic Survey, 153–154
  Toronto Observatory, 151, 156–159, 162
Lequeux, J., 284
Lévis-Mirepoix, Gaston-Pierre-Charles, 287
Lillie, Dr., 95–96
Linnaeus, Carl, 18, 24, 287, 296–297
Literary and Historical Society of Quebec, 54, 66, 70, 90–92, 136, 162, 178, 280, 283
Little Ice Age, 40–44, 88, 92, 237
Ljungqvist, F. C., 41–43
Lloyd, Humphrey, 154, 275
Locher, F., 303
Locke, J., 145
Logan, William, 67, 280
London Meteorological Society, 70, 72
London Observatory, 288
London Times, 129
"Long Winter" of 1708-1709, 39
Loomis, Elias, 144, 162, 299

Lorenz, Edward Norton, 9, 129, 300
Louis XIV, 9–10, 12, 168, 283
Louis XVI, 168
Lunt, J., 282
Lusignan, Paul-Louis de, 23
Lyell, Charles, 8, 68, 280, 293–294
Lyon Observatory, 289

MacKenzie, Alexander, 53, 279
MacKenzie, John, 185
magnetic declination, 127–128, 141–145
magnetism, 70, 127–128, 136, 198–199, 275–276, 284, 302
    Lefroy and the Northwest Magnetic Survey, 153–154
    magnetic crusade, 138, 141–164
Mahoney, Gunner, 157
Malaria, 168, 287
Maline, Thomas, 155
Manley, G., 252
Mann, Theodore Augustin, 33–40, 42, 44–46, 91–93, 301
Mannheim Meteorological Society, 12, 34
Maraldi, Giacomo Filippo, 127, 284–286
Maraldi, Jean Dominique, 12
Maraldi, Jean Philippe, 11–12
Marsden, W., 76
Marshall, J. S., 170, 273
Mascart, Jean, 264, 289
Mason, John, 29
Mason, Joseph, 188–189
Massachusetts Institute of Technology, 30
Maury, Lieutenant, 167
McCave, I. N., 42
McClellan, J. E., III, 51, 168–169
McCord Family Papers, 59, 69
McCord Museum of Canadian History, 57, 67, 69, 272
McCord, Anne, 81
McCord, David Ross, 57, 68–69, 272
McCord, John, 57–58
McCord, John Samuel, 6, 8, 30, 33–34, 50, 53–54, 57–58, 183, 192, 199, 207–210, 212, 214, 217, 217f, 218, 222f, 224, 228f, 234–235, 237, 271–272, 274, 282, 293, 301–302
    and the Natural History Society, 65–82
    climate change theory, 90–96
    contributions to science, 68–70
    education, 59–60
    influence of, 60–61
    influences on temperature, 76–77
    on climate, 74–76
    on instruments, 73–74
    second weather journal, 68–73
    Toronto Observatory, 146–147, 152, 160–161
    weather journal, 58–59, 125–137, 168, 172, 174, 177, 255, 257, 259t, 263
    winters of 1841–1842 and 1842–1843, 260–261
McCord, Thomas, 6, 50, 54, 57–58, 214, 220, 222f, 234, 270
McCord, William King, 58, 61, 69
McCulloch, I., 270
McDermott, F., 41–42
McDougall, E. A. K., 279
McGill College, 178, 271
McGill Medical School, 273
McGill Observatory, 165, 186, 197–207, 223f, 272–273
    after Smallwood, 201–203
    Archives, 170
    expansion into meteorology, 199–200
    problem of Smallwood, 200–201
    timekeeping, 197–199
    transition to professional meteorology, 203–204
McGill Stormy Weather Group, 171
McGill University, 70, 76, 81, 170, 177, 179, 194, 200, 271–273, 279, 281
    Archives, 56–57, 92, 170, 169, 179, 273
McGill, James, 81, 278–279
McGuire, S., 271
McKenzie, R., 282–283
McLeod, Clement Henry, 179, 201–203, 273
McNaught, John, 156
Mechanics' Institutes, 65–66

medical meteorology, 165–181
    Archibald Hall, 177–178
    atmospheric ozone research, 172
    Charles Smallwood, 169–177
    Joseph Henry, 177
    storms and storm warnings, 166–168
    weather and health, 168–169
    William Sutherland, 179
medicine, 19–20, 277–278, 281, 285–287, 289, 291, 296–297
Medieval Warm Period, 40, 44
Mémoires and Histories de L'Académie Royal des Science, 9, 14, 30, 79, 90, 92
Menzies, Thomas, 154, 156–157, 159–160
Menzies, William, 160
Meteorological Service of Canada, 147, 151, 183–195, 197, 201–202, 204, 207, 275, 277
    international cooperation, 187–194
    Kingston and, 186
    networks and standards, 184–186
meteorology
    Corps of Royal Engineers, 135–138
    Herschel's instructions, 128–131
    Humboldtian science, 126–128
    military and, 125–140
    professionalization of, 197
    Saint Helen's Island experiment, 131–135
Meyer, W. B., 220
Middle Ages, 42
Middleton, W. E. K., 14
military, 131–135, 165, 184–185
    Royal Artillery, 125, 133, 136, 144, 146, 149, 153, 159, 161, 167, 169, 185, 275–276, 294–296
    Royal Engineers, 125, 129, 133, 135–138, 148, 166, 169, 184–185, 270, 275, 281, 295–296
    Toronto Observatory, 146–162
Military College of West Point, 298
Miller, D. P., 291
Miller, P., 59
Ministére de La Marine, 7
MMA Collection, 217*f*

Moberg, A., 252
Montague, Mary Wortley, 291
Montcalm, Louis-Joseph de, 7, 288
Montenegro, A., 96
Montpellier, 285–286
Montreal, 208–212
    agricultural indicators of autumn, 230–232
    agricultural indicators of spring, 227–228
    apple blossom dates, 228
    autumn weather, 229–230
    cold summers, 251–258
    first autumn frost, 229*f*, 230*f*
    first snow, 231*f*
    first summer thunder, 226*f*
    freeze and first river crossing, 214, 214*f*
    frost in July, 256*t*
    frost-free season, 232*f*, 233*f*
    harvest dates, 235*f*
    ice melt and road breaks, 214–215
    last spring frost, 222*f*, 223*f*
    last spring snow, 225*f*, 226*f*
    other seasons, 236–238
    shipping open, 215*f*
    spring sowing dates, 228*f*
    strawberry maturity dates, 228*f*
    summer of 1753, 247, 249
    summer of 1859, 257–258, 259*t*
    summers of 1807 and 1808, 251–254
    weather observers, 270–273
    wheat and hay crops, 233–236
    winter roads, 215
    winters, 258–264
Montreal General Hospital, 178, 272
Montreal International Airport, 208
Montreal Natural History Society, 125, 171–173, 178, 198–199, 271–273, 279–280
Montreal School of Practical Medicine and Surgery, 179
Montreal Star, 78, 80
Montreal's North West Company, 53
Moore, Charles, 7, 166

Moore, J.-D., 260
Moore, P., 73, 167, 296
Morrison-Low, A. D., 303
Mount Pinatubo eruption, 254
Mount Tambora eruption, 65, 165, 254–255
Mountain, Jacob, 177
Murdoch, P., 50, 136

Napier Shaw, William, 169, 197, 283, 289–290, 292–293, 295–296, 298–300, 302
Napoleonic Wars, 8, 12, 54, 56, 65, 87, 125, 141, 212–213, 270, 281–282, 294
National Institute of the United States, 200
Natural History Society of Montreal, 54, 65–82, 162
    establishment of, 66–68
Naturalist's Journal, The, 291
Naval College, Quebec City, 276
Naval College, U.K., 276
Negretti, Enrico Angelo Ludovico, 172, 191, 303
Netherlands Meteorological Institute, 302
New France, 6–8, 17–19, 23–24, 95, 168, 219, 277–278
New Republic, 143
New World Encyclopedia, 128
Newman, John, 74, 187, 303
Newton, Isaac, 8, 144, 284–285, 289, 290, 293
Nicolet College, 76
Noble, Andrew, 136
North-West Company, 274, 278–279
North-West Mounted Police, 188
Novascotian, 87–88

O'Connor, J. J., 129, 284, 300
Observatoire de Paris. *See* Paris Observatory
Oldfield, Colonel, 148, 150
Olson, S., 169
Ørsted, Hans Christian, 144
Ovid, 30–31, 35–36, 42
Oxford University, 66, 69, 293

Oxford University Press, 283–285, 288–289, 297–302
ozone. *See* atmospheric ozone research

Pacific Railroad Survey, 300
Palatine Society of Mannheim, 301
Paris Exposition, 280
Paris Observatory, 11–14, 18, 142–143, 259, 284–285, 287–289, 295
Parker, D. E., 252
Parry, Admiral, 294
Pascal, Blaise, 9, 283
Pasteur, Louis, 8, 168, 289
Patriot Rebellions, 69, 81, 131–132, 149, 235, 272, 281–282
Patterson, W. P., 42
Payne, B., 286
Peninsular War in Spain, 281, 295
Périer, Florin, 9, 283
Peter the Great, 14
Petronius, 35–36
Pfister, Christian, 43–44, 250
Philippe, Jean, 11
Philosophical Transactions of the Royal Society, 12, 29–30, 50, 136, 289
phlogiston theory, 288
Picard, Jean, 284
Pictou Academy, 280
Piery, M., 169
Plackett, R., 9
Pliny the Elder, 35
Pliny the Younger, 36
Poincaré, Henri, 300
Pomponius-Mela, 35
Ponari, Federico, 79
Pontarron, Monsieur de, 22
Pontus, 30
Poole, Henry, 185
poudrerie, 236
Pouillet's scale, 77
Presbyterian Church, 66–67, 80–81, 270, 279–280
Priestly, Joseph, 288
Prince of Wales, 290

Princeton University, 70, 154, 161, 298–299
Puy de Dôme exercise, 9

Quakers, 291–292
Quebec Citadel, 5, 282
    Observatory, 186, 207
Quebec City, 208–212, 215, 218
    1740s weather, 244–246
    agricultural indicators of autumn, 230–232
    agricultural indicators of spring, 227–228
    apple blossom dates, 228f
    autumn weather, 229–230
    cold summers, 254–258
    first autumn frost, 229f, 230f
    first snow, 231f
    first summer thunder, 226f
    founding, 277
    frost in July, 256t
    frost-free season, 232f
    harvest dates, 235f
    last spring frost, 222f, 223f
    last spring snow, 225f
    other seasons, 236–238
    shipping open, 217f
    spring season, 220–227
    spring sowing dates, 228f
    strawberry maturity dates, 228f
    summer of 1754, 246–251
    summer of 1859, 257–258, 259t
    summers of 1807 and 1808, 251–254
    weather observers, 269–270
    wheat and hay crops, 233–236
    winter roads, 215
    winters, 258–264
Quebec City Airport, 225
Queen Anne's War. *See* War of Spanish Succession
Queen's College, Kingston, 275

Raudzens, G. K., 281–282
Ray, John, 290

Réaumur, René Antoine Ferchault de, 10–11, 13–14, 20–22, 92, 285–286, 290
Red River Rebellions, 149
Redfield, William, 73, 135, 161, 166, 295, 298
Regourd, F., 51, 168–169
Reid, William, 135, 166, 295–296
Reign of Terror, 288
Repcheck, J., 290
Richmann, Georg Wilhelm, 172
Riddell, Charles, 6, 70, 74, 168, 184, 199, 276
    Toronto Observatory, 146–147, 155, 159, 161, 275
Rideau Canal, 137, 281
Rioux, C., 270
Roberts, E. F., 284
Robertson, Dr., 177
Robertson, E. F., 129, 300
Robertson, J., 290
Robertson, William, 73
Roberval, Jean-François, 277
Robinson anemometer, 187
Robinson, Emily, 160
Robinson, John, 160
Roland, C. G., 273
Roman Empire, 30–32, 34–38, 41–44
Roman Warm Period, 41–43
Rose, Alex, 50, 136
Ross, Anne, 69
Ross, Catherine, 297
Ross, Huw, 152
Ross, John, 303
Rothman, 40
Rousseau, J., 23, 56, 278, 288
Royal Academy of Brussels, 301
Royal Academy of Painting and Sculpture, 284
Royal Academy of Science. *See* Académie Royale des Sciences
Royal Artillery, 125, 133, 136, 144, 146, 149, 153, 159, 161, 167, 169, 185, 275–276, 294–296
Royal Astronomical Society, 283, 293
Royal Canadian Mounted Police, 188
Royal College of Surgeons, 178
Royal Grammar School, 271

Index 321

Royal Institution, 303
Royal Meteorological Society, 66, 187, 292
Royal Military Academy, 275–276, 281, 294–295
Royal Naval College, 296
Royal Observatory, 9, 51
Royal Sappers and Miners, 295
Royal Society in London, 52–53, 66, 68, 72, 133, 135, 142, 145, 154, 274–276, 280–281, 285–287, 289, 291–292, 294–295, 301, 303
  Geomagnetic and Meteorological Committee, 146, 167–168
Royal Society of Medicine, 51
Royal Swedish Academy of Science, 18
Rudwick, M., 294
Ryerson, Egerton, 276

Sabine, Edward, 125, 133, 136, 198, 275–276, 294–295, 302
  Toronto Observatory, 144–147, 150, 152–154, 159–161
Sabine, Mrs., 160
Saint Andrew's in Scotland, 291
Saint Andrew's Presbyterian Church, Quebec City, 55, 252, 270, 279
Saint Gabriel Street Church, 66–67, 279
Saint Helen's Island experiment, 131–135, 147–149, 272
Saint Helen's Observatory, 153
Saint Lawrence River Valley Region, 208–241
  agricultural indicators of autumn, 230–233
  agricultural indicators of spring, 227–228
  agriculture in colonial Canada, 218–220
  autumn weather, 229–230
  average summer maximum temperature, 210f
  average winter minimum temperature, 211f
  cities and the river, 208–211
  cold summers, 251–258
  frost in July, 256t
  ice melt and road breaks, 214–215
  other seasons, 236–238
  shipping season, 216–218
  spring in Quebec, 220–227
  St. Lawrence River, 213–218
  summer of 1753, 247, 249
  summer of 1859, 257–258, 259t
  summers of 1807 and 1808, 251–254
  transport and communication changes, 211–213
  wheat and hay crops, 233–236
  winter roads, 215–216
  winters, 258–264
Saint Petersburg Observatory, 286
Sarrazin, Michel, 11, 23, 277–278
Saul, J. R., 132, 149
Saunders, R., 50
Schaffer, S., 52
Schönbien, Christian, 173, 175–176
Schott, Charles A., 62, 78–79
Scott, Robert Henry, 167–168, 187–192, 294
Scottish Enlightenment, 54–55, 66–67, 145, 270, 279–280, 290
scurvy, 277
seasonal weather prediction, 128–131
Seneca, 35–36
Senior, E. K., 271
Seppä, H., 42
Seven Years' War, 6–8, 50, 57, 65, 274, 286–287
Seymour, A. A. D., 281
Shapin, S., 52
Sheets-Pyenson, S., 68, 281
Shelley, Mary, 254
Sheppard, Mrs., 66
Shey, Dr., 93
Sicre, M.-A., 42
Simpson, George, 72, 154
Siveright, John, 6, 53, 67, 72–73, 76, 274
Skakel, Alexander, 6, 55, 58–61, 66–68, 74, 135, 145, 165, 177–178, 210, 212, 222f, 271–272, 280
  association with John Samuel McCord, 59–60
  weather journal, 78–80, 255, 256t

Skakel, William, 3, 6, 59, 145, 178, 210, 222*f*, 271
   weather journal, 78–80
Slonosky, V. C., 67, 75, 92, 209
smallpox vaccine, 291
Smallwood, Charles, 3, 6, 8, 68, 76, 80, 135, 161, 171–172, 178–179, 183–186, 204, 210, 213, 222*f*, 272–273
   at McGill Observatory, 197–200
   medical meteorology, 165, 169–179
   problem of, 200–201
   snows of 1869, 261–264
   weather journal, 253–255, 257, 271–272
Smallwood, Mrs., 177
Smith, Adam, 5, 290
Smithsonian Institution, 70, 78, 154, 161–162, 172, 183, 193, 198, 201, 272, 297–299, 303
   Archives, 177
   Meteorological Project, 183, 185
   Smallwood and Henry at, 177
Smolenaars, M., 290
Société Météorologique de France, 199
Société Royale de Médecine, 168
Somerville, James, 66, 279
space–time relationship, 143–144
Spark, Alexander, 6, 50, 54–57, 66–67, 69, 73, 75, 78, 81–82, 88, 90, 92, 145, 165, 199, 210, 212–213, 221, 222–223, 222*f*, 226*f*, 229*f*–232*f*, 234, 270, 279
   association with John Samuel McCord, 59–60
   influence of, 60–61
   weather journal, 55–57, 243, 251–255
spatial mapping, 141–145
Stanley, Frederick Arthur, 203
Star/L'étoile, 279
Statius, 35
Stehr, N., 96, 264, 303
Stern, R. T., 276
Stevenson, Robert, 215, 303
Stevenson, Robert Louis, 303
Stevenson, Thomas, 303
Stockholm Academy, 297
Storch, H. von, 96, 264, 303

"storm controversy," 73, 135, 166, 295–296
storm warnings, 5, 126, 166–168, 204, 213, 277, 296
Strabo, 31, 35–36, 42
Strachan, Bishop, 271
Strachan, John, 80
Sunderland, William. *See* Sutherland, William
Sutherland, William, 6, 169, 179, 212–213, 216, 222*f*, 237, 253, 255, 256*t*, 273
Sweden, 296–297
Swindon, Jean Henri van, 142
synoptic maps, 167, 298

Taillemite, É., 278, 288
technological changes, 3–6, 303
   airplanes, 204
   automobiles, 204
   barometers, 127–128, 130, 172
   "blackbulb" thermometers, 137
   Enlightenment science, 8–12
   magnetic measurement, 142
   maps, 50–52
   McCord on instruments, 73–74
   steam trains, 204
   steamships, 204, 213
   telegraph, 5, 172, 198, 204, 213, 298
   thermometers, 12–15, 20–22, 39, 73–74
   timekeeping, 197–199
   transport and communication, 211–213
   under Kingston, 187–189
   weather maps, 126
   wind rose, 51
   wireless communication, 204
terrestrial magnetism. *See* magnetism
Terricelli, Evangelista, 9, 300
Tesio, S., 17
Theodore, Karl, 34
Theophrastus, 28
thermometers, 12–15, 39, 73–74, 172, 187, 285–286, 290, 297, 300–301, 303
   "blackbulb," 137
   Delisle's, 13–14, 20–22
   Réaumur's, 14–15, 20–22

Thiessen, A. D., 152, 159–160, 162
Thomas, M. K., 153–154, 162, 277
Thompson, David, 52
Thompson, James, 270
Thompson, Kenneth, 29, 96
Thornton, P., 169
Thury, César François Cassini de, 12
timekeeping, 197–198, 273
Tomkins, H., 53
Toronto Magnetic and Meteorological
   Observatory, 5, 71–72, 74, 94, 125,
   133–134, 136–137, 185–186, 200, 270,
   275–276, 282
  at King's College, 152
  Charles Riddell and, 147–149
  Charles Younghusband and, 152–153
  founding years, 146–164
  John Henry Lefroy and, 159–161
  Lefroy and the Northwest Magnetic
    Survey, 153–154
  observers and their families, 149–150,
    154–159
  on Bathurst Street, 149–152
  transition from military to civil
    institution, 161–162
Toronto Public Library, 276, 277
Toronto weather observers, 275–277
Torricelli, 283
Tournefort, Joseph Pitton de, 11, 278, 284, 285
*Transactions of the American Philosophical
   Society*, 145
*Transactions of the Literary and Historical
   Society of Quebec*, 67, 131, 283
Transactions of the Royal Society, 52
Treaty of Paris, 8, 49
Trudel, M., 277
Tunis, B. R., 281
Turcotte, Benoit, 209
Tweedale, Corporal, 133
two-pole magnetic theory, 144–145
typhus, 165, 168–169, 255

U.K. Meteorological Office, 126, 136–137,
   184, 187–189, 294

U.S. Army Air Corps, 300
U.S. Army Medical Department, 169
U.S. Coast Guard, 204
U.S. Coastal Survey, 299
U.S. Congress, 297
U.S. Geodetic Survey, 204
U.S. Navy, 297
U.S. War Department, 199, 297
U.S. War Office, 177
United States, 297–300
Université Laval, 7
University College of London, 272
University of Åbo, 297
University of Edinburgh, 178, 280, 303
University of Glasgow, 278, 279
University of Göttingen, 301
University of Pennsylvania, 298
University of Toronto, 151–152, 276
University of Utrecht, 302
Upminster in Essex, 289
Uppsala University, 297
urbanization, 208–211

Vachon, C., 270
Vaillant, Sébastien, 11, 278, 285, 286
Vetch, R. H., 276
Victoria Bridge, 215
Victoria, Queen, 213
Vienna Conference on International
   Meteorology, 187
Virgil, 31, 35–36, 42
Vogel, Brant, 29

Wales, William, 52, 273–274
Walker, James, 154, 159–160
War of 1812, 8, 49, 54, 56, 65, 67, 144,
   148–149, 169, 174, 183, 282, 294–295
War of Austrian Succession, 7
War of Spanish Succession, 7
Ward, William Cuthbert, 3, 138, 148, 270
Watson, George, 154–155
Watson, Mrs., 155
Watts, Mr., 77, 282

weather and health, 168–189
weather forecasting, 5, 126, 297–298
weather journals
    Bethune's, 80–82, 255, 256*t*, 257, 258*t*–259*t*, 279
    Gaultier's, 243-251, 253-254, 258-260
    Glackmeyer's, 254–255, 256*t*, 257, 258*t*–259*t*
    Lambert's, 251-254, 253*t*
    McCord's, 58–59, 68-73, 125–137, 168, 172, 174, 177, 255, 257, 259*t*, 263
    Royal Engineers', 125
    Skakel brothers', 78-80, 255, 256*t*
    Smallwood's, 253-255, 257, 271-272
    Spark's, 55-57, 243, 251-255
    storm warning systems, 126
Westfall, R. S., 289
Wetter, O., 250
Whewell, William, 143
White, Gilbert, 291
White, Sam, 29, 33
Whitehouse, Richard, 29
Whitfield, C. M., 276
Wien, T., 17, 218–219
Wilkie, Daniel, 58–59, 66–68, 74, 78, 92, 279–280
Williams, G., 274
Williams-Mystic, 298

Williamson, Hugh, 32–33, 36–38, 40, 42, 44–46, 297
Williamson, James, 4, 275
Wilson, A., 281
Wilson, Cynthia V., 33, 44, 52–53
Winder, G. Gordon, 280
Wolfe, James, 7, 49, 234
Woodward, F. M., 137
World War II, 194, 300
Wulf, A., 52, 145, 274, 302

Yakir, D., 41, 43
Yale University, 299
"year without a summer," 65, 211, 221, 251, 254–255
Young, Brian, 58–59, 68, 71, 82, 272
Young, G., 88
Younghusband, Charles W., 6, 152–153, 184, 275–276
    Toronto Observatory, 155–159

Zambra, Joseph Warren, 172, 191, 303
Zeller, Suzanne, 68, 145, 153, 162, 280
Zilberstein, Anya, 11, 29–30, 52, 65, 87
Zoology, 66, 68, 272

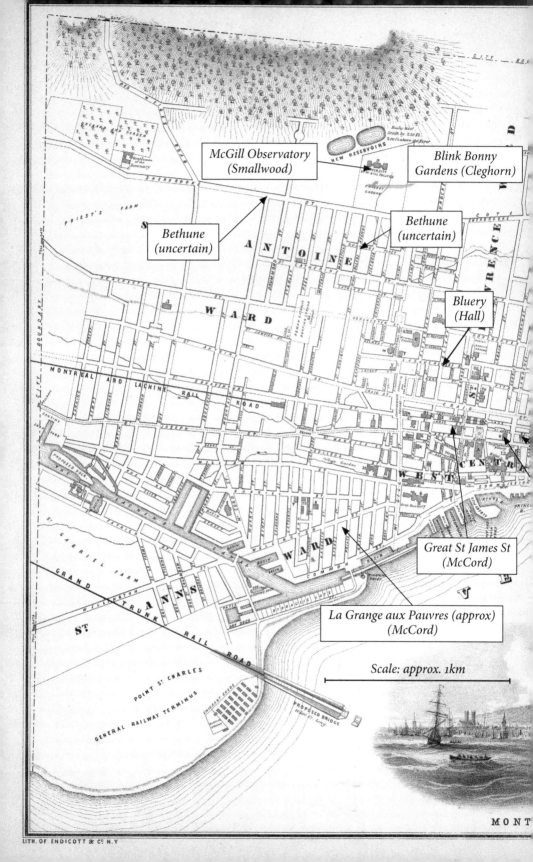